THE HENRY HOLT GUIDE TO

ASTRONOMY

DAVID BAKER
ILLUSTRATED BY DAVID A. HARDY

An Owl Book
HENRY HOLT AND COMPANY
NEW YORK

Acknowledgments Appreciation is extended to Carol Oliver for much good advice on the material structure of the book and for offering useful comments on the general subject areas, to M. Wilson and E. W. W. Baker for support with information on illustrations and aspects of archaeoastronomy respectively and to T. Ainsworth for his comments on the preliminary text. Finally, a very special mention goes to my dear wife Ann for her support and forbearance throughout the period of preparation for this book and the long hours spent in writing the text.

Photographs Ron Arbour, Bishopstoke 267 bottom, 273 bottom; Arizona Office of Tourism 225; Aspect Picture Library, London 220, 249, 261 top; T. P. Byatt, Moulton 22; Jet Propulsion Laboratory, Pasadena, California 199; Los Alamos Scientific Laboratory, New Mexico 218; Movie Newsreels, Hollywood, California 168 top; NASA, Washington D.C. 168 bottom, 169, 170, 173 bottom, 185 top, 185 bottom, 186–187 top, 186–187 bottom, 188, 191 bottom, 194, 197 bottom, 201, 202, 203, 207, 263, 271 top, 271 bottom, 272, 273 top; Novosti, London 179 bottom; Photri, Alexandria, Virginia 174, 179 top, 181, 217, 235 bottom, 241 bottom, 251, 261; Royal Observatory, Edinburgh 219; Space Frontiers, Havant 242; University Museum, Oxford 227 top; ZEFA-Photri 68–69, 189; NASA/Perkin-Elmer Corp. 25.

The photographs on pages 64–65, 66–67, 70–71, 235 top, 240, 241 top and 243 are copyright of the California Institute of Technology and Carnegie Institution of Washington and are reproduced by permission of the Hale Observatories, Pasadena, California.

The publishers gratefully acknowledge the contribution made by Storm Dunlop during the preparation of this book.

First published in the United States in 1990 by
Henry Holt and Company, Inc., 115 West 18th Street,
New York, New York 10011.
Originally published in Great Britain under the title *Astronomy* in 1978 by
The Hamlyn Publishing Group Limited.

Library of Congress Cataloging-in-Publication Data
Baker, David, 1944–
[Astronomy]
The Henry Holt guide to astronomy / David Baker ; illustrated by David A. Hardy. — 1st American ed.
p. cm.
"An Owl book."
Originally published in Great Britain in 1978 by the Hamlyn Pub. Group under the title: Astronomy.
Bibliography: p.
Includes index.
ISBN 0-8050-1197-8 (pbk.)
1. Astronomy. I. Title.
QB45.B145 1990
520—dc20 89–11018
CIP

Henry Holt books are available at special discounts
for bulk purchases for sales promotions, premiums,
fund-raising, or educational use. Special editions
or book excerpts can also be created to specification.

For details contact:

Special Sales Director
Henry Holt and Company, Inc.
115 West 18th Street
New York, NY 10011

First American Edition

Produced by Mandarin Offset
Printed and bound in Hong Kong
10 9 8 7 6 5 4 3 2 1

Contents

Introduction 4

Telescopes
- Introduction 6
- The reflector and the refractor 6
- Light 8
- Types of reflector 10
- Instrument performance 14
- Spectroscopy 16
- Telescope mountings 20
- Large telescopes 20

Stars
- Stellar motion and distance measurement 26
- Stellar classes 32
- The Sun 34
- The Hertzsprung-Russell diagram 38
- Formation of heavy elements 44
- Multiple star systems and novae 46
- Stellar evolution 52
- Giant stages and beyond 54
- The question of life 62

The Constellations
- The celestial sphere 72
- The constellations 76
- The constellations: a description 86
- Constellation notes 144

The Planets
- Properties of the solar system 146
- The Moon 150
- Index to Moon maps 160
- Mercury 170
- Venus 174
- Mars 180
- Jupiter 190
- Saturn 198
- Uranus 204
- Neptune 208
- Pluto 210
- Minor planets 212
- Evolution of the solar system 213

Comets and Meteorites
- History and designation 214
- Structures 215
- Orbits 218
- Meteors 222
- Meteorites 224
- Tektites 226

The Galaxies
- The Milky Way 228
- Structure of the galaxy 230
- The spiral arms 230
- Galactic nebulae 232
- Extragalactic nebulae and classes 234
- Galactic evolution 244
- The redshift 246
- The universe – evolutionary or steady state? 252
- Curvature of the universe 260

The Development of Celestial Observation
- Historical preface 264
- Telescopes – on earth and in space 268

Tables 274

Glossary 277

Greek alphabet 280

Further reading 280

Index 281

Introduction

Although the science of astronomy has been developing for about 4000 years we can be sure that the structure of the universe has intrigued Man for the entire duration of his existence on this planet, perhaps for more than two million years. Today, the constellations no longer represent ideological principles, mythological characters or valiant deeds of war. We no longer wonder at the celestial vault surrounding Earth and place ourselves at the centre of everything, nor do we regard comets as portents of disaster, the firebolts of the gods that rain upon the planet as retribution for misguided deeds. We think we are perhaps wiser; too knowledgeable to allow ourselves the indulgence of mythological conjecture.

However, the immensity of the visible universe still inspires us with awe, and we feel the same great affinity for the wonders of the universe of which we are so small a part. This book, though, is not concerned with the aesthetic and philosophic appeal of astronomy but rather with presenting a brief summary of the universe – its content, make-up and evolution. It must, by necessity, be incomplete and many of the ideas and explanations set out in the seven chapters reflect the current view or the most scientifically acceptable theory and can in no way be taken as unequivocal dogma. There are many things we have yet to learn about the universe, and what we know today is primitive conjecture to the view future generations will have of space and astronomy. It has been said that the only truly factual statement one could make about the universe is that there are a lot of lights in the sky! We have come a long way from recognizing that as a fact but the bases of our assumptions are still highly debatable and we must be continually aware of our own scientific fallibility when dealing with astronomy and cosmology.

With that in mind, it may assist the reader to obtain full value from the book by summarizing the reading programme. For the potential observer, it is recommended that chapters one, seven and three are studied in detail, followed by a general reading sequence of chapters four, five, two and six. The chapters on telescopes, constellations and the history of observing will provide information necessary for an introduction to practical observation but a background knowledge of the physical processes at work in the universe is essential as an incentive for observational activities.

For the armchair theorist, studious attention to chapters two, four, five and six will provide a basic working knowledge of the universe, followed by chapters one, seven and three. For persons with a general interest in the practice and mechanics of astronomy a sequential reading programme will be appropriate. To the armchair theorist and the general reader alike it must be emphasized that this book can only be a prelude to the sense of achievement derived from actually taking a telescope outside on a starlit night and sampling the magnificent array of celestial phenomena at first hand. Nothing can compensate for this rewarding participation in the subject.

Due to the necessary limitations imposed by the size of this book, the aspiring amateur will doubtless feel it necessary to expand his or her horizons still further by consulting other, perhaps more specialized, works; a bibliography is therefore included before the index to assist with selection. This practice is to be encouraged since a thorough appreciation of astronomy must take in the full gamut of subjective thinking and theoretical consideration. However, it is to be hoped that *Guide to Astronomy* has taken an objective view where theory diverges from observed phenomena.

Finally, it should be emphasized that in as rapidly changing a science as astronomy the week-by-week events can quickly obscure, modify or update existing knowledge and it is very important to stress the value of belonging to a recognized astronomical society. Most areas are covered by local groups and the amateur should be encouraged to join these whenever possible. Accurate and timely news can be received through membership of the British Astronomical Association, and information on aspects of space research and space-borne astronomy can be obtained via membership of the British Interplanetary Society.

In addition, the weekly and monthly scientific periodicals carry much of both general and specialized interest to the amateur astronomer. No single book can do more than take a cursory look at such a vast subject and one or other of these activities is essential for keeping up to date with developments.

Telescopes

Introduction

Several times during the last five centuries significant discoveries or inventions have opened up a window on one or more of man's scientific pursuits. One science, astronomy, has been completely revolutionized by the 17th century invention of the telescope, allowing an observer to study the precise configuration of celestial bodies and piece together theories about the structure of the universe in a way previously impossible.

Using the telescope as a light collector and magnifier more precise instrumentation has been developed over the past three centuries. Unlike many scientific instruments, the telescope lends itself to a variety of applications at different levels of complexity and because of this has found favour with amateur and professional alike.

The reflector and the refractor

Although it is possible to collect a variety of radiated energies for measuring celestial phenomena the most common, and the simplest and most adaptable for professional and amateur use, is the form of visible radiation called light. The two most important criteria of the equipment being used are adequate magnification of the object under study and good light gathering characteristics.

There are two main types of instrument capable of achieving these objectives: the reflector and the refractor. Each has unique properties useful for certain tasks and there are a broad range of common specifications against which performance can be evaluated.

The reflector **(Fig. 1a)** was first developed by Isaac Newton in the latter half of the 17th century and because of its robust flexibility it has been favoured by amateurs ever since. When light is reflected from the main concave mirror, it is intercepted by a smaller, secondary mirror mounted at an angle of 45° to the axis of the primary. When the light reaches the secondary mirror it is turned through 90° to an eyepiece where the image can be viewed outside the tube of the telescope. It would, theoretically, be possible for an observer to situate himself directly in front of the telescope, but for comparatively small instruments the amount of light obscured by the physical mass of the human body would render it inoperative. The Newtonian design is only one possible configuration using the principle of reflection and other variations are detailed on p. 10.

Large reflectors occasionally enable an observer to be placed at the prime focus since the amount of light obscured by a supporting cage is only a small fraction of the total amount of light falling on the primary. This position is usually used for photographic purposes only.

The refractor is simpler in concept, more compact in design but less

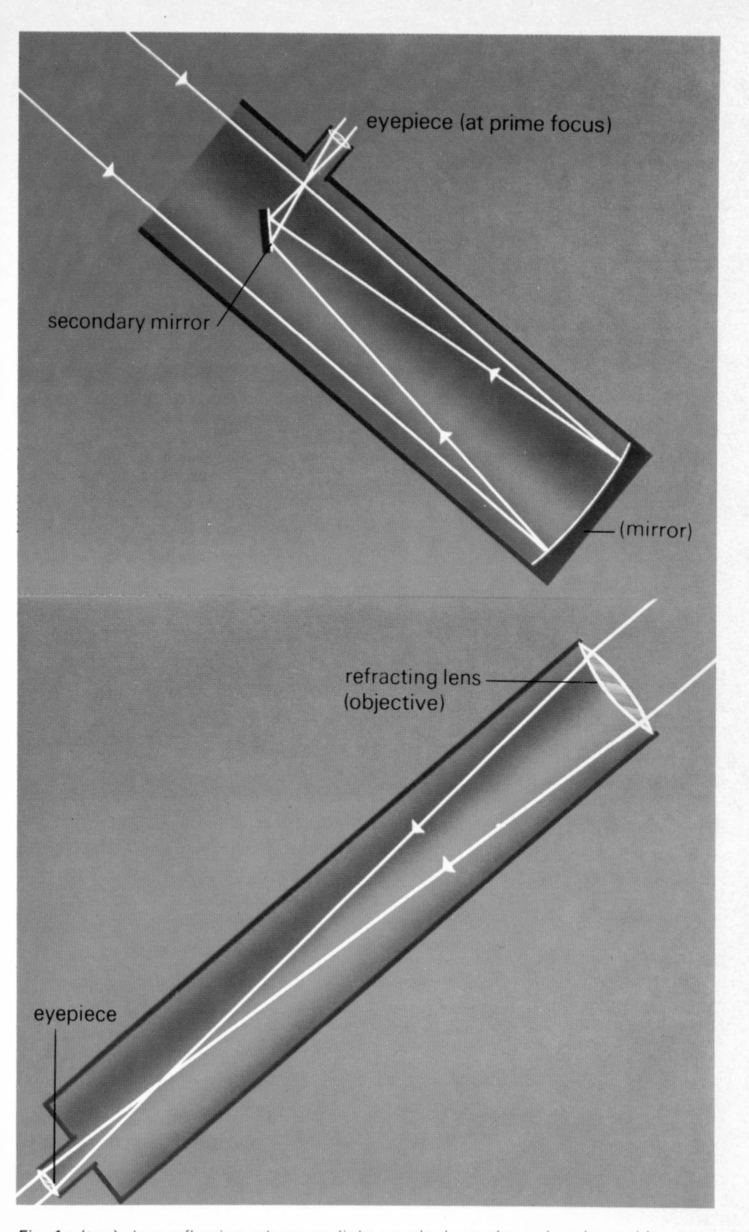

Fig. 1a (top) In a reflecting telescope, light travels down the main tube and is turned back by a concave mirror to a secondary mirror where it is directed through the magnifying lens in the eyepiece. **b (above)** A refracting telescope brings the image to a focal point and on to a magnifying lens; no mirrors are used.

adaptable in application. **Figure 1b** shows the basic operating principle of a typical refractor, first used astronomically by Galileo in 1609 when he became the first person to view the sky through a telescope. Instead of reflecting light with a mirror the telescope incorporates a main lens which refracts light to a specific plane, the focal plane, from where a magnifying lens takes the refracted image and presents it to the eye.

Light

Any examination of the merits of the reflector and the refractor must depend upon their detailed characteristics and the methods adopted to counteract the distortion of light. The general principles discussed below will clarify this problem, which besets amateur and professional alike.

Figure 2 shows the effect of a beam of incident light falling upon a solid block of glass, when the light is at an appreciable angle to the perpendicular. The refracted ray is bent towards the normal and a high proportion of light is reflected back at the same angle as that of the incident beam. If the incident beam deviates only slightly from the perpendicular, the reflected light is substantially reduced, with more light being refracted through the glass block.

Figure 3 shows the situation where rays of light falling on a convex lens, of the type forming the main objective of a refracting telescope, are made to converge at a point known as the focus. The distance from the objective to the focus (or focal length) is important in determining magnification. At this distance from the lens an image of any object is formed in the focal plane.

In the same way as light rays passing through a prism are split into their component colour bands, the convex lens will disperse the incident rays with a shorter focal length at the blue end of the spectrum and longer at the opposite, or red, end as shown in **Fig. 4**. Light falling along a line perpendicular to the centre of the convex lens will not be dispersed. If the lens is concave the dispersion will be in the opposite direction, blue light being refracted more than red.

In the case of a telescope made from a simple convex lens, at any one position only a single colour will be sharply imaged. To avoid this, a second lens of the same glass as the objective can be used **(Fig. 5)** to correct the dispersion and bring all colours to the same focal point.

However, if two carefully selected but different types of glass are brought together **(Fig. 6)** the combined compensation of convex-concave lenses provides the same effect and dispersion is eliminated. This type of dual lens is known as an achromatic objective. The effect of colour dispersion, or chromatic aberration, also applies to the eyepiece, which is similarly shaped to the convex objective but in reality consists of two lenses and provides a built-in correction. Since the focal length of the eyepiece is only a small fraction of the focal length

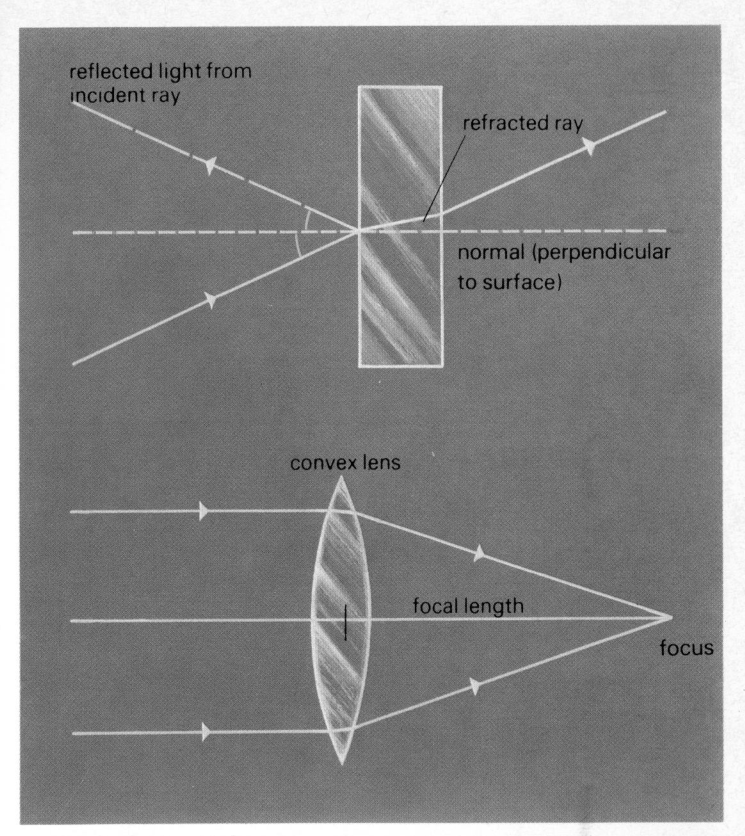

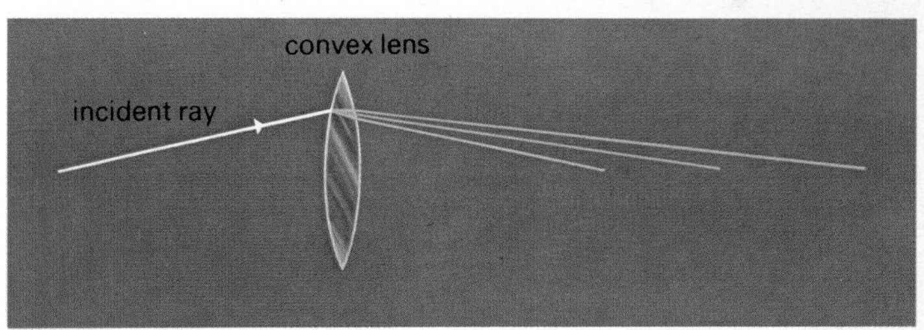

Fig. 2 (top) An incident ray of light will pass through a solid block of glass at a unique angle depending on the angle of the incident ray from the perpendicular and will leave the glass at the same angle as the initial incident beam. A portion of the light falling on the block will be reflected back at a similar angle to that of the initial, incident ray. **Fig. 3 (centre)** A convex lens will bring refracted light to a point known as the focal position, thus determining the focal length. **Fig. 4 (above)** Light rays passing through a prism are split into their component colour bands. This is also true with a convex lens, an effect known as chromatic aberration.

of the objective, the colour separation is minimal even for a single lens eyepiece.

The effect of chromatic aberration hampered development of the refractor late into the 18th century partly because of difficulties in the manufacture of the different lenses. Despite a revival in astronomical use of refractors, imperfections in the optical qualities of the object glass limited the theoretical size. Also, the lenses in a refractor can only be supported at their circumference, and distortion produced by changes in temperature and the sheer weight of the glass limited objectives to a diameter of about 1·2 m.

A different form of image distortion, spherical aberration, is experienced by the reflector telescope. Spherical mirrors bring rays reflected from the centre to a different focal plane to those from the edge, and the corrective procedure is simpler and cheaper than that required for compensating the effect of chromatic aberration in a refractor. When the mirror is ground from a single block of glass or pyrex the surface must be contoured so as to present a parabolic profile in cross-section. This provides a compensatory factor and brings light from all areas of the mirror to a common focal plane.

Types of reflector

The principle of the reflector lends itself to many applications. The most common form of reflector, the Newtonian already shown in **Fig. 1a**, is the type widely accepted by amateurs. Simple to set up, it is reliably designed for minimum attention with ease of access for component elements or parts replacement.

Figure 7 shows four other variations of the basic reflector and these represent the degree of versatility applied to the concept. The Herschel reflector incorporates an angled parabolic mirror which reflects the image through an angled tube along one side of the telescope so that the observer sits with his back to the object under study. This type is no longer used since it was unsuitable for prolonged observation; the observer had to be situated in a very awkward position when observing objects at high altitudes.

The Cassegrain reflector uses a concept much favoured today by removing the eyepiece from a position upstream of the incoming light. It also has the added advantage of increasing the prime focal length by replacing the plain secondary mirror of the Newtonian reflector with a small convex one that accepts the image from the primary and reflects it back through a hole in the main mirror. Minimal loss of light-gathering power is experienced, and the light reaches prime focus behind the primary at which point the eyepiece provides magnification. Here the observer actually faces forwards as with a refractor, and the Cassegrain is the only optical arrangement which allows a reflector to be used in this way.

The principle of the coudé telescope affords even greater opportunity for extending the focal length of the main mirror. After reflecting from the primary the light is returned by the convex secondary to a third mirror, which feeds it to a fixed position outside the telescope. The third mirror tilts to compensate for changes in the telescope's elevation. Depending upon the design, a fourth mirror is sometimes required to bring the light to a fixed focal point where heavy or cumbersome equipment can be mounted without affecting the telescope.

Generally, chromatic aberration is eliminated by an achromatic objective but this only applies to the naked eye interpretation of the most sensitive visual portions of the spectrum: green and yellow. Parts of the far blue or red ends of the spectrum are not as accurately brought to prime focus as the eye would instruct the brain to believe.

Fig. 5 (below) A concave glass placed some distance behind the convex lens will compensate for aberration and bring all the spectral colours to the same focal point. **Fig. 6 (bottom)** Two different types of glass can be brought together in the same mounting to achieve the same corrective effect for chromatic aberration. **Fig. 7 (following page)** Four common types of reflector (Herschel, Cassegrain, coudé and Schmidt) use various combinations of mirror and lens to achieve specific tasks.

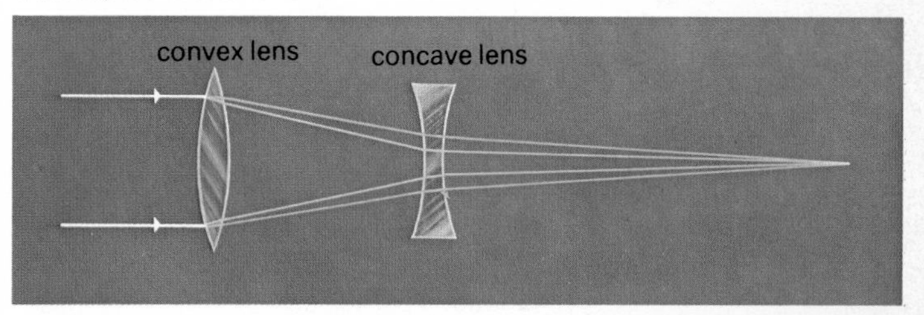

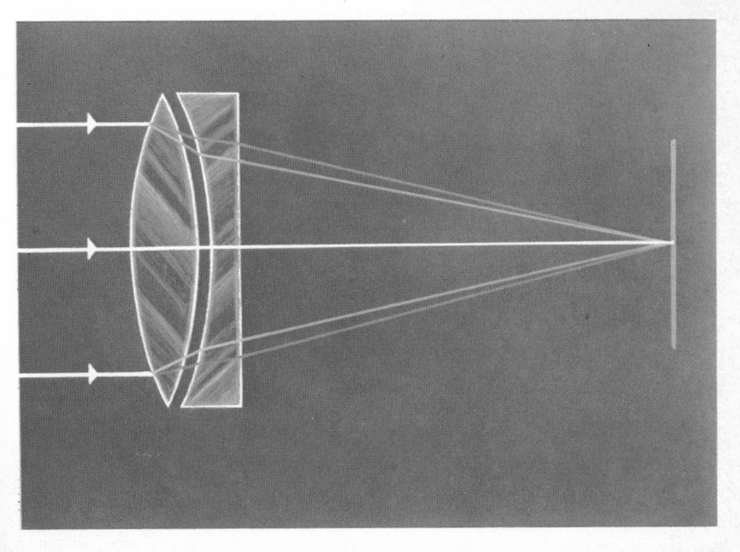

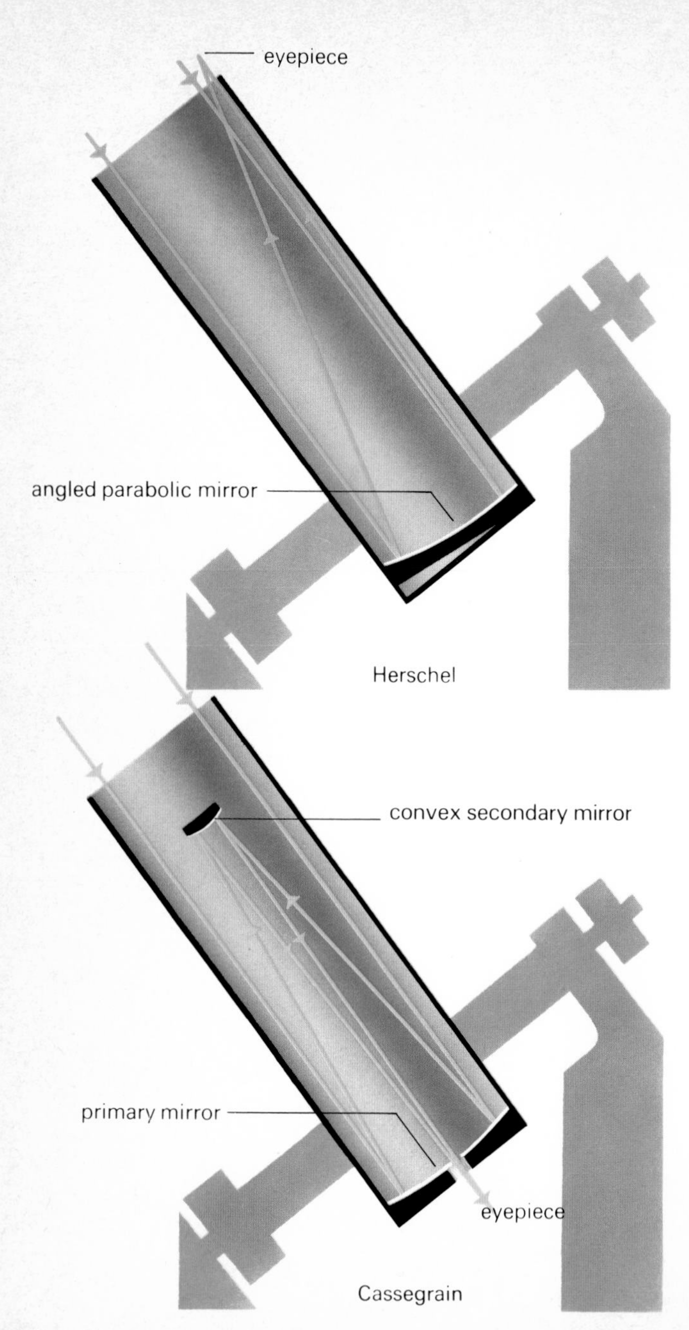
eyepiece
angled parabolic mirror
Herschel
convex secondary mirror
primary mirror
eyepiece
Cassegrain

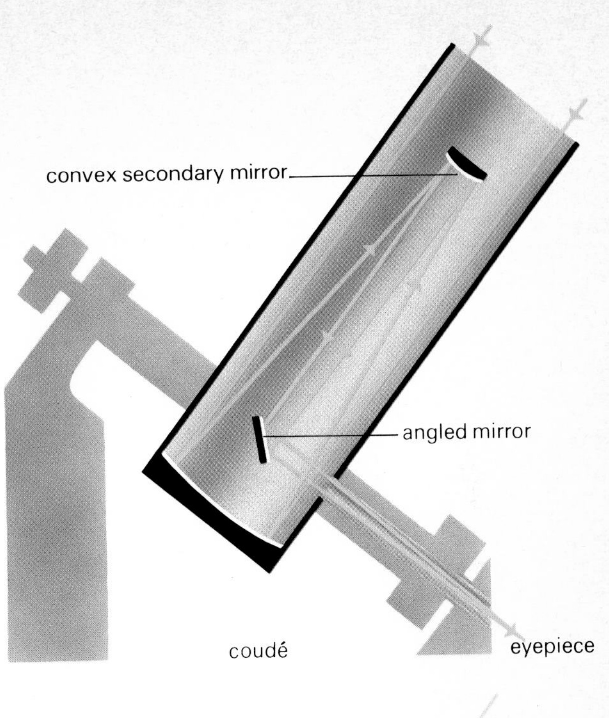

coudé

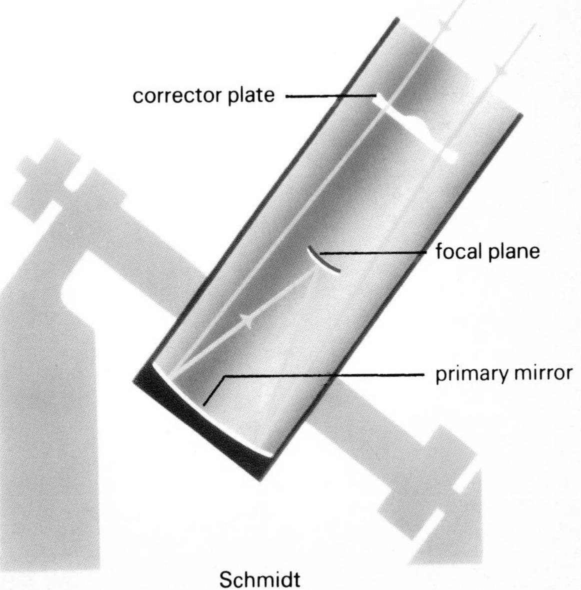

Schmidt

For photographic purposes, where the blue end of the spectrum is most sharply perceived it is necessary to attach a special type of objective to the photographic plate. However, a problem arises due to the optical characteristics of camera systems. A fast camera obtains a limited useable field of view while good definition over a wide area requires very long exposures.

The problem is solved by using a Schmidt camera **(Fig. 7)**, named after the Estonian instrument maker who first designed this type of telescope in 1930. The telescope places a correcting plate in front of the spherical mirror so that the combination of convex and concave surfaces exactly compensates the spherical aberration caused by the main objective, distorting the beam as it passes through the correcting plate. The reflected image is brought to a focal point between the correcting plate and the main objective so that a very wide field of view is maintained free of any distortion, unlike the conventional telescope fitted with a photographic objective. The Palomar instrument has an effective field of view measured in arc minutes, whereas the Schmidt has a theoretical field of view extending to more than 20°. This ensures maximum efficiency when photographing faint objects covering a large area of the sky but is useless for visual observation.

Instrument performance

There are three primary constraints to be considered when selecting a particular telescope: magnification, light gathering properties and resolving power.

Magnification depends upon the ratio of primary to secondary focal lengths and is obtained by dividing the focal length of the eyepiece into the focal length of the main objective, a mirror in the case of a reflector and an achromatic lens in a refractor. For example, if the focal length of the eyepiece is 5 cm and the focal length of the objective is 100 cm, the telescope is set up to magnify the object under study 20 times (since 100/5 = 20). **Figure 1** shows the position of respective focal planes for reflectors and refractors. Increase in magnification can be effected by replacing the 5 cm eyepiece with one of, say, 2 cm, which (using the same formula) provides the telescope with a magnification of 50 times.

There is obviously a lower limit to the size of an eyepiece which can be of practical use, and the only option for increased magnification with most telescopes is to increase the effective focal length of the objective. This can be achieved with a Barlow lens **(Fig. 8)**. This consists of a tube containing an achromatic diverging lens which is placed in the eyepiece holder with the normal eyepiece inserted in the top of the tube. The upper limit here is set by the increased atmospheric distortion, as magnification increases, and the reduced amount of light gathered, as the field of view gets smaller.

Generally, the theoretical limit is considered to be 35 times the magnification for every centimetre of objective diameter; a 15 cm reflector would have a magnification limit of 525 times. Larger instruments are not subject to the same ratio and the 508-cm Mount Palomar telescope has an effective magnification limit of 2750 times, or 5·4 per centimetre. Light gathering power is directly related to the surface area of the objective and can be determined by comparing this with the surface area of the pupil in the human eye, which is usually about 0·25 cm^2. A telescope which carries an objective with 300 times the surface area of the observer's pupil will gather 300 times the amount of light captured by the naked eye.

These two parameters are of utmost importance in any telescope; magnification is determined by the ratio of prime to secondary focal lengths and light gathering power is directly proportional to the surface area of the objective.

Increased resolving power is rather more dependent upon the diameter of the objective than the degree of magnification since resolution for a given objective will not be improved by continued magnification; distortion will increase and the larger the image becomes the more useless it is for observation. Resolution, defined as

Fig. 8a (below) The Barlow lens increases the focal length of the objective, and the eyepiece (with its own lens) is carried on the extremity of the Barlow. **b (bottom)** A 90° adaptor turns the image in a refracting telescope to a convenient position.

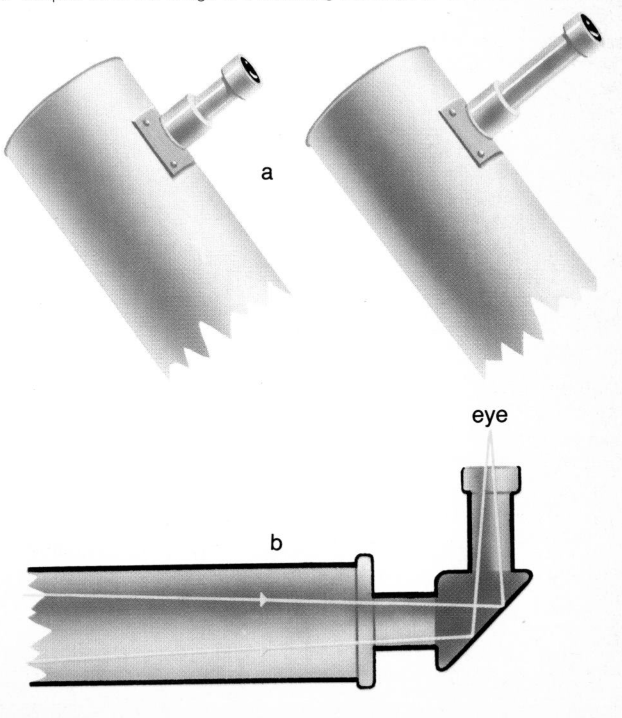

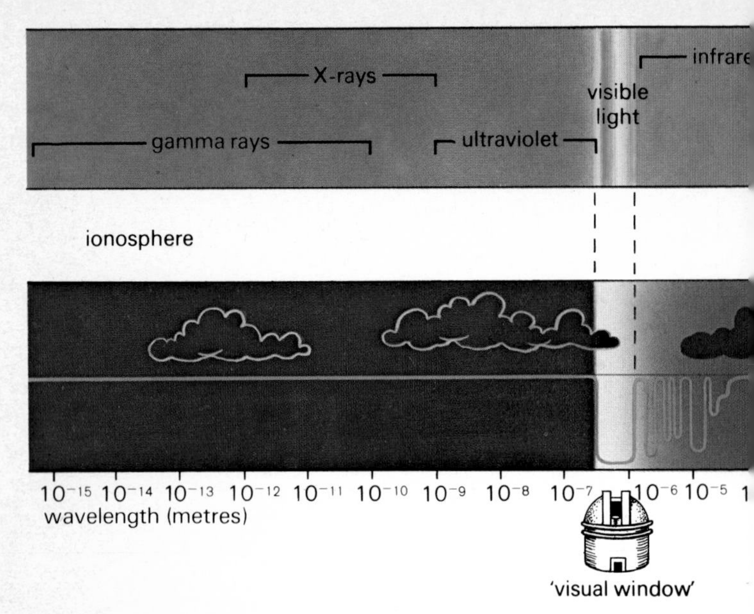

the ability to resolve two separate stars of equal brightness, can be calculated by dividing 12·7″ (seconds of arc) by the diameter of the objective in centimetres. This means that a 15 cm reflector will have a resolution of 0·85″ and a 7·5 cm refractor will show a resolution of 1·69″. The only way to increase resolving power, and improve the amount of information available for magnification, is to increase the diameter of the objective. The amount of detail visible directly depends upon the telescope's aperture.

In selecting instruments, the reflector has decided advantages in that for a given cost, its considerably greater aperture will give very much enhanced resolving power. The advantage of the refractor is that it is simpler to set up and carries its lenses in fixed mountings which should need no further adjustment. The inconvenient arrangement of the eyepiece can be a problem when tilting the refractor to high elevation if only because the observer has to physically locate his eye towards the object in view and for this reason an adaptor is usually provided which turns the image through 90°. Also, a reflector needs occasional attention due to atmospheric or weathering effects on the main mirror and a silvered mirror will degrade more quickly than an aluminized mirror.

Spectroscopy

Although the optical telescope is primarily concerned with gathering light from celestial sources for magnification and visual observation, the basic concept has opened up a wide range of applications. It is

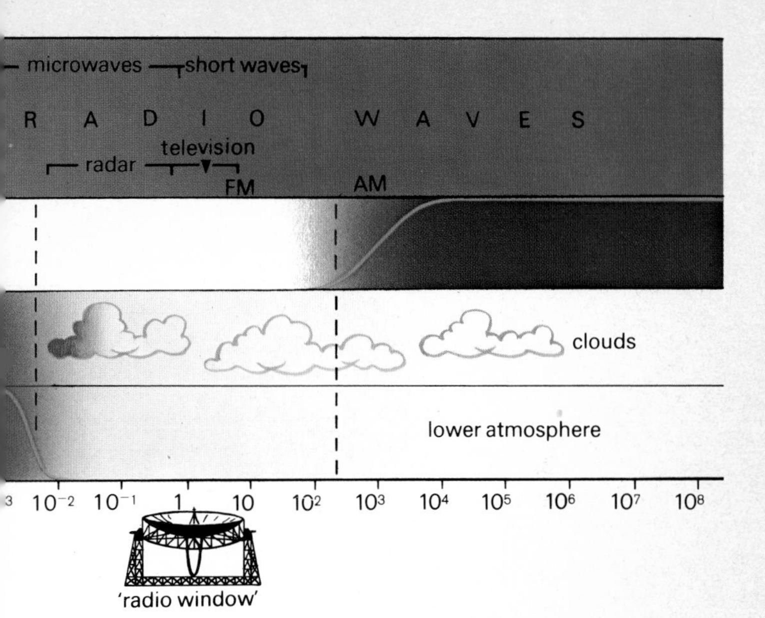

Fig. 9 The atmosphere is transparent to some bands on the electromagnetic spectrum and opaque to others. Visible light and radio waves can be measured from the surface of the earth, other bands can only be detected from high altitude.

less than 400 years since Galileo first saw objects in space invisible to the naked eye and less than 100 years since Hertz first demonstrated the presence of other forms of radiation by producing radio waves.

On the electromagnetic spectrum **(Fig. 9)** radio waves occupy just one band of possible frequencies. In 1895, Röntgen discovered X-rays at the other end of the spectrum, and less than five years later Villard found even more energetic radiation, now called gamma rays.

Figure 9 shows that radiation is measured in wavelengths, the distance between successive points where the fluctuating intensity repeats itself like crests of a wave, and the frequency, a measure of the time elapsed between the passage of successive crests across a fixed point. Since all forms of radiation propagate through a medium at the same velocity ($2{\cdot}9979 \times 10^8$ m/s) the wavelength is determined by dividing the velocity by the appropriate frequency and is expressed in metres.

Obviously, as the frequency increases the wavelength decreases and thus microwaves are referred to as short waves since they are the shortest form of radio wave exhibiting the highest frequency. However, there is no sharp line of demarcation between various forms of radiation and the names applied to different bands of the electromagnetic spectrum are arbitrary and more a matter of convenience rather than related to specific properties.

Radiation in the infrared, visible, ultraviolet and X-ray bands is sometimes expressed in a unit of measurement called the ångström, after the Swedish scientist Ångström. One ångström is equal to 10^{-8} cm and visible light is found to lie within the range of 3850–7600Å, ultraviolet between 3850–100Å and X-rays less than 100Å. To assimilate as much information as possible about the object in view, it is often necessary to use a telescope for gathering light in a manner which provides a broader view of the subject.

Early in the 17th century it was discovered that a beam of sunlight could be split into its component colours by passing it through a fine slit and on to a prism **(Fig. 10)**. The spread of colours thus obtained contained dark lines, or absorption bands, called Fraunhofer lines after the German optician who first recognized the phenomenon. These lines are caused by chemical absorption of light in the relatively cool outer layers of the Sun and can be used to determine the nature and comparative abundance of elements found in that region.

The instrument developed to study the arrangement of the spectral lines is called the spectroscope **(Fig. 11)** of which certain types can be modified to record a photographic image of the spectra. Such spectrographs are used in conjunction with a telescope.

A prism spectrograph uses a reflector or a refractor to form a concentrated image of the Sun; the light passes through a narrow slit and a collimating lens which directs it into a parallel beam and through a prism. Glass can be used to form the lenses and prism for visible light, but it impedes the transmission of radiated light for wavelengths below about 3800Å. For observations through the ultraviolet portions of the spectrum it is necessary to use quartz which is transparent down to about 1800Å, and for wavelengths down to 1100Å lithium fluoride must be used. Above 7600Å (in infrared bands) rock salt prisms are necessary.

Fortunately, such a design of spectroscope has high resolving power with good separation between spectral lines and adequate definition of narrowly spaced wavelengths, but there is a limit to the resolution; enhanced resolution is achieved by increasing the size of the prism and this has certain practical limitations. To surmount this problem a second type of instrument, the diffraction-grating spectroscope is used.

Instead of permitting the light to fall on a prism, the beam is directed against a grating engraved with numerous parallel grooves on a polished metal surface which acts in the same way as a mirror. The light falling upon the grating is reflected back in the form of a spectrum and photographed in the usual way. For spectrographs from ultraviolet through to infrared regions the grating contains between 4000–12 000 lines per centimetre with resolution directly proportional to the number of lines used. Careful selection of incident radiation angles directed against the grating permits a wide range of wavelengths to be examined.

The most sophisticated use of a spectroscope is that adopted in the spectroheliograph, used for studying the absorption bands in the atmosphere of the Sun. In principle, the instrument focuses a beam of light through a high-dispersion spectroscope located at the prime focal plane of the telescope's objective, and then through a collimating lens. Instead of passing directly on to a photographic plate, the light first goes through a screen containing a narrow slit. The screen blocks most of the solar spectrum and only allows a single

Fig. 10 (below) A beam of sunlight can be separated into its component colours by passing it through a slit and directing the rays through a prism. **Fig. 11 (bottom)** A conventional spectroscope takes light directed through the lens of a telescope, places it through a slit and collimating lens to a prism which separates the colours and reveals the absorption bands, or Fraunhofer lines.

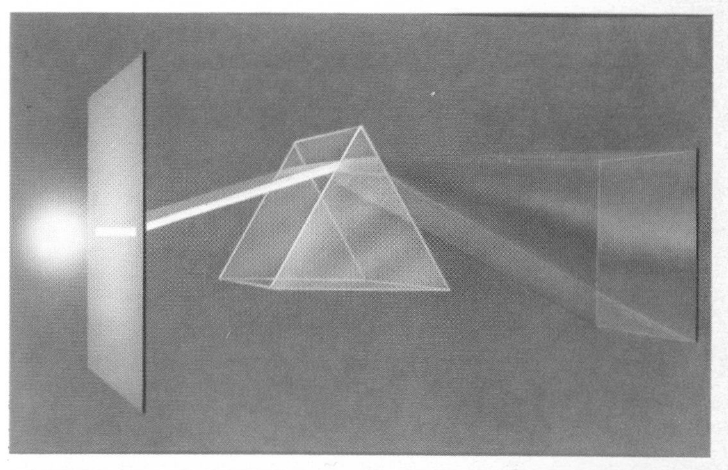

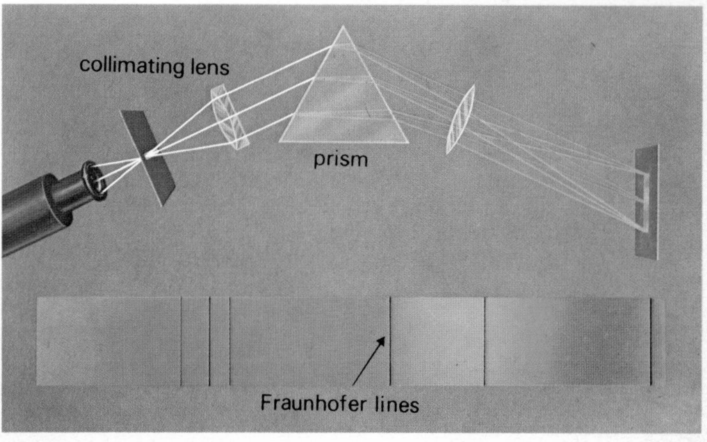

band to pass through.

Scanning the Sun in this way permits a record to be built up showing the solar disc in just one selected wavelength. These measurements enable specific solar features, such as flares and spots, to be analysed in greater detail than would be possible with a normal prism spectrograph. Several important discoveries have been made using this equipment. Although spectroscopic analysis of solar phenomena is the professional's prerogative, many amateurs are finding that the effort required to perform these measurements is well justified in view of the increased rewards. Complementary use of the spectroscope and telescope, therefore, broadens observation programmes by providing a useful adjunct to visual activities.

Telescope mountings

For all practical purposes the only types of mounting that need to be considered are the altazimuth and equatorial types, either of which can be set upon a portable tripod for small instruments or a fixed pedestal for larger telescopes. The altazimuth .(altitude-azimuth) mounting permits movement around vertical and horizontal axes. This mounting would be suitable for observation of objects moving directly up and down or from left to right, but the tilt of the Earth's axis creates the impression that stars and planets move in sweeping arcs across the sky, so motion has to take place around both axes at the same time.

To track a celestial object effectively, the observing instrument must be made to combine both vertical and horizontal motion which is made possible by fixing two axes at right angles to each other. The equatorial mounting shown in **Fig. 12** is such that the telescope can be aligned with one axis, the polar axis, pointed directly at the north celestial pole. This means that the angle of the polar axis to the horizon is the same as the latitude of the observer's site. A typical amateur mounting is shown in **Fig. 13**.

The second, or declination, axis is generally attached to the upper end of the polar axis. One single movement of the telescope is sufficient to track a star or planet on its arcing motion across the sky. Measurement of the telescope's field of view can be made by aligning with a star near the celestial equator so that when the telescope is held stationary the star will move across a central line bisecting the eyepiece. The time taken by the star (in seconds) to move fully across the field of view will, when multiplied by 15, give the diameter of the field of view in arc seconds.

Large telescopes

The application of the telescope to 20th century astronomy has led to a significant increase in the amount of information acquired by the scientist. Therefore, it is fitting to conclude this section with a brief look at a large observatory in use today and a preview of future trends.

Although the Hale Observatory at Mount Palomar, USA **(Fig. 15)** is no longer the largest telescope in the world, it has undoubtedly made a major contribution to astronomy over the past three decades since its mirror was completed in 1948. Set on a mountain 1·7 km above sea-level, the Hale reflector sits in a dome 41 m high and consists of a tube in a 'horseshoe' mounting, with a total weight of about 500 tonnes. The primary is 508 cm in diameter, 60 cm thick, and weighs 14·5 tonnes, while the central hole is 102 cm across.

Three observational modes are possible within the one instrument: when used at the prime focus (see page 6) the mirror has a focal length of 1676 cm; the central cut-out in the main mirror affords use as a Cassegrain with a focal length of 8138 cm; and in adopting the

Fig. 12 The equatorial mount provides two fixed axes at right angles with one axis aligned with the celestial north pole. The angle of the polar axis to the horizon is the same as the observer's latitude.

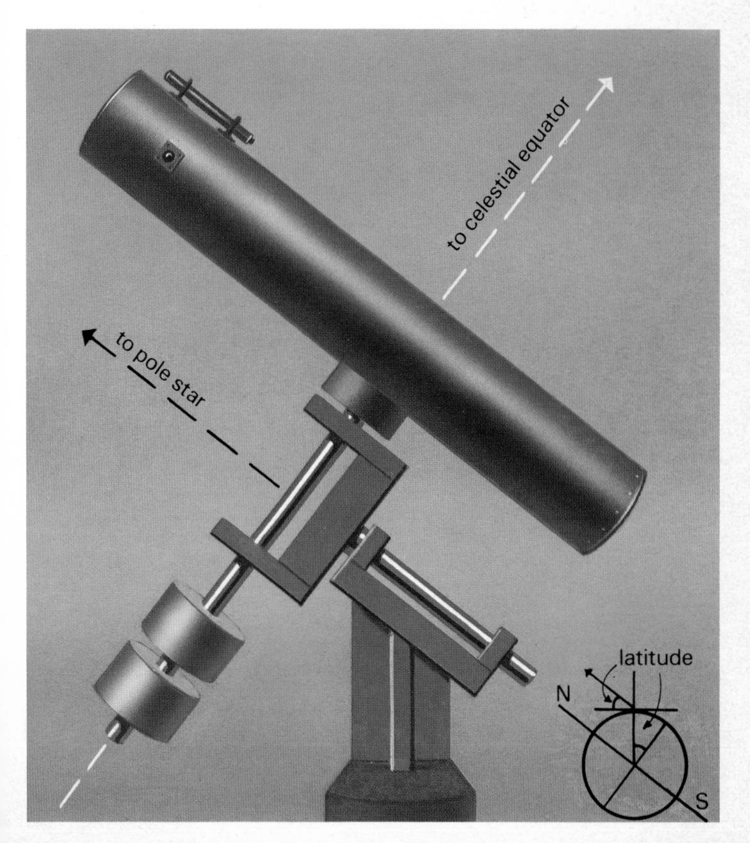

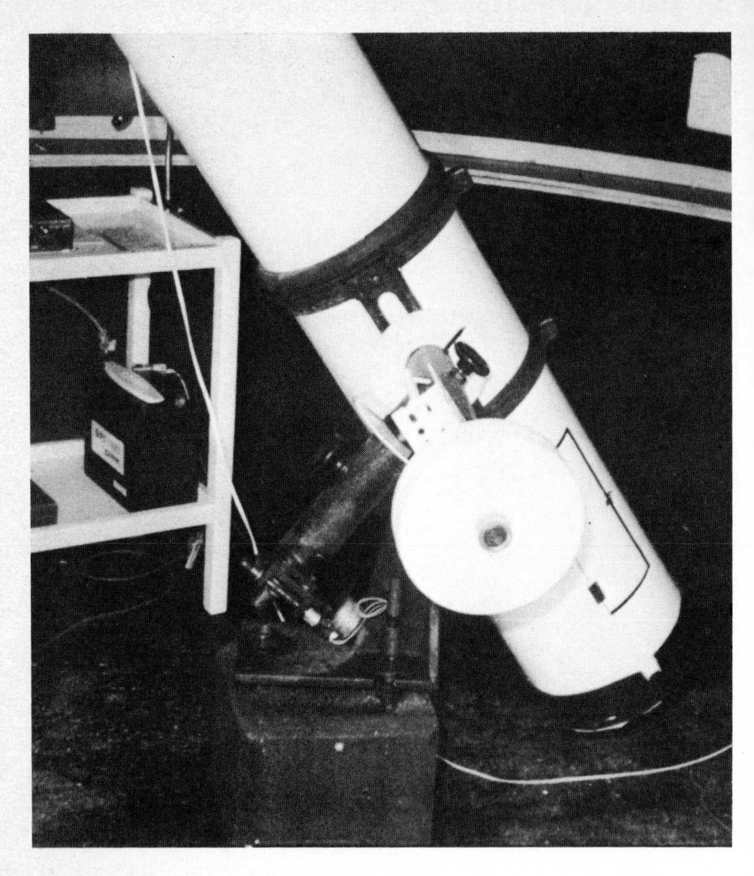

Fig. 13 (above) A typical mounting for an amateur telescope enthusiast. The 21-cm reflector is driven by an electric motor controlled by a variable frequency oscillator. **Fig. 14 (right)** The only safe way to view the sun is by using a white card upon which to display the image. It must never be viewed directly through an eyepiece. **Fig. 15 (following page left)** The Hale Observatory at Mount Palomar. **Fig. 16 (following page right)** The Hubble Observatory which is being launched by the Space Shuttle in late 1989. With its large mirror of 2·4 m diameter it should be able to detect objects about 4 magnitudes (roughly 40 times) fainter than any Earth-based telescope.

coudé principle, the mirror has an effective focal length of 15 240 cm.

The massive telescope is supported in a large yoke assembly which allows the tube to turn as far as the north pole, and a large motor drives the instrument on a system of bearings and oil pads.

Despite the remarkable performance of the Hale telescope, theoretical limits of refined observation are severely hampered by the disturbing effect of the Earth's atmosphere. Astronomers today place considerable emphasis on the increased efficiency gained from a telescope positioned in space beyond the degrading influence of the atmosphere. Towards that end the USA is putting a major optical observatory in orbit in late 1989.

The Hubble Observatory **(Fig. 16)** employs a main mirror 240 cm in diameter within a basic structure about 15·2 m long and 4·3 m in diameter. Weighing 6800 kg, the instrument will be powered by 'wings' of solar cells which convert sunlight into electricity and the whole assembly will be stabilized to a pointing accuracy of about 0·01" (1/360 000th of 1°). Operating at the very limit of theoretical performance the telescope will observe the sky across a spectral range of 1 mm–1150Å, covering infrared, visible light and a portion of the ultraviolet bands of the electromagnetic spectrum.

The disturbing effect of the Earth's atmosphere can most readily be understood by considering that the 508-cm Hale telescope on Mount Palomar is, at best, equivalent to only a 38-cm reflector in space, and for most of the year no better than a 10-cm space-borne instrument due to various atmospheric phenomena. As a result, the 240-cm Space Telescope will observe objects 50 times fainter across a frequency spectrum 10 times greater than that observed to date from the surface of the Earth. Capable of seeing seven times as far as the Hale instrument, the Space Telescope will explore a region of space about 350 times the volume so far observed.

a secondary mirror
b declination axis
c primary mirror
d Cassegrain focus
e skeleton tube
f 'horseshoe' mounting
g north pier

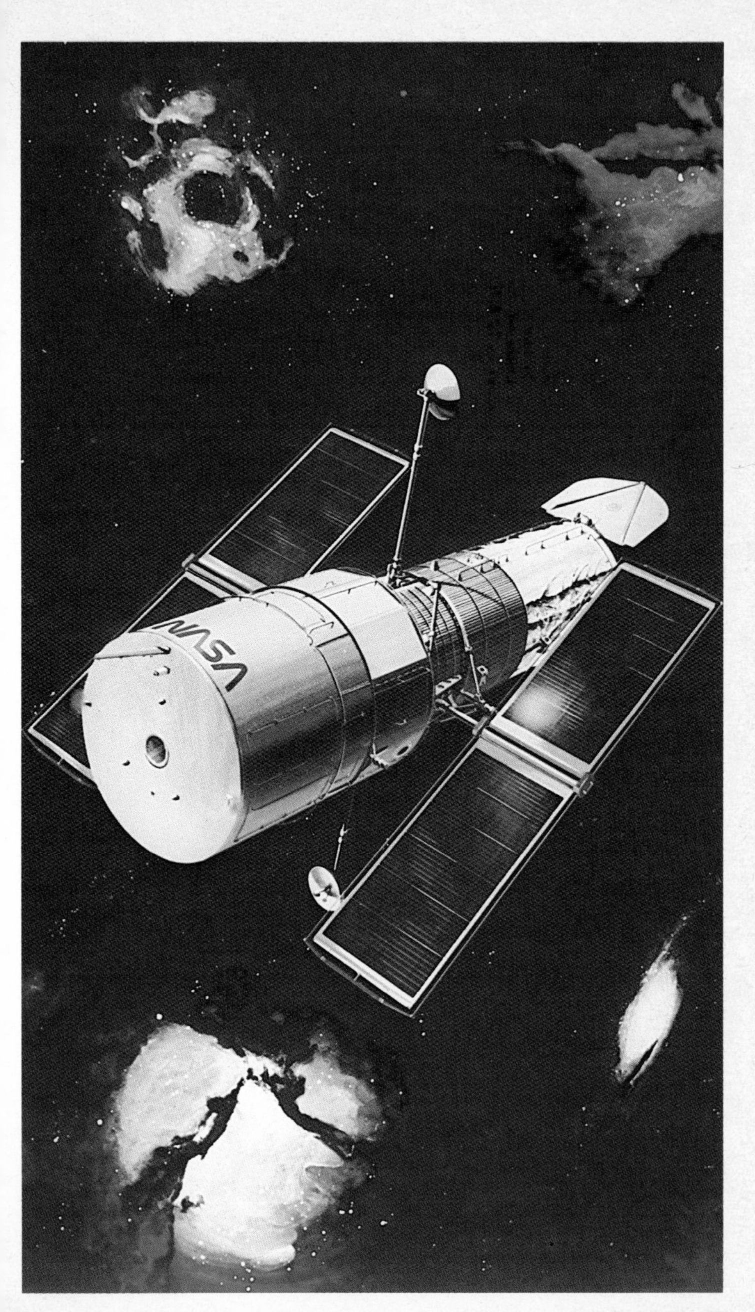
NASA

Stars

Stellar motion and distance measurement

Movement dominates the universe; planets move round the Sun, stars are observed to move across the sky and grouped patterns are seen to change with time. This perplexity of systematic motion delayed satisfactory models of planetary orbits for several centuries. It was under the aegis of men like Galileo and Kepler, who began to replace chaotic theory with scientific fact in the early part of the 17th century, that planetary and stellar motion came to be applied to many fundamental problems in astronomy.

Paramount was the question of stellar distance, a fundamental uncertainty that prevented the ancients establishing a baseline for their deliberations on the 'physics' of the universe. Two methods of distance measurement will be considered here and a third deferred to p. 30.

The first, called trigonometric parallax **(Fig. 17)**, uses the diameter of the Earth's orbit as a baseline from which to construct a triangle with the star under observation. Two measurements six months apart ensure maximum displacement of the star so that the angles of the triangle can be determined. The calculations envisage the Earth on a circle and the star at the centre with the radius determined by the distance between the two.

Calculation of stellar distances is simplified by the concept of radians. A full 360° = $2\pi r$ radians, r being the radius, so 1 radian (when the arc of the circumference of the circle equals the radius) is about 57·3°, 3 438 minutes, or 206 265 seconds of arc. For very small angles the length of the arc is regarded as directly proportional to the angle. In the case of stars, the arc, or Earth–Sun distance, is constant, and the distance Sun–star is inversely proportional to the angle. Dividing one radian by the measured parallax gives the Sun–star distance in AU. Thus if the parallax is 0·443″: 206 265/0·443 = 465 609. As 1 AU = 149×10^6 km, the star's distance equals $6{\cdot}97 \times 10^{13}$ km or about 7·36 light years, where one light year is the distance traversed at a speed of nearly 3×10^8 m/s (see page 17). Therefore the motion of the Earth in its orbit provides a basis for establishing the distances of the nearer stars.

Another form of movement has important repercussions for distance measurement **(Fig. 18)**. Although stars appear to be stationary,

Fig. 17 (above right) Seen against a distant stellar background, star S is observed to move due to the motion of the Earth in six months (E1 and E2). This provides sufficient information for the measurement of trigonometric parallax, or angular displacement. **Fig. 18 (right)** The actual motion of a star relative to the Earth is shown by line S–S1. Radial motion is the component of the actual motion away from, or towards, the Earth while proper motion is that observed to carry the star through the background field (S–X and S–Y respectively).

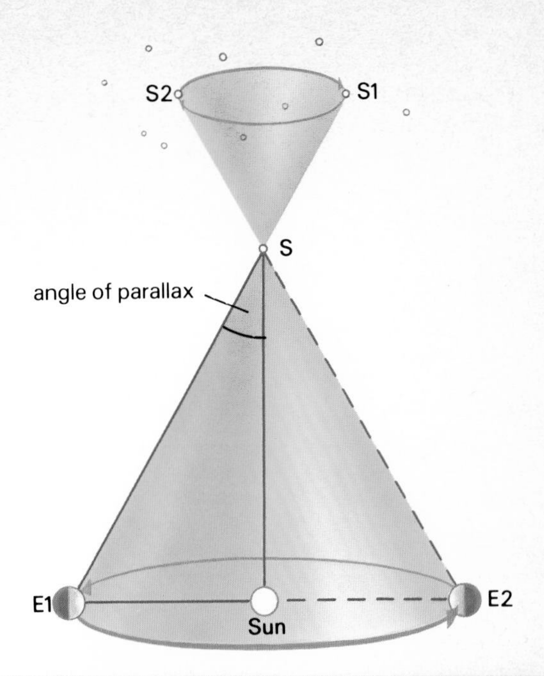
S2
S1
S
angle of parallax
E1
Sun
E2

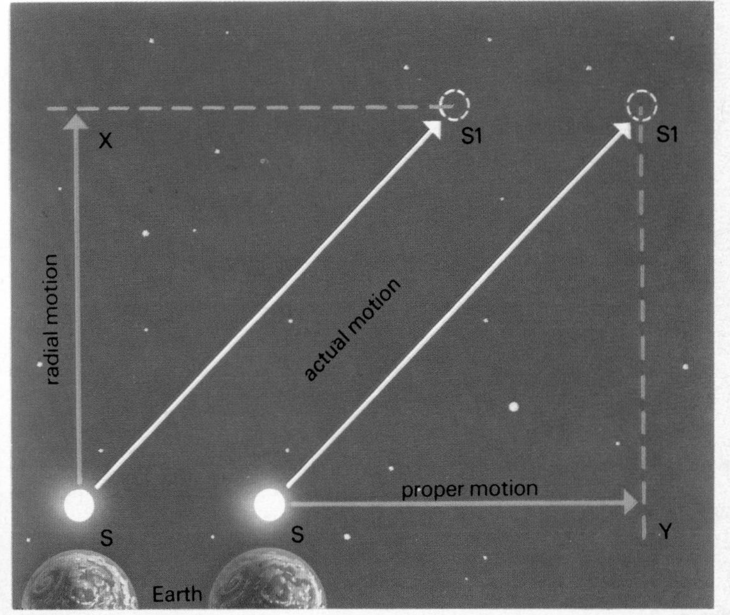
X
S1
S1
radial motion
actual motion
proper motion
S
S
Y
Earth

there is a slight angular displacement with time which may consist of radial motion, transverse motion or coupled motion. The transverse, or proper, motion is measured in units of arc per year but this, of course, is only the apparent speed with respect to the Earth.

Radial motion can only be determined by analysis of its spectrum (see p. 16) which exhibits a shift in the spectral lines proportional to the velocity of the star. If the star is approaching Earth, the waves will be compressed moving the spectral lines towards the blue end of the spectrum, and if it is receding, the waves will be extended moving the lines towards the red.

If it is possible to measure the distance between the Earth and a comparatively near star, and also obtain the object's radial and transverse velocities, the actual motion may be determined by combining these three pieces of information. This provides the true space velocity of the star with respect to the Earth.

Distances are often referred to in light years rather than kilometres due to the enormous values involved. However, using the principle of trigonometric parallax, a third unit of measurement is sometimes employed called the parsec. One parsec is equal to the distance set up by one parallax second of arc and is roughly 3·26 light years, less in fact than the distance to the nearest star (except the Sun).

Trigonometry can only be used for comparatively close stars and increased distance obviously means increased error factors in the calculations; the small angular shift in stellar parallax is only good for distances up to 150 light years from Earth. To understand the second, more widely available, method of measurement it is necessary to introduce an important factor inherent in the properties of light transmission, which is that light diminishes on the inverse square law. Therefore, two units must be incorporated to equate distance with the amount of light measured for any particular star – absolute magnitude, or luminosity, and apparent magnitude.

Absolute magnitude is the amount of light received from a star at a standard reference distance of 10 parsecs (32·6 light years) which means that a comparative chart of stellar luminosity is a true measure of the different levels of brightness exhibited by many stars at this standard distance. However, not all stars are at a fixed distance from Earth and careful measurement of the amount of light actually received (where the absolute magnitude is known) enables the star's distance to be calculated. Apart from correcting for the inverse square law, allowance may also have to be made for light absorption in dust clouds and other factors. This also applies to the alternative procedure, where the absolute magnitude may be calculated if the apparent magnitude is known and the distance may be determined from parallax measurements.

A most interesting group of stars, known as Cepheid variables, very greatly assist stellar distance measurements by relating absolute magnitudes to periodic variations in brightness. These stars fluctuate between fixed levels of maximum and minimum brightness in precise relation to absolute magnitude; the brighter the star the longer the period of oscillation. This means that if a group of Cepheid variables were all found to be lying at the same distance from Earth, careful measurement of the period would determine the absolute magnitude of each star. By comparing this known luminosity value with the light received on the Earth, the apparent magnitude, it would be possible to calculate how much light had been lost to the inverse square law factor. This would then reveal how far the light had travelled and, consequently, the distance of the Cepheid cluster.

The first variables of this type were observed in the constellation Cepheus, hence the name Cepheid for the group and the prefix Delta for the prototype. Many Cepheids have been observed in regions of our own and other galaxies and their accurate measurement has refined the astronomer's knowledge of distances in the universe. Leavitt first brought the Cepheids to fame in 1911 and for more than 40 years variables (later known as Population II Cepheids) in the nearest galaxy of comparable size to our own, Andromeda, were used to obtain a distance estimate of 800 000 light years. This changed to

Fig. 19 Cepheid variable stars have been observed to vary in brightness in precise relation to absolute magnitude. Period (of brightness) is shown against magnitude for types I and II.

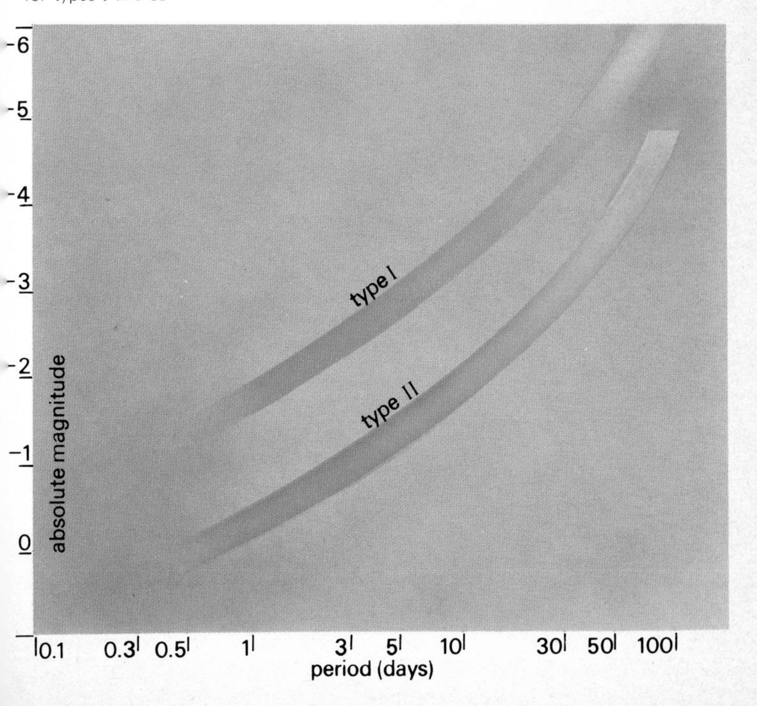

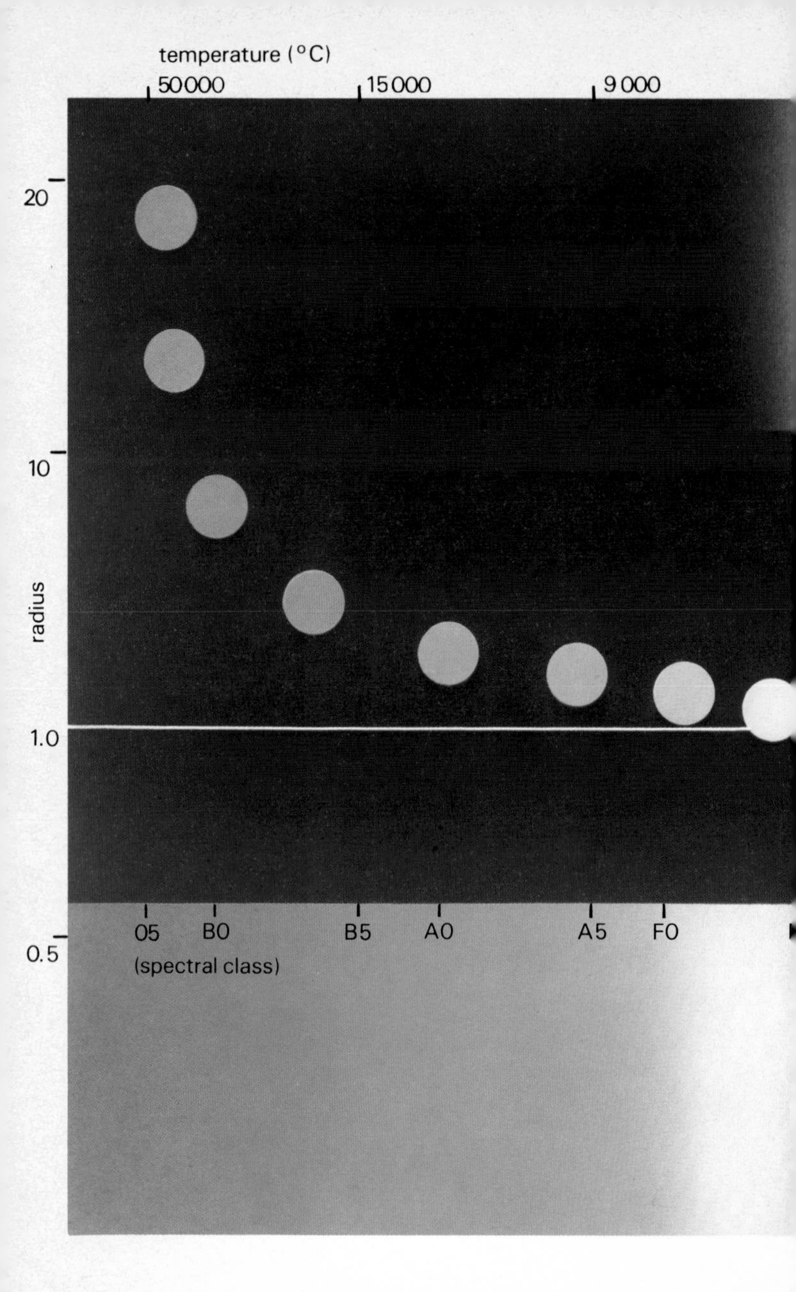

Fig. 20 Physical characteristics of stellar classes O–M provide identification of the properties inherent in different stars.

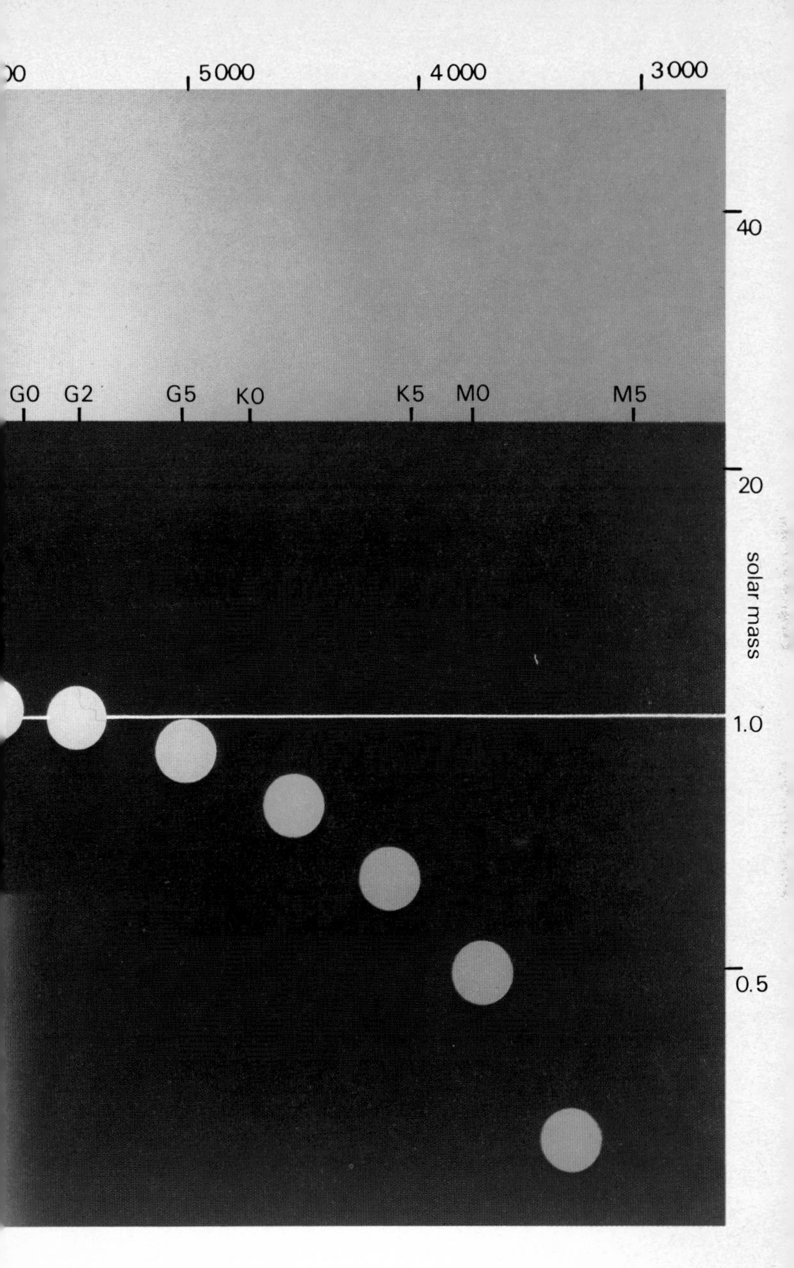

00
5000
4000
3000
40
G0
G2
G5
K0
K5
M0
M5
20
solar mass
1.0
0.5

2·2 million light years when a more luminous class (known as Population I Cepheids) was discovered by Baade in 1952.

Period-luminosity curves are shown for both types in **Fig. 19**. It had been assumed that the Cepheids measured in the Galaxy had the same period-luminosity relationship as those in far-off galaxies. In 1952, the Population I type were distinguished from the Population II group necessitating the increased distance estimate between the galaxy and Andromeda. This had a profound effect on restructuring the size of the universe. It is interesting to note that the first distance measurement of Cepheids, vital for setting up the scale of absolute magnitude, was derived from the observed proper motion of local stars of this type.

Stellar classes

All stars in the universe radiate light and other forms of energy as products of the enormous reactive forces at work in their centres, and therefore it is possible to observe the spectral characteristics and obtain useful information of their chemistry. This, and determination of distance and absolute magnitude, lead to a convenient system of stellar classification.

The characteristics determined by examination of stellar spectra can be used to group the stars into categories, and to arrange those groups (or classes) into a specific order. For convenience, the normal classification scheme is based on temperature, with cool stars at one end of the scale and hot stars at the other. It should be emphasized that there is a continuous range of temperatures, so the boundaries of the major individual classes are entirely arbitrary. When more characteristics are taken into account, the classification becomes very complex and sub-groups may include very small numbers of stars.

The major divisions, or groups, are allocated letter designations O, B, A, F, G, K and M, with subdivisions adopting a scale from 0 to 9 for most letter groups. For instance, F9 precedes G0; after this come G1 through to G9, and then K0. Beginning at O5, the sub-divisions give a range of 65 stellar types across the series of temperature classes, arranged in order of increasing spectral colour. Four other groups, W, R, N and S, are minor divisions with W preceding O type stars and R, N and S types following the M stars. These are side branches however, and can be ignored for the moment.

The arrangement of stars in order of spectral colour indicates that the very hot blue stars lie at the O end of the scale with cool red and orange-red stars at the M end. This method of classification provides other interesting scales of comparison and **Fig. 20** reveals the extremes of mass and radius in addition to surface temperature. For example, the Sun is a G2 and its mass, volume and luminosity are frequently used as units for measuring those of other stars.

Figure 21 displays the relationship between stellar luminosity,

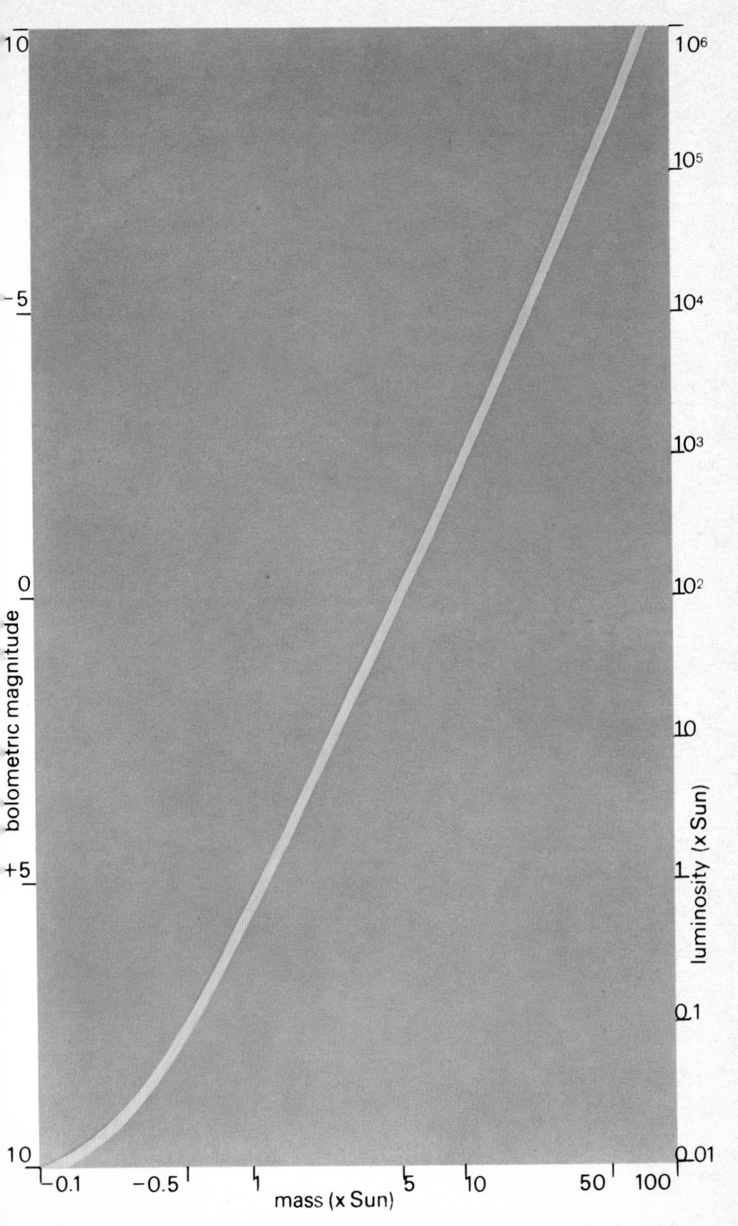

Fig. 21 The combined total of visual and non-visual forms of radiation (known as the bolometric magnitude) plotted against stellar mass.

masses, and bolometric magnitude. Bolometric magnitude measures the total energy emitted by a star, taken throughout the whole electromagnetic spectrum. Some classes of stars emit most of their radiation outside the visible region, so the use of bolometric magnitudes gives a better indication of their true luminosity.

There are three different but related definitions of magnitude: apparent, absolute and bolometric. The scale, or measure, of relating the magnitude of one star to another originated in Babylonian mathematics using six as the base system. It is from a preoccupation with this value (and the 360° circle) that time is measured in the familiar units of 60 (seconds into minutes, minutes into hours) and 24 (hours in a day), both being divisible by six. Thus, the Babylonians established six divisions between the brightest and faintest star visible to the naked eye and this has influenced the present system.

As first-magnitude stars were about 100 times as luminous as those of sixth magnitude, brightnesses were later expressed on a logarithmic scale, where each unit, or magnitude, is 2·512 times the brightness of the preceding value. The divisions, with the zero point at an arbitrary brightness level, indicate increasing (−) or decreasing (+) intensity. A star of first magnitude is exactly 100 times as bright as one of magnitude +6, and a star of magnitude −6 is 251·2 times as bright as a star of zero magnitude. The visual magnitude of the Sun is −26·8 and its absolute magnitude (luminosity at 32·6 light years) is +4·86.

Obviously, any star less than 32·6 light years from the Earth has an apparent magnitude greater than its absolute magnitude. However, the vast majority of stars lie at distances greater than the 32·6 light years used to measure absolute magnitude and, therefore, the reverse situation applies. Betelgeuse, at a distance of 520 light years, has a measured visual magnitude of +0·7 but its absolute magnitude is higher (because it is more than 32·6 light years distant) at −5·6, with a luminosity 13 000 times that of the Sun.

This accepted logarithmic scale is used for recording the apparent, absolute and bolometric magnitudes of all stars and it is common practice to delete the + sign for stars of decreasing brightness in stellar catalogues. For example, Procyon has an apparent magnitude of 0·37 and an absolute brightness of 2·7, whereas Rigel has apparent and absolute magnitudes of 0·15 and −7·1 respectively.

The Sun

Although the Sun is a comparatively ordinary star of G2 stellar class, the observer's unique position enables the varied phenomena associated with this massive thermonuclear reactor to be studied **(Fig. 22)**.

Fig. 22 (above right) Sunspots, faculae and prominences are important features of the Sun's 'weather'. **Fig. 23 (right)** The Sun is thought to comprise an inner core where thermonuclear fusion originates, a zone of radiation and an outer region of convective forces.

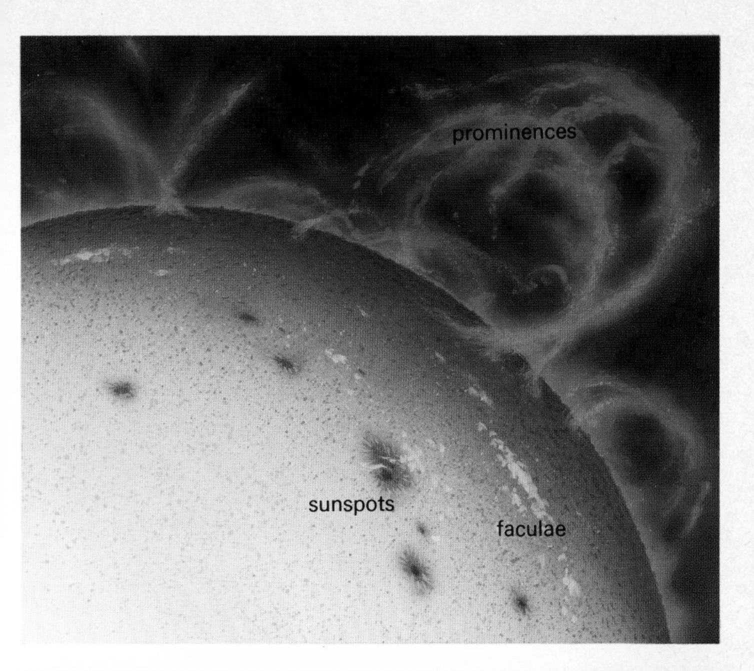
prominences
sunspots
faculae

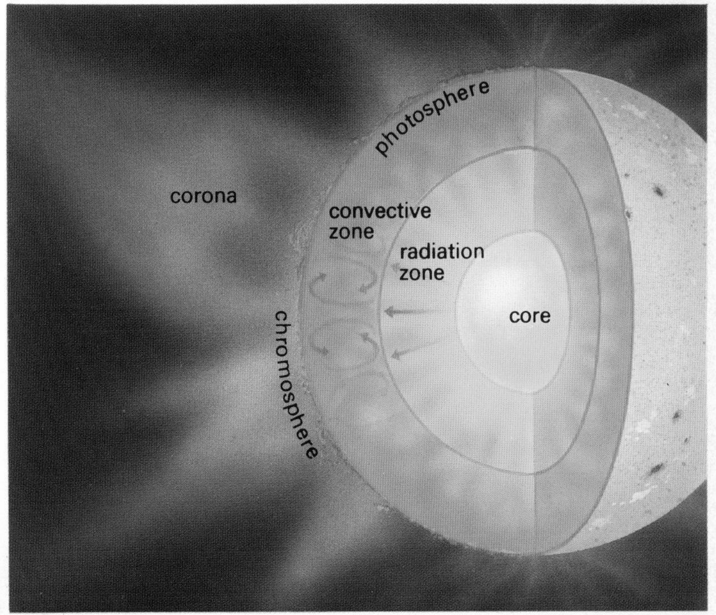
photosphere
corona
convective zone
radiation zone
core
chromosphere

The interior of the Sun is thought to comprise a core where temperatures reach $1 \cdot 4 \times 10^7$ °C radiating heat through a region extending to 470 000 km from the centre. From here convection carries thermal energy a distance of 225 000 km to the surface or photosphere which is less than 5000 km deep and exhibits temperatures ranging between 4500 and 5800 °C. Above this lies the chromosphere from 130 000 km above the surface with a temperature range increasing from 4500 to 50 000 °C. In this zone, the pressure and density rapidly decrease with increasing height and it is here that the spectral, or Fraunhofer, lines originate **(Fig. 23)**.

At the top of the chromosphere temperatures rise dramatically to meet the transition zone of the corona, a diffuse atmosphere of ionized gas where temperatures reach 3×10^6 °C. Therefore, the thermal inversion layer between the inner, convective, region of the Sun and the corona **(Fig. 24)** appears to contravene the second law of thermodynamics which states that heat will not continuously flow from cold regions to hot regions. It has been suggested that friction generated by convection is dissipated through the chromosphere to the corona and that the increased temperature is an energy dump unrelated to thermodynamic laws.

Although the Sun's corona is diffuse and invisible through the intense light radiated from the surface, it can be studied during an eclipse or from a satellite that provides an artificial occultation by placing a disc over the image of the photosphere.

The corona extends several million kilometres from the Sun and sends a stream of particles, protons and electrons into the solar system at 300–600 km/s. This solar wind impinges on planetary bodies or sweeps round the magnetic fields of Earth and Jupiter, the only planets known to have radiation belts locked into intense magnetic fields. The force of the solar wind fluctuates according to the magnitude of emission rates, while the rotation of the Sun creates a spiral effect in the released particles. The wind is thought to extend some 5×10^9 km into the solar system where the heliosphere is no longer of sufficient strength to resist incoming particles from other star sources.

The rotation of the Sun is typical of a fluid body where the equatorial region rotates faster than zones in higher latitudes. Measurement of surface features indicates an equatorial rotation period of 25 days.

Sustained observation of the Sun's behaviour is necessary since the long-term fluctuations are expected to change the balance of temperatures in the solar system; **Fig. 25** displays the gradual warming effect and its repercussions on Earth. The Sun exhibits several repetitive cycles and the most familiar, spanning 11 years, brings increased

Fig. 24 (above right) The temperature of the Sun from the convective zone to the corona, showing the prominent thermal inversion layer between the chromosphere and the corona. **Fig. 25 (right)** The temperature curve on this chart represents one school of thought on the changing thermal environment of the Earth from the formation of the solar system to a period 1000 million years into the future.

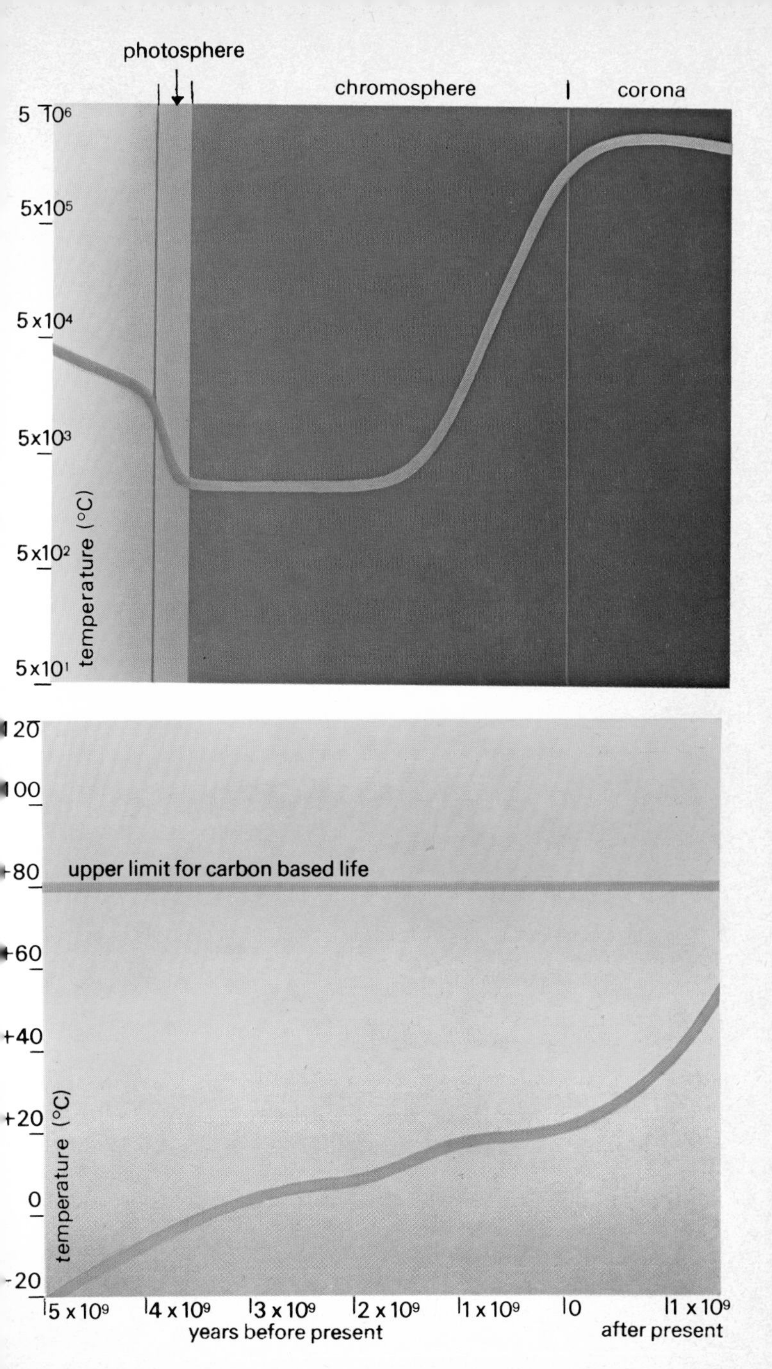
photosphere
chromosphere
corona
5 10⁶
5x10⁵
5x10⁴
5x10³
5x10²
5x10¹
temperature (°C)
120
100
+80
upper limit for carbon based life
+60
+40
+20
0
-20
temperature (°C)
5 x 10⁹
4 x 10⁹
3 x 10⁹
2 x 10⁹
1 x 10⁹
0
1 x 10⁹
years before present
after present

sunspot activity and develops increasingly intense flares which interact with the Earth's magnetic field to perturb the transmission and reception of radio waves. Other, longer-period cycles are suspected which may have influenced the ice-ages on Earth, but the trend is towards an increasingly hot Sun.

The Hertzsprung-Russell diagram

At present there is no possibility of studying other stars in the same detailed manner as that which can be applied to the Sun. Generally, in investigating stars, only major characteristics can be inferred from observed data, and here a Hertzsprung–Russell diagram can be invaluable. Generated by the work of the Dane, Hertzsprung and the American, Russell, in 1912 and 1914 respectively, the Hertzsprung-Russell (H-R) diagram as in **Fig. 26** plots absolute magnitude against stellar class.

The classes discussed on p. 32 are placed along the bottom with the log scale of varying magnitude on the left and the luminosity on the right (where $L_\odot$ equals the luminosity of the Sun). It will be seen that the stars fall on a diagonal band from the upper left to the lower right of the chart on a slope with a variation equal to the temperature at a power of 5·5. This indicates that the radius of a star increases with temperature and satisfactorily endorses the situation presented in **Fig. 20**.

The H-R diagram can be used to determine both distance and mass where efforts fail to secure these parameters by direct observation. If the spectral class (or colour index) of a star can be determined, which will also define the temperature, this will place the star at a single location on the H-R diagram, and thus indicate its true absolute magnitude and luminosity. By comparing absolute and apparent magnitudes the star's distance can be derived by means of the inverse square law. This does, of course, assume that the star in question lies on the band known as the main sequence, but this can also be deduced from its spectrum.

It has been observed that the luminosity of a star varies between the third and fourth power of the mass; where mass is unknown, the stellar spectrum will provide not only the category but also the mass, radius and surface temperature. The mass/luminosity relationship can be applied to the problem of stellar longevity, and derivations of the life-time calculated for each type of star are important considerations for many astronomical questions.

If M is the mass of a star and L is the luminosity, M/L is the duration in time taken for consumption of a certain amount of material in the star. Since luminosity is proportional to M^3 or M^4 (with time varying as $1/M^2$ or $1/M^3$) it will be seen that the more mass a star has the more rapid will be its consumption to produce the additional energy. In fact it is just this very aspect of differential life-time that has provided such a variety of stellar forms, as we shall see later.

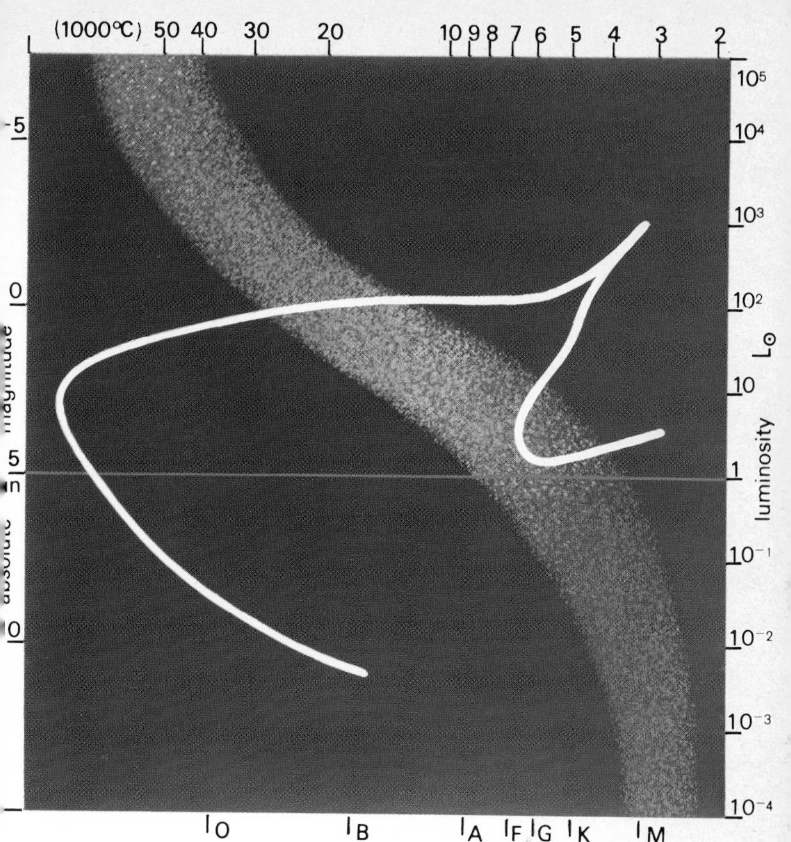

Fig. 26 The H–R diagram plots absolute magnitude against stellar class and shows the evolutionary cycle for stars of solar mass.

The old theories concerning the H-R diagram envisaged a star forming from a nebulous cloud of gas and dust at the upper right, moving slowly across to the left as temperatures rose due to contraction and slowly down to the main sequence as the radius decreased. It was presumed that a star continually contracted down through the measured radii for all classes from O to M until it endured a cooling process at the bottom right, diminishing in size and temperature.

It is now known that the intense temperature and pressure at the centre of a star produce thermonuclear reactions whereby hydrogen is converted to helium **(Fig. 27)**. This cycle is known as the proton-proton reaction and applies to stars approximately the same mass as the Sun. For reasons which will be explained subsequently, stars up to twice the solar mass end their life as white dwarfs, black dwarfs or neutron stars, but in the overall evolution from initial cloud to white dwarf the star spends the majority of its time on the main sequence. Once established in its appropriate stellar class it only moves away during its final evolutionary stages.

The star condenses down to a position on the main sequence determined by the initial mass of the cloud, and when nuclear fusion reaches a critical level the star expands to a red giant and moves to the upper right of the chart. Finally, it condenses down to a white dwarf, following a track to the left and bottom centre of the H-R diagram.

Stars more massive than the Sun follow a different track **(Fig. 28)**. Taking the example of a star of about 15 solar masses, a B type star, it is apparent that contraction will be appreciably more rapid due to the larger mass of the initial cloud and stars of this type will join the main sequence in about 1/200th of the time required for a G2 type star. From its main sequence position, this massive star will convert hydrogen to helium on a carbon-nitrogen cycle at a faster rate than permitted by the proton-proton cycle of cooler stars.

It is believed that a star will move off the main sequence when 10 per cent of its hydrogen is converted to helium which for solar, G2, stars means about 10^{10} years of stable life. Obtaining a value of the mass (M) and the luminosity (L) of a star will enable calculations to

Fig. 27 Two protons collide to form deuterium which then combines with another proton to form a light isotope of helium. In 86 per cent of cases, two light isotopes of helium combine to form stable helium and eject two protons. (Orange = proton; purple = neutron.)

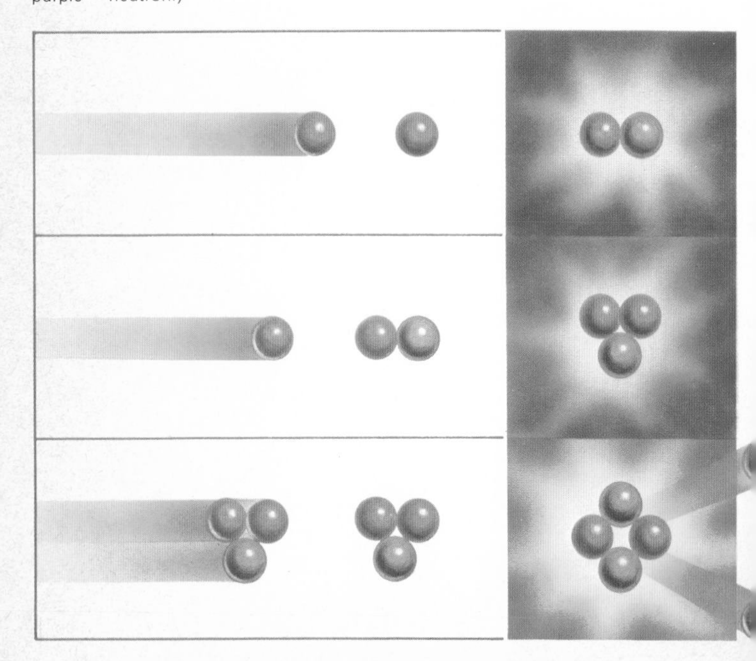

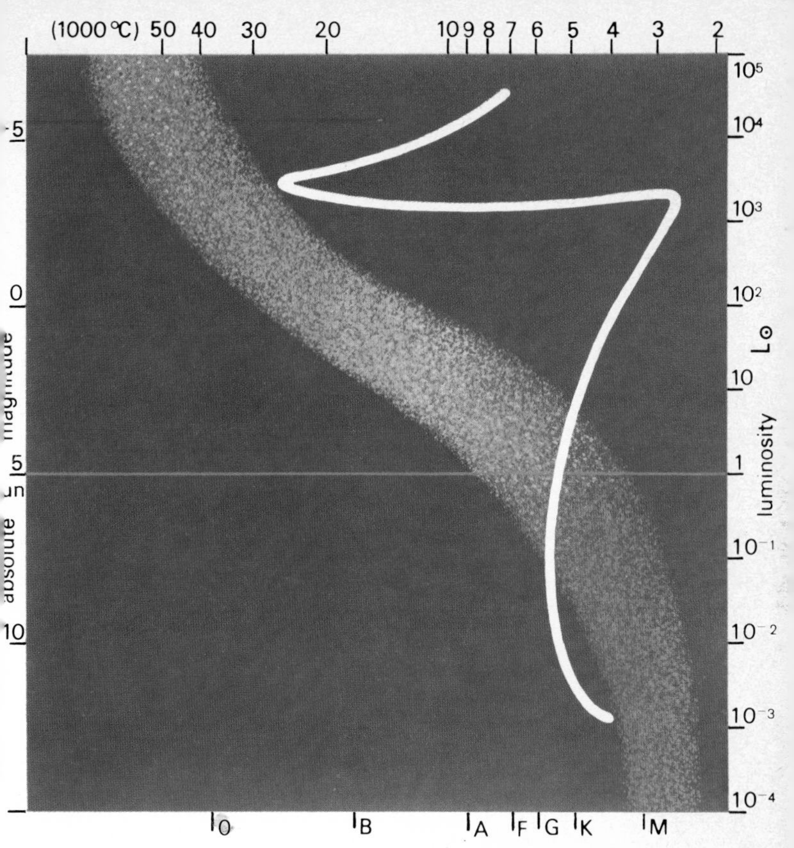

Fig. 28 Stars of greater mass than the Sun follow a different evolutionary path to that characterized by stars of solar mass. The track of a star of 15 solar masses is plotted against the H–R diagram (see **Fig. 26** for comparison).

determine its stable life. For instance, M/L determines the ratio of mass conversion to energy compared with the Sun and since the main sequence time of the Sun is estimated to be 10^{10} years, $(M/L) \times 10^{10}$ years will provide the stable life duration of the star in question.

For a star of 17 solar masses and a luminosity 13 000 times that of the Sun, $(17/13\,000) \times 10^{10}$ displays a predicted stable life of $1{\cdot}3 \times 10^{7}$ years. This formula indicates the maximum possible main sequence time and should be reduced somewhat to allow for late-stage perturbations observed in the majority of massive stars. Stars in the B and A categories will have a stable life of between $10^{7} - 2 \times 10^{9}$ years and at the end of this period they follow the track shown in **Fig. 28**. The star rapidly moves to the right and joins the giant branch before collapsing to a neutron star or collapsar (black hole).

The carbon-nitrogen cycle **(Fig. 29)** applies to stars with a core

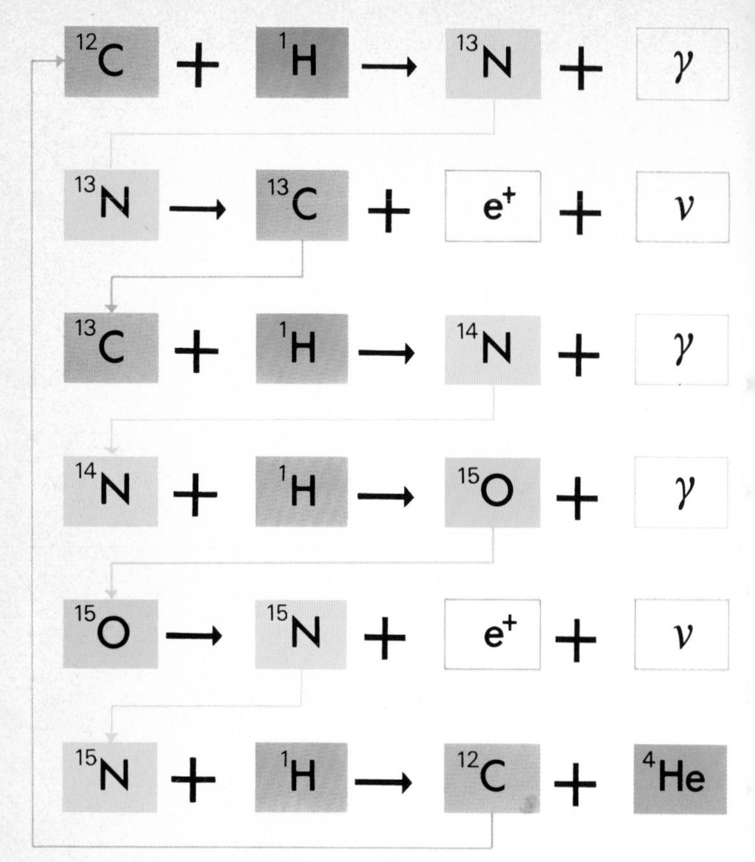

Fig. 29 Very massive stars fuse hydrogen to helium by the carbon–nitrogen cycle. Positrons (e^+) and neutrinos (ν) are produced at the second and fifth stages, while photons (γ) are liberated in the first, third and fourth.

temperature exceeding 2×10^7 °C, with the proton-proton cycle dominant in stars where core temperatures are below this value. All material in the known universe is built from the building blocks of atoms which link to form elements and, in turn, molecules. The reactive processes at work in stars, which provide the energy release necessary to keep the mass inflated and in a state of equilibrium with gravity, are the only means by which most of the elements can be manufactured and re-distributed for subsequent condensation into secondary stars and planets. In fact, the materials from which the planets and even human beings are made were processed in the core reactions of a first generation star.

Figure 30 shows that the neutron, the proton and the electron form a base from which all other materials are made. The process is active for proton-proton fusion when temperatures approach 10^7 °C, and because of this the line between massive cloud condensations and the most diffuse star is difficult to find.

Jupiter has similar relative abundances of hydrogen and helium to the Sun itself. However, its core temperature of 5×10^4 °C at formation has cooled to 3×10^4 °C, which in either case is clearly too low a value to trigger nuclear fusion. Since it has only 1·4 per cent of the mass necessary to raise gravitationally the temperature and pressure to fusion point, it is destined to remain a non-luminous body. How-

Fig. 30 All elements and molecules are built from protons, neutrons and electrons, seen here in comparative size scales.

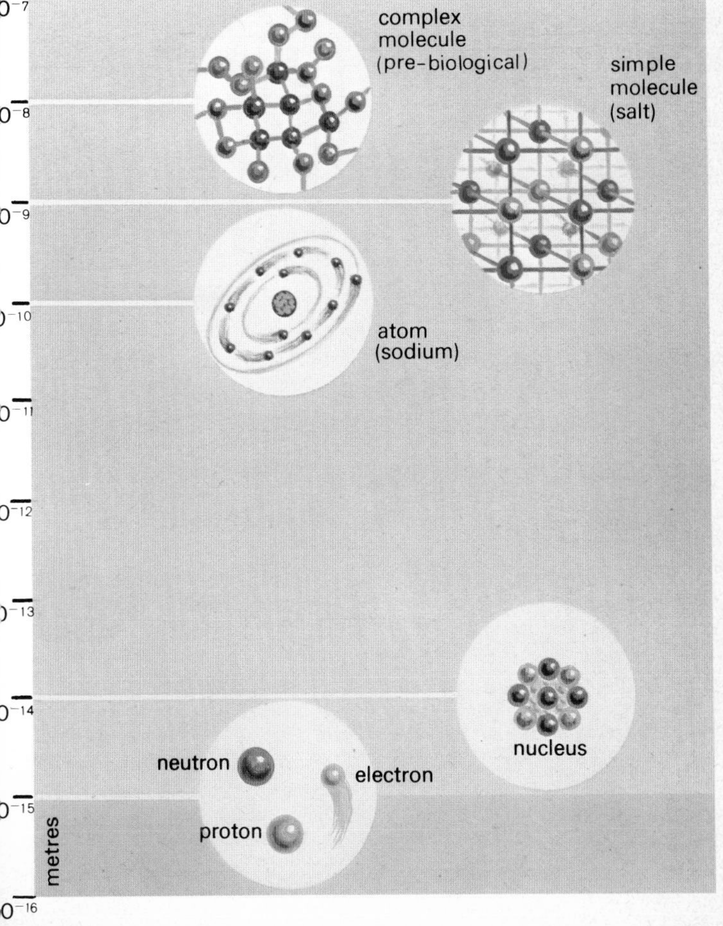

ever, there may be many cases where condensed clouds of gas are close to becoming stars in their own right.

Concerning the relative abundance of elements denser than hydrogen and helium account must be taken of the limitations imposed by the energy yield profile. Here, if the binding energy per nucleon is plotted against mass number (the total mass of protons and neutrons in the nucleus) it is found that for elements up to Fe^{56} (iron, with 56 particles in the nucleus) energy is liberated by the fusion of nuclei and this is the source of all stellar energy.

At the lowest fusion temperature (around 3×10^6 °C) hydrogen (protons) is formed into helium on the proton-proton cycle, while at about 18×10^6 °C the carbon–nitrogen cycle predominates. At higher temperatures in massive stars other processes take over building heavier elements. These may themselves be burnt in yet further reactions, still liberating energy, building elements up to iron. However, beyond iron energy must be added to the reactions in order to obtain the heaviest elements. Furthermore large quantities of neutrons must be available to stand a chance of building up the heavy elements before fission decay predominates. The known presence of elements very much heavier than iron confirms the existence of a mechanism where abundant neutrons can fuse with existing nuclei and also which subsequently disperses them into space, later to be incorporated into other bodies.

Formation of heavy elements

Stars can build nuclei with 56 particles in the fusion reactions during main sequence life, a long way from known elements containing more than 200 particles. The heavier elements are thought to form when a massive star explodes, and in this event the prolific abundance of neutrons would permit the elements with a large mass number to develop and become distributed within the inter-stellar medium.

Such an event is called a supernova. The star manufactures the elements from which secondary proto-stars accrete their compositional material so that dense terrestrial-type planets can form. It is worth reflecting for a moment on the fragility of events which leads to the chain of circumstances that supports the tenuous platform of biological evolution. The condensations that form new stars *must* contain a certain percentage of heavier elements since without them only gaseous, Jupiter-like, bodies will form. In this way material is continually re-cycled, and careful observation of the frequency rate of supernovae in our galaxy indicates that less than 2 per cent of the galaxy has been moved through the stellar process.

Fig. 31 In a scale of comparative abundance, hydrogen is seen to be the most common element in the universe.

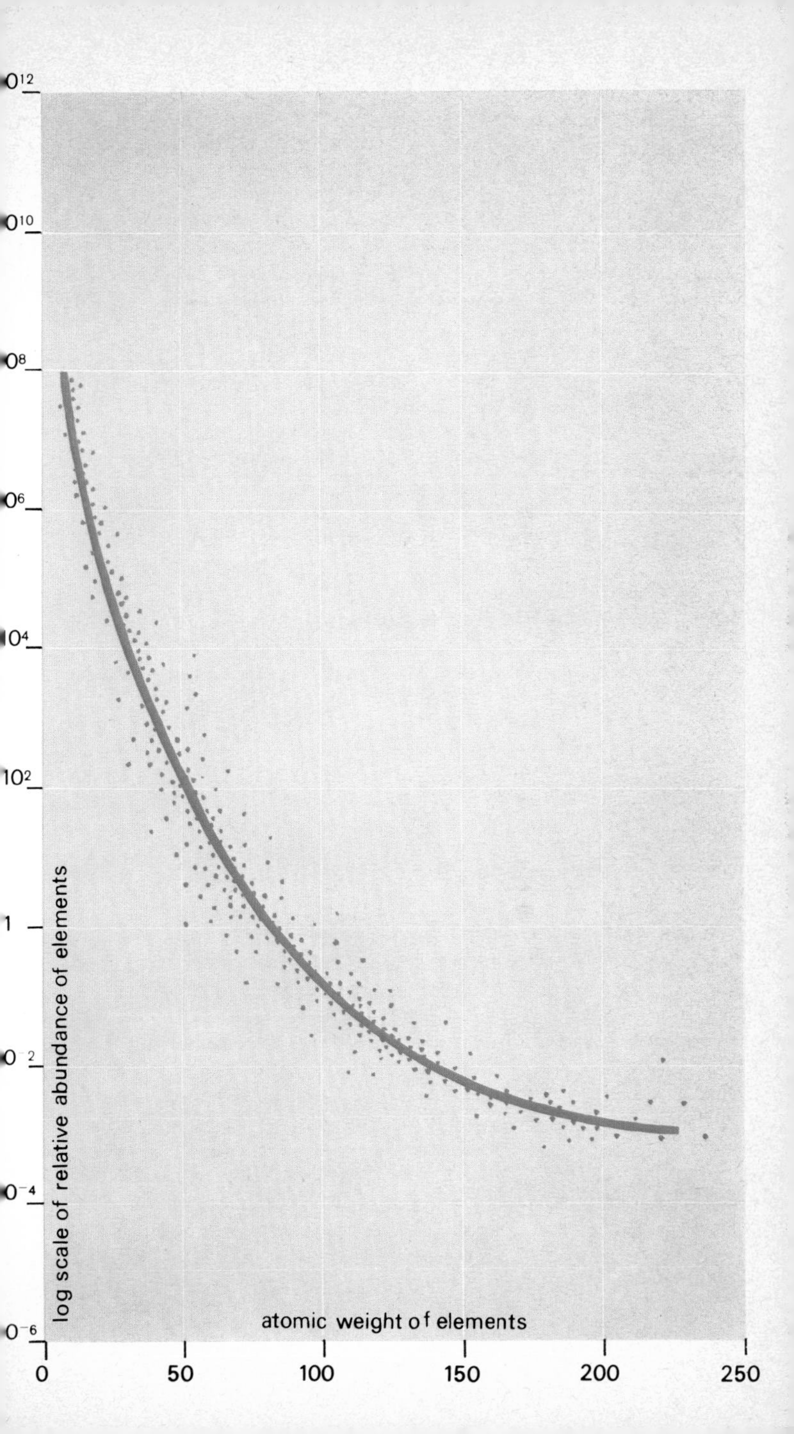
log scale of relative abundance of elements
atomic weight of elements
0¹²
0¹⁰
0⁸
0⁶
10⁴
10²
1
0⁻²
0⁻⁴
0⁻⁶
0
50
100
150
200
250

This is in agreement with the theory of elementary abundance at the beginning of the universe and the relative quantities of heavier material can be accounted for by the measured ratios present today. The galaxy, containing stars totalling about 10^{11} solar masses, generates nearly 6×10^{33} kW in radiated energy from the conversion of hydrogen to helium at the rate of 10^{25} g/s. The oldest stars so far dated in the galaxy are about 10^{10} years old and from that time to the present the current measured output should have remained constant.

The total mass of the galaxy (3×10^{44} g) divided by the mass of hydrogen converted to helium in 10^{10} years (3×10^{42} g) confirms that about 1 per cent of the galactic mass has moved through the fusion cycle. However, the relative abundance of helium, the most simple atom formed from hydrogen nuclei, is too high to allow it all to have been manufactured in stellar interiors **(Fig. 31)**.

There is some evidence to suggest that about one-third of the galaxy's mass is helium, and since the galaxy is thought to have converted 3×10^{42} g of hydrogen into helium since its formation 10^{10} years ago, the calculated abundance of helium exceeds the theoretical value by 30:1. This may indicate a higher abundance of helium at the beginning of the universe with hydrogen not playing the dominant role once proposed, or that galactic energy output in the past has been considerably more prolific than at present. It should be emphasized that the accepted method of energy production in stars may be only approximately correct, and that the theories of stellar fusion processes may require very considerable revision. For example, there is a very large discrepancy between the calculated rate at which the Sun should produce neutrinos – massless particles which are released in certain individual fusion reactions – and that which has been observed. For this to be explained it may be necessary to alter many theories about stellar interiors.

Multiple star systems and novae

Stars have so far provided an insight to the basic processes at work in the universe by presenting a mechanism for stellar re-incarnation and the manufacture and distribution of heavy elements. They can also provide information on the early history of the galaxy.

During the 1940s, Baade collected spectral details of many stars in the nearby Andromeda galaxy. **Figure 32** displays the result of this work. The stars are clearly arranged in two populations: one runs down the main sequence, the other remains below a centre line on the H-R diagram. Baade suggested that these two populations represented collective groups of stars marking distinct phases of galactic evolution. The Population I groups were predominantly blue and occupied positions on the arms and within the disc of the galaxy while Population II types were older, giant red stars lying at the centre of the galaxy and in clusters high above the plane of the arms.

This reveals the early phases of galactic evolution (see below).

Fig. 32 A temperature/luminosity diagram of stars in the Andromeda galaxy indicates two distinct population types: type I, or second generation stars, and type II, the early components of the galaxy.

When the galaxy contracted down from a much larger cloud of constituent matter which formed stellar clusters, stars emerged occupying a much larger orbital radius to the centre of the cloud than is represented today by the extent of the spiral arms. Population I stars (second generation stars like the Sun) appeared later, with older Population II stars occupying a halo around the flattened galaxy.

The sharp line of demarcation between Populations I and II is now absorbed within a new five-tier structure. Extreme Population II stars represent the oldest group and are to be found in dense clusters forming a spherical halo encompassing the galaxy. The intermediate Population II stars are slightly younger, although still 75 per cent as old as the galaxy, found in a slightly flattened halo with velocities which are not as high as the extreme Population II types. This means that their orbits round the galaxy, although highly inclined to the plane of the arms, are more elliptical.

The vast majority of stars in the galaxy are of disc population, occupying positions in loose clusters within the dust-free regions between the galactic arms. They appear to have been formed from the debris, albeit essential for forming Earth-like planets, ejected from first generation stars and consist of stars no older than the Sun ($\sim 5 \times 10^9$ years) and frequently younger. They contain a high relative abundance of elements heavier than helium and this signifies an origin ensconsed in the swirling clouds of supernovae residue.

Younger stars belong to the intermediate Population I type and are characterized by open clusters in close proximity to the central plane of the galaxy. Ranging from about $10^8 - 2 \times 10^9$ years of age they are in an early phase of main sequence life as expressed by the H-R diagram. Youngest of all are the emerging stars of extreme Population I type, condensations which have emerged as stars in their own right within the past 10^8 years. Much of the condensing dust and gas in the spiral arms of galaxies belongs to this group.

This five-tiered sequence of stellar population levels has enabled observation of the unique phases of galactic evolution. The oldest, extreme and intermediate Population II types, are more commonly called globular clusters, and each contains up to 10^5 stars in tightly bound colonies far above the galactic equatorial plane. Open clusters of more youthful stars were formed when the galaxy had reached its present, flattened profile and are more susceptible to the gravitational perturbations of the central regions of the galaxy. Consequently, unlike globular clusters, they are more easily separated and are less easy to define since escaping stars of one cluster mingle with wanderers from neighbouring clusters.

It is quite common to find two stars which are gravitationally bound and rotating about their common centre of mass. There are also many multiple systems consisting of three or more components. Such systems may arise from splitting of late-stage stellar condensations, or from a group of stars formed within the same dust and gas cloud. The Sun is rather unusual in being unattended, although it is a member of a group of local stars, which formed from the same cloud. This may be of considerable significance to the possible methods by which the planetary bodies could have been formed (see page 146).

Figure 33 shows the relative motion of component stars in a binary system with a large massive star close to the centre of mass of the coupled unit and a smaller, less massive, star orbiting further out. The barycentre is found by drawing a straight line between the component stars and defining the centre of mass, always at some point on the line, from the respective intrinsic masses of the two bodies.

It may be that the two components subtend so small an arc of triangulation from Earth that it is impossible to separate them visually. In this case, a spectrographic record of the dual system reveals radial motion towards (blue shift) or away from (red shift) the observer for each star, thus enabling the different velocities to be established. Since the barycentre is adequately defined from the relative motion it is possible to determine the mass ratio from the measured velocity of the two components.

Where M_1 and M_2 equals mass and A_1 and A_2 equals distance of the two components from the barycentre, the relation $M_1A_1 = M_2A_2$ is fixed. Taking a hypothetical example of a binary system with components 20·5 AU apart (1 AU = Earth–Sun distance, or $1{\cdot}49 \times 10^8$ km) and an orbital period of 50 years it is necessary to use the equation:

$$M_1 + M_2 = \frac{a^3}{p^2}$$

where a is the distance separating the two bodies in AU and p is the

Fig. 33 The relative motion of two stars in a binary system. CG marks the barycentre. When stars are at A2/B2 and A4/B4 there is no separation in the spectral lines, because transverse motion is not appreciably changing the relative distance of each component to the Earth; when stars are at A1/B1 and A3/B3 the lines will separate.

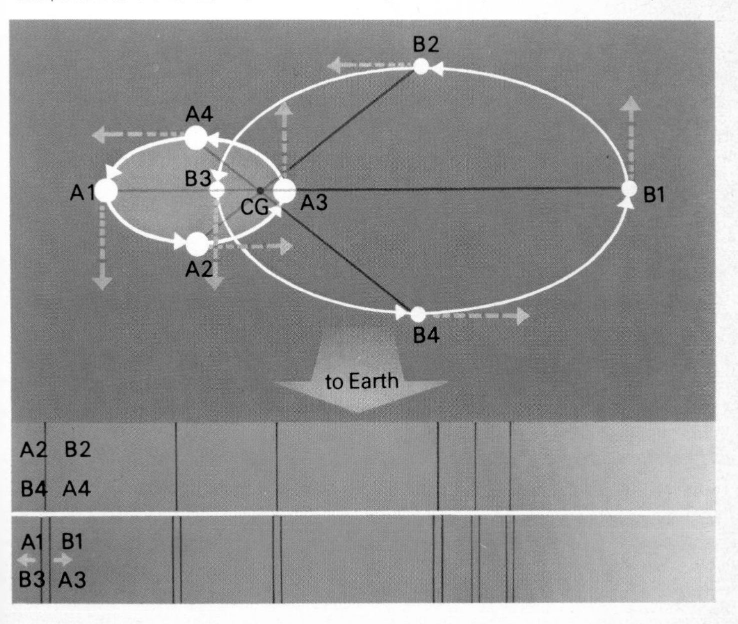

period of each component. Using the values known for the system it is now possible to derive the equation:

$$M_1 + M_2 = \frac{20 \cdot 5^3}{50^2} = 3 \cdot 4 \text{ solar masses.}$$

From the inverse ratio of the respective distance of the two components from the barycentre it can be seen that $M_1 = 2 \cdot 4$ solar masses and $M_2 = 1$ solar mass.

In reality, the requirement that the line of sight from Earth lies in the orbital plane of the two components is not wholly met and it is essential to derive the angle of orbital inclination with respect to the Earth to obtain a compensating factor and thereby establish the values necessary to derive the above equation. But, it is possible to confirm a minimal angle of inclination by seeking eclipsing binaries.

In this situation, where the line of sight is virtually in the plane of the orbits about the barycentre, the two components of unequal luminosity will cause a double decline in the observed magnitude of the total system. In **Fig. 34**, the faint star passes in front of the bright component on the Earth side of its orbit and the bright star passes in front of the faint star as the bodies rotate.

It is comparatively rare for binaries to be exactly aligned with the line of sight, and even if they are the amount of the light changes during the eclipses depends greatly upon the luminosity of the two components. Sometimes one component is very dim and no secondary (lesser) eclipse is seen when the bright star passes in front of the fainter member. However, it is possible to detect the presence of a companion by careful measurement of the perturbed orbit of the brighter, visible component since the dim star will still possess sufficient mass to exert a reasonable influence on the system **(Fig. 35)**.

Otherwise invisible components may make their presence known in other, more dramatic ways. Binary systems exist in which one star is losing material, which falls towards the other, white dwarf, companion. A disk of gas forms around this smaller star, and in some systems, known as dwarf novae, periodic brightenings of this disk occur. However a much more dramatic event happens in novae, where the gas accumulates on the surface of the white dwarf, building up for hundreds or thousands of years until its temperature rises high enough for hydrogen fusion to begin in the layer. The sudden outburst of energy lifts the outer envelope from the star and blows it out into space. The amount of material transferred and lost in an outburst is small, perhaps 10^{-3} solar masses, but the increase in luminosity is usually around 11 magnitudes. The stages of this process are shown in **Fig. 36**.

Until recently it was thought that novae resulted from the normal evolution of single stars. However it is now generally accepted that they are all binary systems and that the basic process which has been

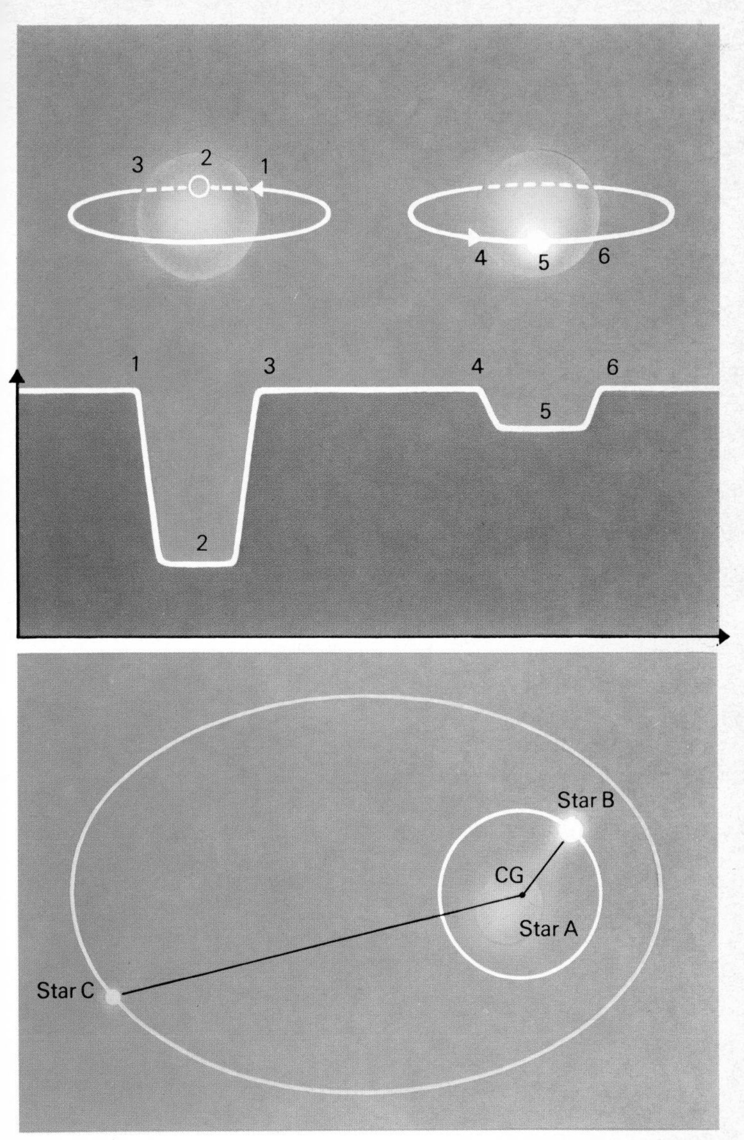

Fig. 34 (top) In an eclipsing binary, greater loss of light is observed when the faint companion passes in front of the bright star. The numbers on the magnitude scale equate to the position of the bright star orbiting the larger and cooler primary. **Fig. 35 (above)** The relative motions of stars in a trinary system are seen in this example of the Zeta Cancri group. Note that all three components move round a common barycentre.

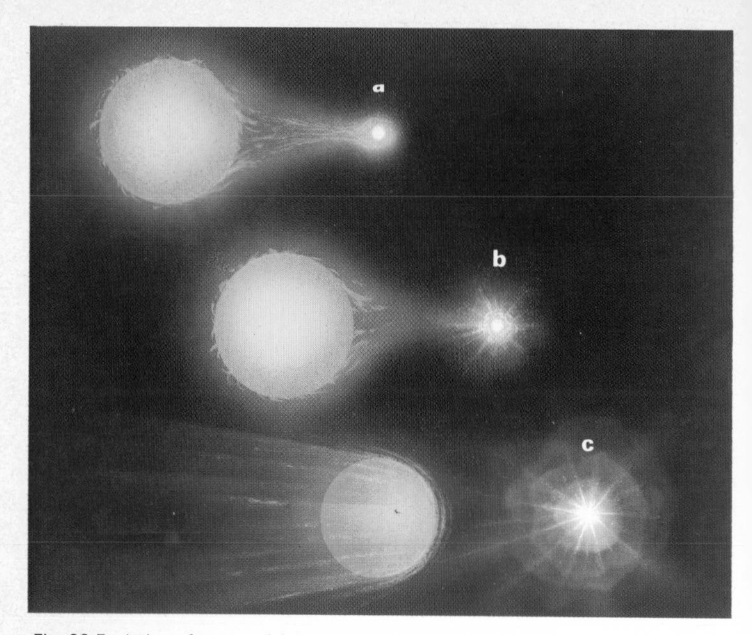

Fig. 36 Evolution of a nova: (a) tenuous outer layers of a large star are drawn towards the companion in a binary system; (b) more material is added as an external envelope to the smaller component; (c) sudden, explosive burning sheds material into space.

described is substantially correct, although there is some disagreement over the exact details. The stars which are losing material to the white dwarf components have not necessarily reached the red giant stage, although in some types of system they are believed to have done so. Obviously, quite complicated interchange of material is possible in binary and multiple systems, as each star goes through its various evolutionary stages, especially as mass exchange can substantially alter the stars' characteristics, as can the loss of material from the system.

Stellar evolution

It is very unlikely that a single star could contract from a cloud of dust and gas, and the mechanism which initiates stellar formation is more likely to give birth to several hundred stars rather than a unique few.

The process begins when some disturbing influence compresses a vast cloud of gas and dust to the point where gravitational forces can take over and complete the contraction to a stellar object. This influence may be the explosion of a supernova where the resulting shock waves compress adjacent material. There is some evidence that this does actually happen in the fact that some supernova remnants appear to be surrounded by shells containing a significant number of young stars. Another mechanism which may initiate stellar formation is the compression wave associated with the arms in spiral galaxies. Certainly both in our own and other spiral galaxies young objects are strongly concentrated along the arms. Once a certain degree of compression has occurred, gravitational and magnetic forces come into play and tend to fragment the cloud into individual aggregations which will later become stars.

With the cloud broken up into several detached condensations, gravity accelerates the collapse and pressures begin to build in what are now emergent proto-stars (**Fig. 37**). A balance is established between the rising temperature and increasing pressure whereby radiation is gradually lost to space which in turn eases the internal pressure and provides for further contraction. Under these conditions the surface temperature tends to remain constant at around 4×10^3 °C, although the core temperature rises towards that at which nuclear reactions can begin. At this stage the protostar is highly luminous and is likely to form a strong infrared source.

With continuing contraction there will come a time when instabilities begin to occur, and such variable partly condensed objects are known to exist. When nuclear reactions finally take place within the core, the star becomes even more variable, emitting a very

Fig. 37 Stellar birth: (a) gravitational compression produces stellar condensation, separating localized whirlpools of matter due to asymmetric magnetic forces; (b) gravity draws material down towards the nebula which spins more rapidly in the conservation of angular momentum.

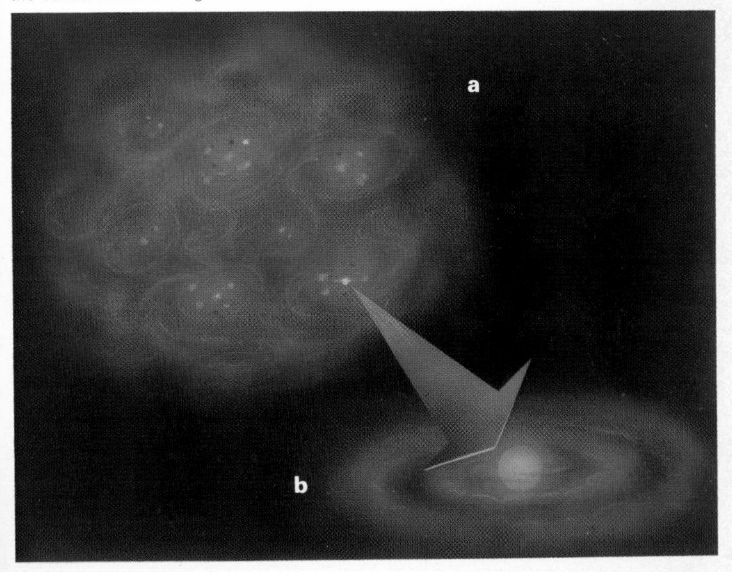

strong stellar wind and blowing a large amount of residual material away into space.

Many stars have been observed to go through this phase and T-Tauri variables exhibit classic examples of this phenomenon. Several times during the staggered development of the emerging star, prodigious quantities of material are ejected in a violent wind capable of ripping through attendant planets with remarkable efficiency. This T-Tauri phase signals the end of the collapsing oscillations although the star may go on expanding and contracting radially for some time.

Thermonuclear stability is reached when the star radiates an equal amount of energy to that received from the fusion cycle in the core. If energy production exceeds the rate of radiation, the star will inflate and cool down so that gravity dominates collapsing the star and increasing temperatures to restore balance and reverse the process. These radial oscillations would be almost undetected by the naked eye; they are so small that prolonged damping is almost inevitable.

The precise point on the H-R diagram where the star joins the main sequence is determined by the mass of the post-T-Tauri product, and the total radiated energy of the sphere must always equal the energy liberated in the core. To achieve this balance the star must deplete its abundance of hydrogen nuclei which inevitably means it will always possess a finite life on the main sequence.

In general, condensations resulting in stars less than 0·1 per cent of solar mass have insufficient pressure to initiate fusion reaction, while at the other end of the scale, stars of more than 60 solar masses are so short-lived that they have difficulty in settling down. Also, very massive stars do not have time to leave the very dense clouds from which they originated, and thus hamper good observation and limit measurement of their characteristics.

Throughout this phase, from initial contraction to main sequence, the density increases from about 10^4 atoms/cm^3 to between 10^{17}–10^{31} atoms/cm^3. The star will continue to occupy its main sequence position until about 10 per cent of the hydrogen has been converted to helium. It then changes to less efficient burning cycles as temperatures rise and moves across the H-R diagram, trying to reach stability.

Giant stages and beyond

Successively heavier elements will build as the temperature increases with the helium-carbon burning cycle introduced at 2×10^8 °C; carbon → oxygen, neon and magnesium at 8×10^8 °C; oxygen → silicon and sulphur at $1{\cdot}5 \times 10^9$ °C; and perhaps finally, the silicon → nickel reaction at $3{\cdot}5 \times 10^9$ °C. Many of these reactions can occur simultaneously, particularly in massive stars, with spherical shells of decreasing temperature from the core to the surface.

When this group of reactions begins, and thermonuclear fusion moves towards the surface of the star, there is less resistance to its expansion and the star moves into the giant phase, following a course to the upper right of the H-R diagram. Stars of solar mass build carbon and oxygen at the central core until pressure from radiation discharges unburnt hydrogen and helium in expanding puffs of matter somewhat akin to a mini-supernova but with appreciably less mass involved in the release. Several examples of 'doughnut-like' rings are observed and these planetary nebulae are the result of stellar smoke-rings blown from the diffuse outer regions of giant stars.

In the giant phase of its evolution the outer shell of the star is very diffuse; expansion has increased the surface area to an extent where temperatures are very low and luminosity is high since brightness depends on R^2T^4, where R and T are radius and temperature respectively. When the core represents about half the total mass of the star, gravitational compression becomes so great that gas degeneracy sets in and reaches unstable equilibrium; as the temperature increases, however, the core expands, releasing the pressure and removing degeneracy.

The final stage begins when the star starts to contract under the

Fig. 38 Stellar evolution: (a) Contraction continues as more material is added to the proto-star; (b) massive quantities of material are blown back into space by the T-Tauri phase when thermonuclear reactions begin; (c) at the end of its life, the star expands into a red giant

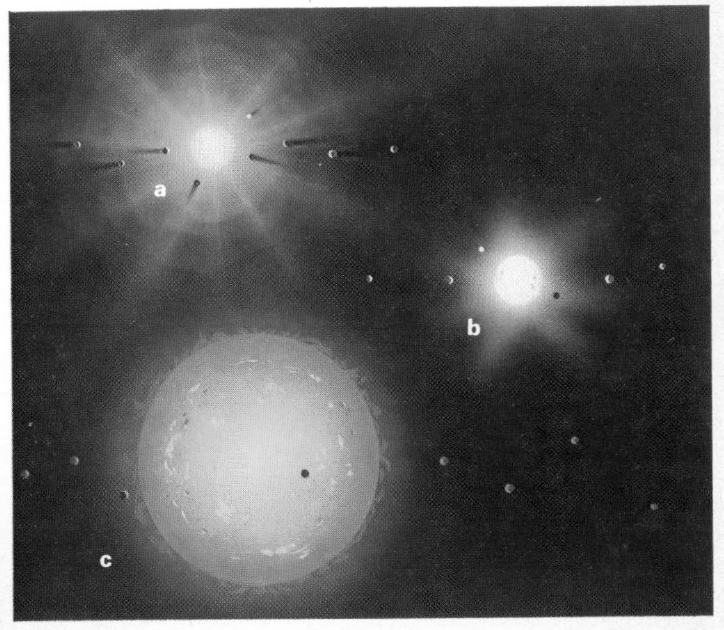

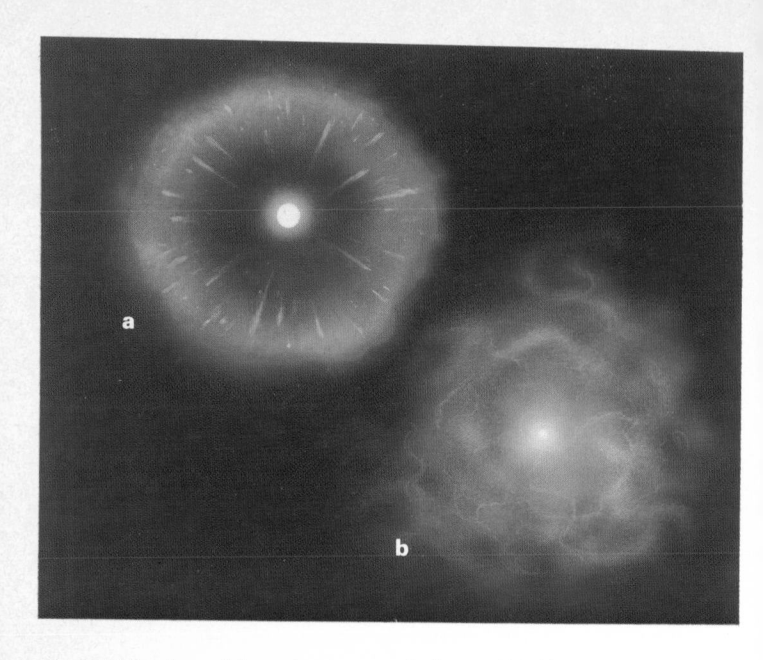

Fig. 39 Stellar decay: (a) massive stars can shed a considerable portion of their mass by moving in to the supernova phase; (b) ejected material moves away from the star at high speed but the nucleus remains as a neutron star.

influence of gravity due to reduced efficiency from diminishing nuclear reactions, and luminosity increases as the radius is reduced. This is only temporary and the loss of outer shells in the planetary nebula phase causes a dramatic loss of energy production. Since the star is no longer in a position to resist gravity, it entirely collapses to a degenerate state – the white dwarf stage (**Fig. 39**).

The pressure at the central region depends on the mass of the collapsed star, but for stars up to about 1·4 solar masses (that is the total mass remaining after the star has moved through the planetary nebula phase) a density of about 10^6 g/cm^3 is not uncommon. The radius of the collapsed star will decrease with greater mass because the magnitude of gravitational compression is related to the available mass.

This is the reverse of the mass-radius relationship that governed both the initial position and main sequence stages where increased mass produced a greater radius due to the higher levels of thermal radiation from the more prolific nuclear reactions. A white dwarf of 0·3 solar masses would collapse to a 10^4 km radius, while a degenerate star of 1·4 solar masses contracts to a radius of 3×10^3 km. Equivalent radii for stars of 0·3 and 1·4 solar mass on the stable main sequence would be $1{\cdot}5 \times 10^5$ and 10^6 km respectively.

It is very difficult to observe the physical characteristics of a white

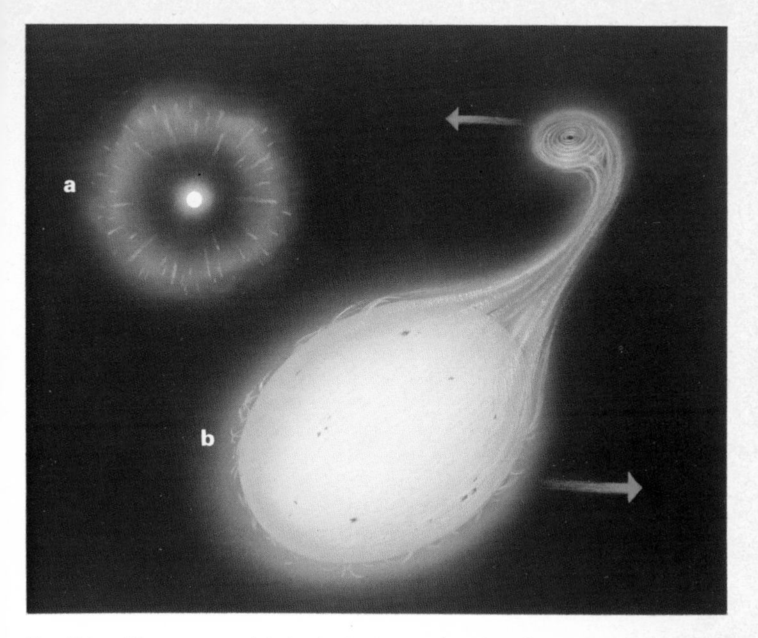

Fig. 40 In a binary system: (a) the death of a massive, short-lived star sometimes results in collapse to a black hole state; (b) the black hole is sucking material from a large companion.

dwarf due to the very low luminosity and radius of the collapsed star. However, theoretical calculations suggest that the dense degenerate core is surrounded by a thin gaseous atmosphere. Inside the core the star is supported by the resistance of the electrons to any further contraction.

Above about 1·4 solar masses, gravity overcomes this barrier and the exceptionally high density, about 10^{12} g/cm^3, combines protons and electrons into neutrons. Like the electrons in white dwarf stars, and as a result of the principles of quantum mechanics, the neutrons also exert a force which resists further compression and leads to a radius of a few tens of kilometres. A stellar mass of between 1·4 and 2·2 solar masses will finish its life as a neutron star.

With a solar mass greater than 2·2 a star should, theoretically, end its life as a collapsar or black hole **(Fig. 40)** whereby the original star material would contract to zero volume with infinite pressure and density. Cursory observation of the stellar classes will reveal the abundance of star types containing main sequence masses considerably in excess of this value and for a long time it was thought that stars appreciably more massive than the Sun avoided total collapse by shedding much of their mass in a supernova phase.

This does indeed appear to happen in many cases, and probably most supernovae leave neutron stars at the centres of their expanding shells of ejected material. (The phenomena seen in many active galaxies, and even our own, appear to require the existence of massive black holes at their centres. However, these are likely to be either of primordial origin, or formed rather differently, not from single stars.)

Several types of supernovae are recognised from the differences between their light-curves, but only two are reasonably well studied. Those known as Type II appear to result from the explosion of a single star of 8 solar masses or more, when it has built up a core of iron (Fe^{56}) which can no longer provide energy to resist gravitational contraction. An initial rapid collapse is followed by the star's explosive disintegration, while the high neutron flux present for a short time gives rise to the heavy elements (see page 44). Supernovae of Type I may arise from the explosion of a component in a

Fig. 41 (below) A neutron star may generate a directional beam of radiation and as the collapsed star rotates the beam sweeps past Earth, appearing to switch on and off, thus forming a pulsar. **Fig. 42 (right)** The moment of eruption of a Type I supernova as seen from one of its planets. The red giant component of the original binary star system can be seen above the clouds of vaporized seas and rock from the 'day' side of the planet. The molten core begins to burst through them and mountains glow and melt; within minutes the scene will be entirely molten, and soon after the planet will be converted to a wisp of gas.

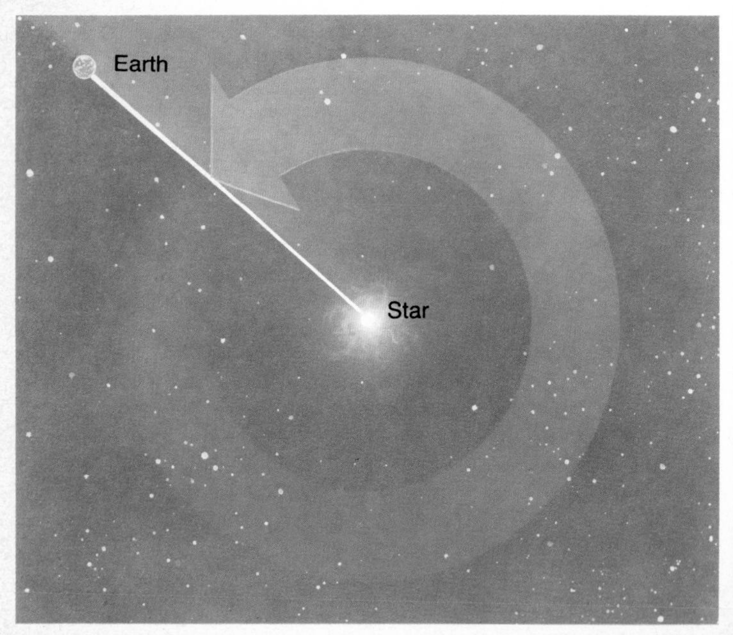

binary system which has accreted hydrogen and helium from its companion, in a way similar to that of the novae (page 50).

Type II supernovae in particular may eject material at speeds in excess of 10^4 km/s, and leave behind a dense neutron star, which may sometimes be observed as a pulsar.

The Crab Nebula is an excellent example of the remnant of a supernova, which was actually observed in AD 1054. The central neutron star is rotating rapidly, giving rise to pulsations at 0·033-second intervals at radio, infrared, optical, X-ray and gamma-ray wavelengths. It is losing energy in the form of energetic electrons, causing it to slow down (by about 10^{-5} s/year). The electrons interact with the very strong magnetic field and give rise to radiation, the ultraviolet component of which excites the visible nebula. Because of its young age the Crab pulsar is very strong and very fast, but nevertheless more than 200 pulsars have been detected, with typical periods of 0·3–3 seconds.

When a central remnant has a mass greater than about 2·2 solar masses, it cannot be sustained by thermal or degeneracy pressures, and collapses to form a black hole. Associated with any such object there is a theoretical sphere with a radius dependent upon the object's mass, from within which nothing, not even light, can escape. This is known as the 'event horizon' and is calculated from: $r = 2M\,(6{\cdot}67 \times 10^{-8}\ \mathrm{cm^3/g/s^2})/c^2$, where M equals the mass of the star and c is the speed of light ($2{\cdot}99 \times 10^{10}$ cm/s).

The implications of this are profoundly staggering for students of classical physics. Light emitted at some point just beyond the event horizon will tend to fall back towards the gravitational radius as the event horizon is approached until, on the 'surface' of this imaginary sphere, the emitted photons will hover for eternity. However, any measurement of photons emitted at the speed of light would have to be made from the event horizon and because physical implosion is occurring at the speed of light, the photons, moving radially from the sphere, would retain their normal velocity of light speed **(Fig.43)**.

In reality, the events would be moving towards the centre of the black hole, with no contravention of the laws of physics, and the beam of photons would be moving radially outward also at the speed of light. The result of this situation is not that one event is moving at 2c relative to the other, but rather that the event horizon appears to remain frozen in space – for an external observer. The implosion of matter within the event horizon is separated from the outside in both space and time.

This means that the only physical record of a black hole is the gravity, or its influence, of the collapsing star as it shrank through its own event horizon. Since there can be no communication from inside the gravitational radius, the field must remain in the position it occupied at the instant of collapse. Inside the event horizon the gravitational tides have awesome implications.

Since the force of gravity varies as the inverse square of the distance in the 'normal' world the tidal forces in the black hole reach extreme proportions, rising to infinity at the centre, and space-time curvature reaches a maximum for which there is no satisfactory formula. It would seem that towards this point, the singularity, matter is broken down into elements, then into atoms and even nuclear particles are destroyed when the radius of space-time curvature becomes smaller than that of the component particles from which the original star was built.

The only way to observe the presence of a black hole is to find a binary system where an invisible component is orbiting the common centre of mass with a larger, visible, companion. Using the equations outlined on page 50 it would be possible to determine the relative mass of the invisible component and, by defining the class of the visible star thereby establish the mass of the suspect black hole. This has been achieved and there seems no doubt that total collapse does occur in the universe, confirming the view that not all stars shed sufficient mass to escape this ultimate fate.

In binary systems where one component is a black hole and the other is a large or giant star, matter will be accreted by the black hole in a similar way to white dwarfs in novae as described on p. 48. In this case, captured matter will be gradually totally absorbed by the black hole with the highly accelerated gases broadcasting power-

Fig. 43 Beams of light sent past a black hole would be deflected by the intense gravity field. At a critical distance, the beam would orbit the photon sphere (b) while a beam directed toward the event horizon (a) would vanish into the singularity.

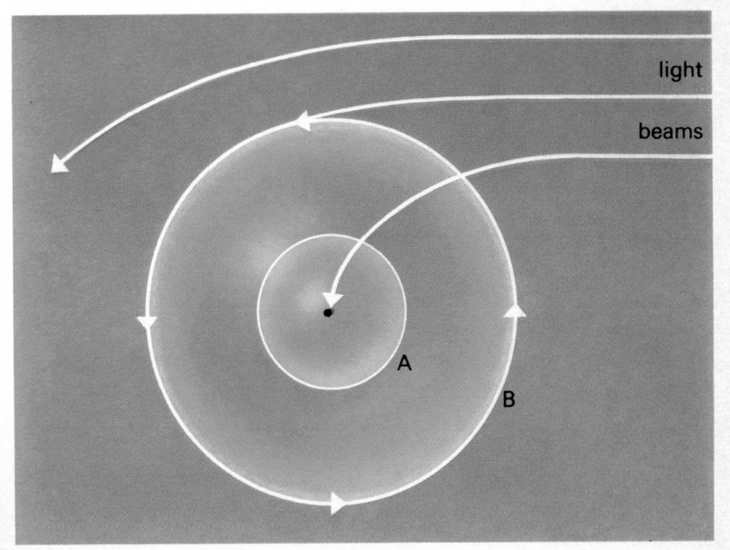

ful X-rays from just outside the event horizon. Several such sources have been observed and the emphasis on X-ray astronomy will afford greater opportunity for confirming black hole theories.

The question of life

In considering the question of intelligent, carbon-based, life in the universe, the stellar classes provide insight to the limitations theoretically imposed by main sequence time and luminosity. Since life as we know it requires at least 2×10^9 years to evolve intelligent creatures (probably $3{\cdot}5 \times 10^9$ years for Earth) it is possible to impose a limitation on the stellar classes likely to support such development.

Figure 44 shows that only the F to M stars are stable for a duration required for emergence of intelligent life and this eliminates classes O, B and A as candidate hosts to life-bearing planets. Also, if planets are too close to the parent star, like the Moon is to the Earth, they will experience captured rotation and provide a very unsatisfactory thermal balance where one side of the planet is too hot and the other too cold. Since carbon-based life requires certain temperature limits, the distance of potential life-bearing planets must always lie within a certain range that is amenable to the evolutionary cycle.

Each star has a different ecospheric distance depending on temperature which in turn is a product of the stellar mass. It will be seen from **Fig. 44** that K and M stars are so cool that their ecospheres lie within the radius of captured rotation. It is possible to conclude, therefore, that Earth-type, carbon-based, intelligent life would probably only develop on planets around F and G stars.

Extreme caution should be applied to sweeping conclusions in this direction since Earth is the only planet on which life has provided a practical model from which to judge the various cycles of biological evolution. Countless permutations of carbon-based life may be possible and as yet there is a great amount of uncertainty concerning the alternate, silicon-based, evolutionary processes.

(Figures 45–48 display phenomena in various stages of stellar evolution as visible in our galaxy.)

Fig. 44 Stellar class. Hatched areas indicate the zones of captured rotation; single colour blocks show the size of the ecosphere for respective stars.

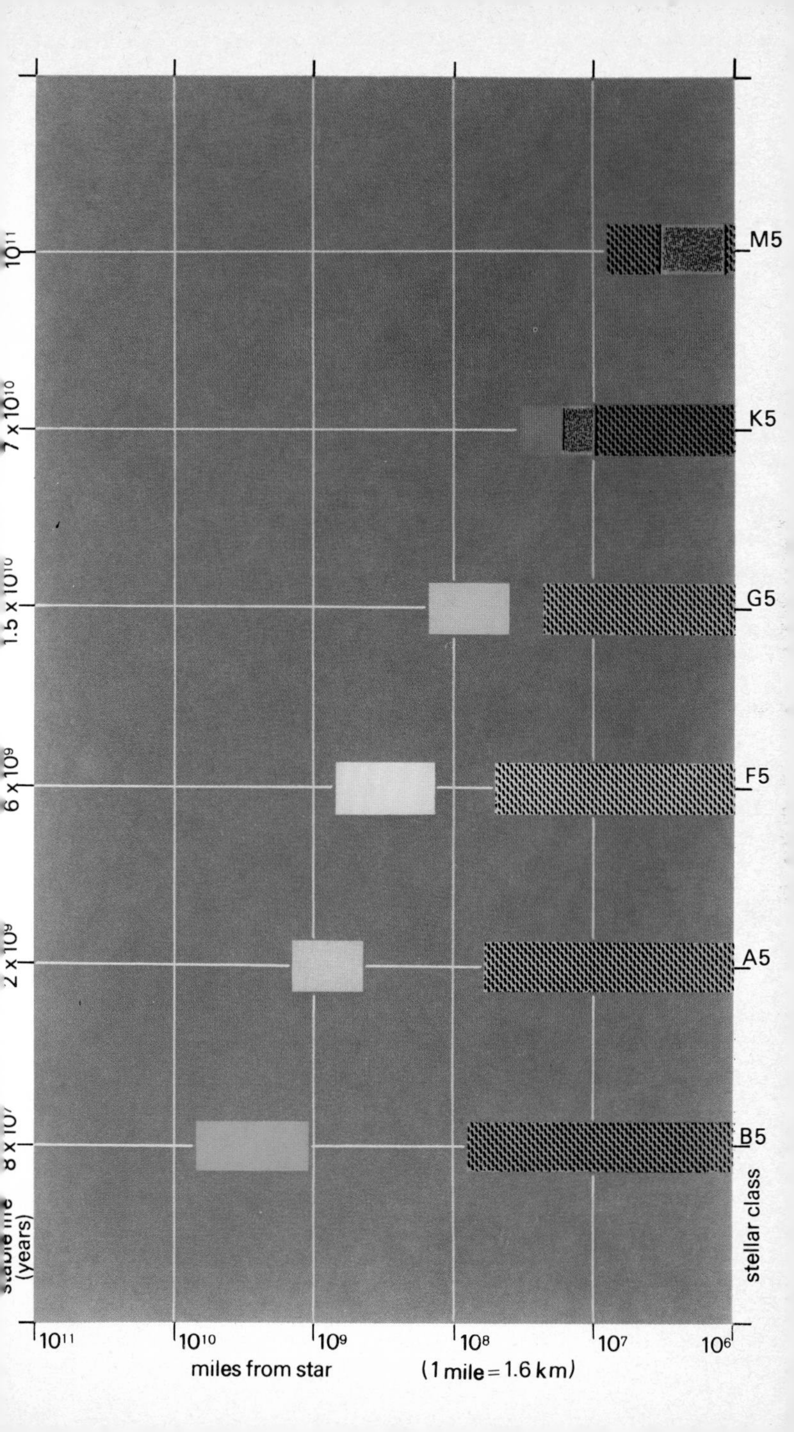

(years)
10^{11}
7×10^{10}
1.5×10^{10}
6×10^{9}
2×10^{9}
8×10^{7}
M5
K5
G5
F5
A5
B5
stellar class
10^{11}
10^{10}
10^{9}
10^{8}
10^{7}
10^{6}
miles from star
(1 mile = 1.6 km)

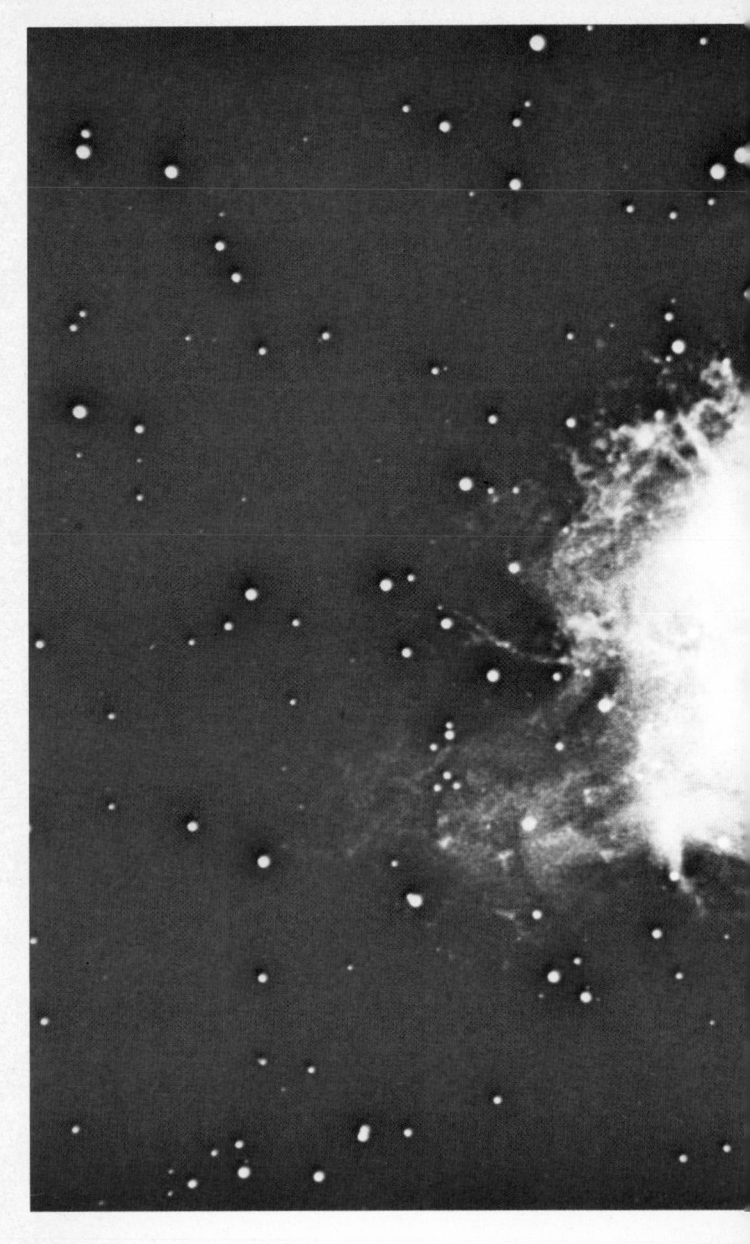

Fig. 45 The Crab Nebula. This is the remnant of a supernova explosion.

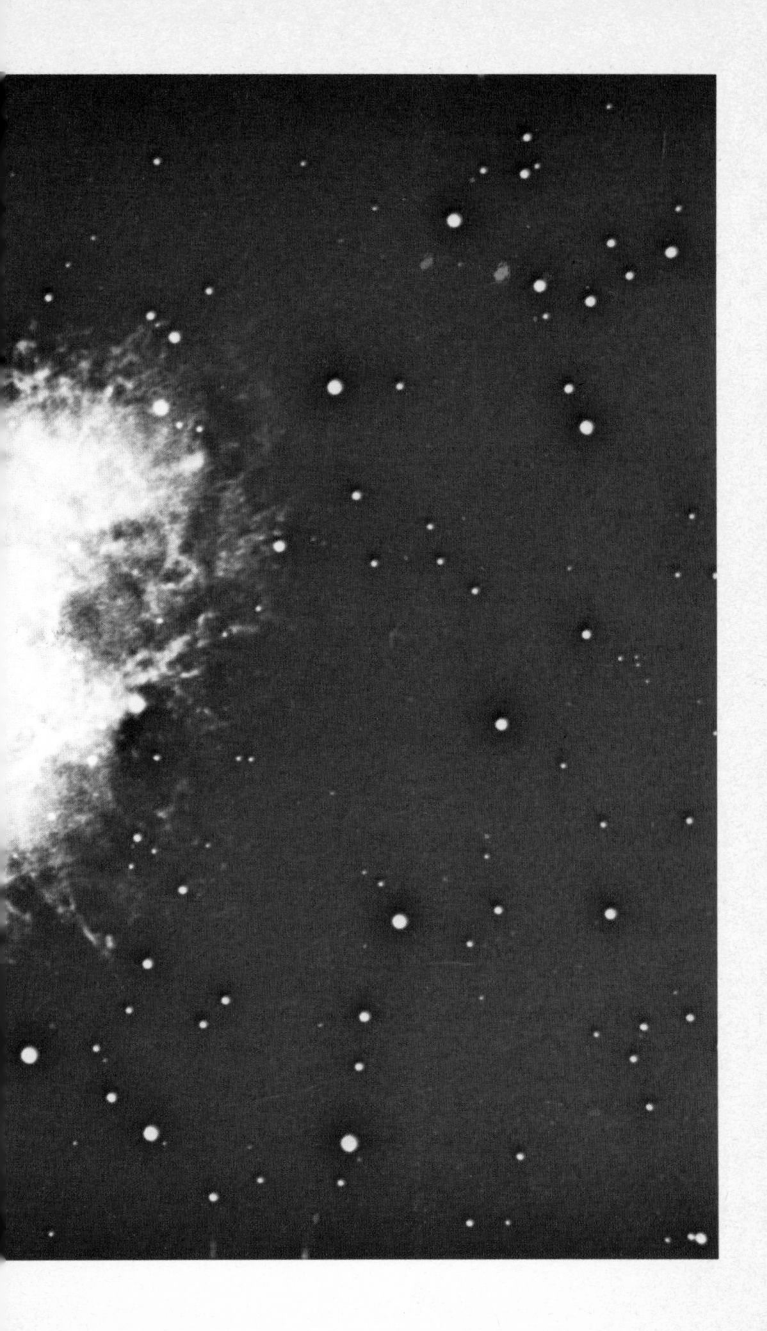

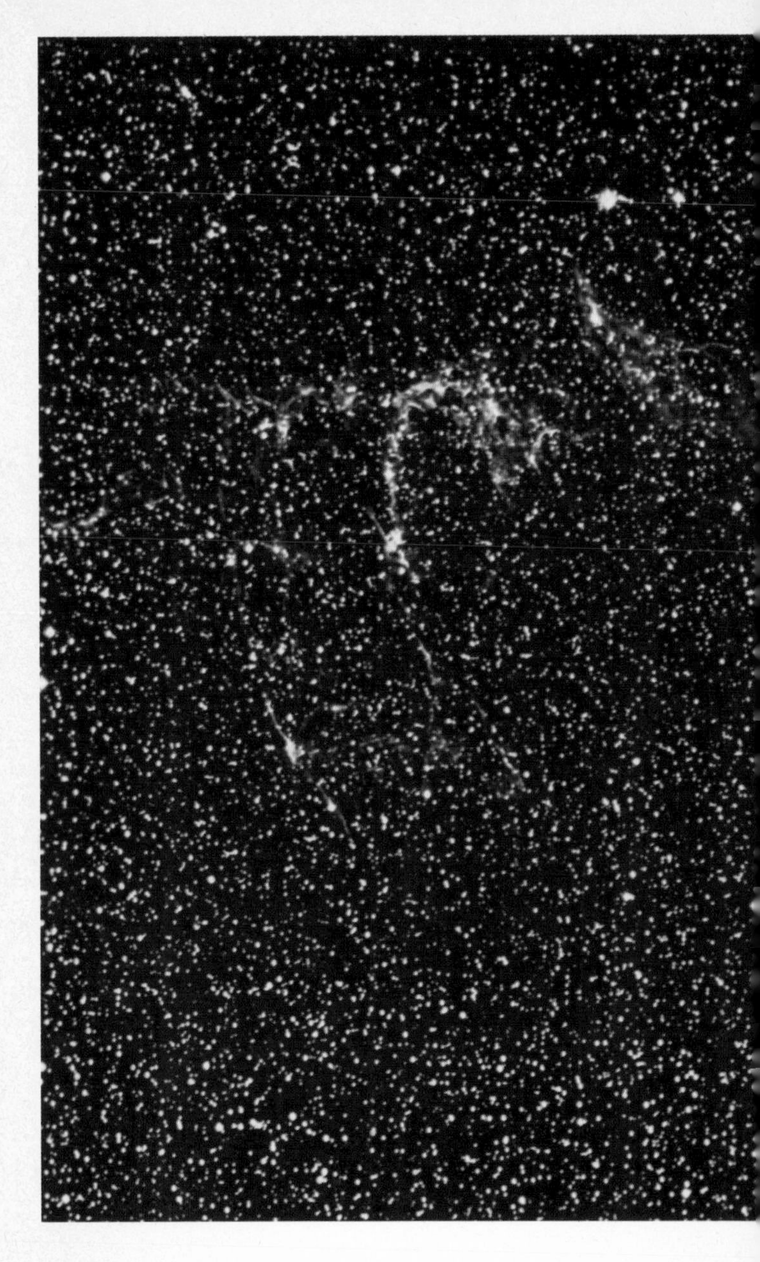

Fig. 46 Part of an old supernova remnant, the Veil Nebula in Cygnus.

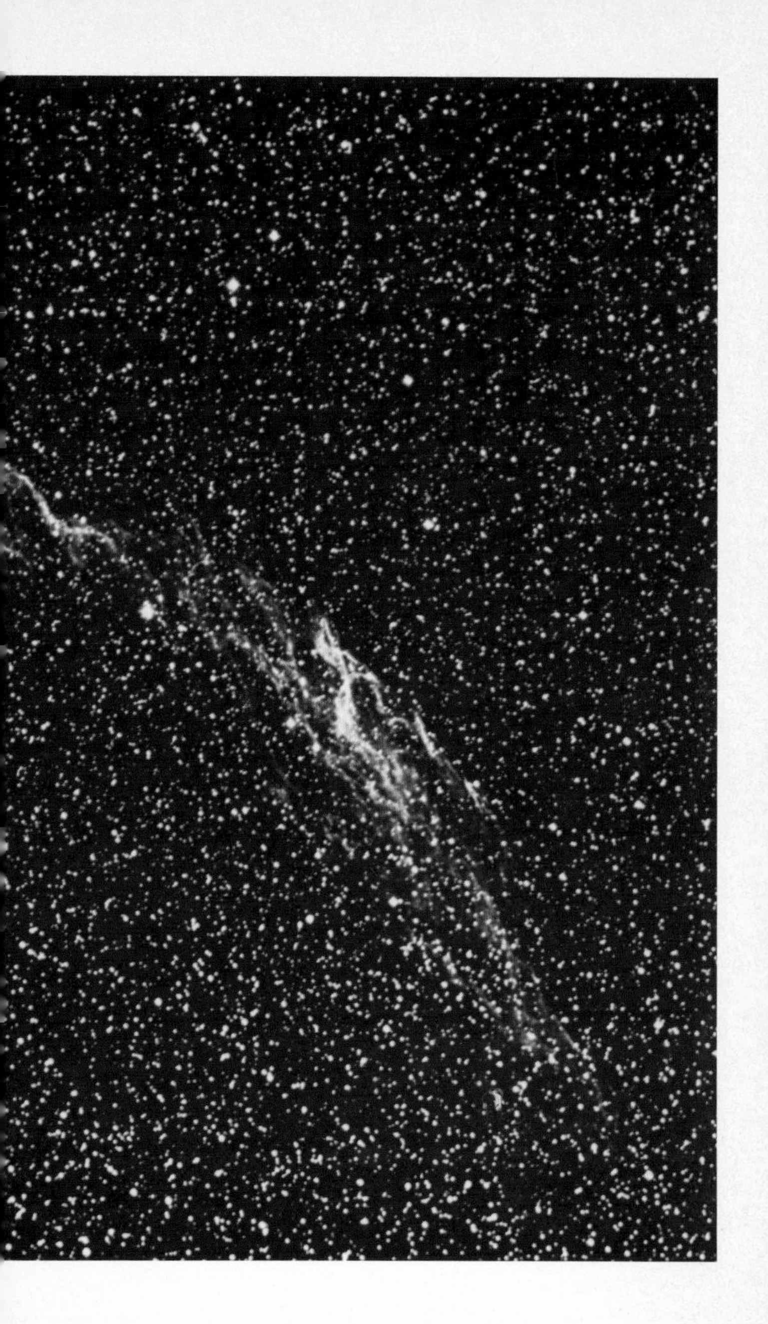

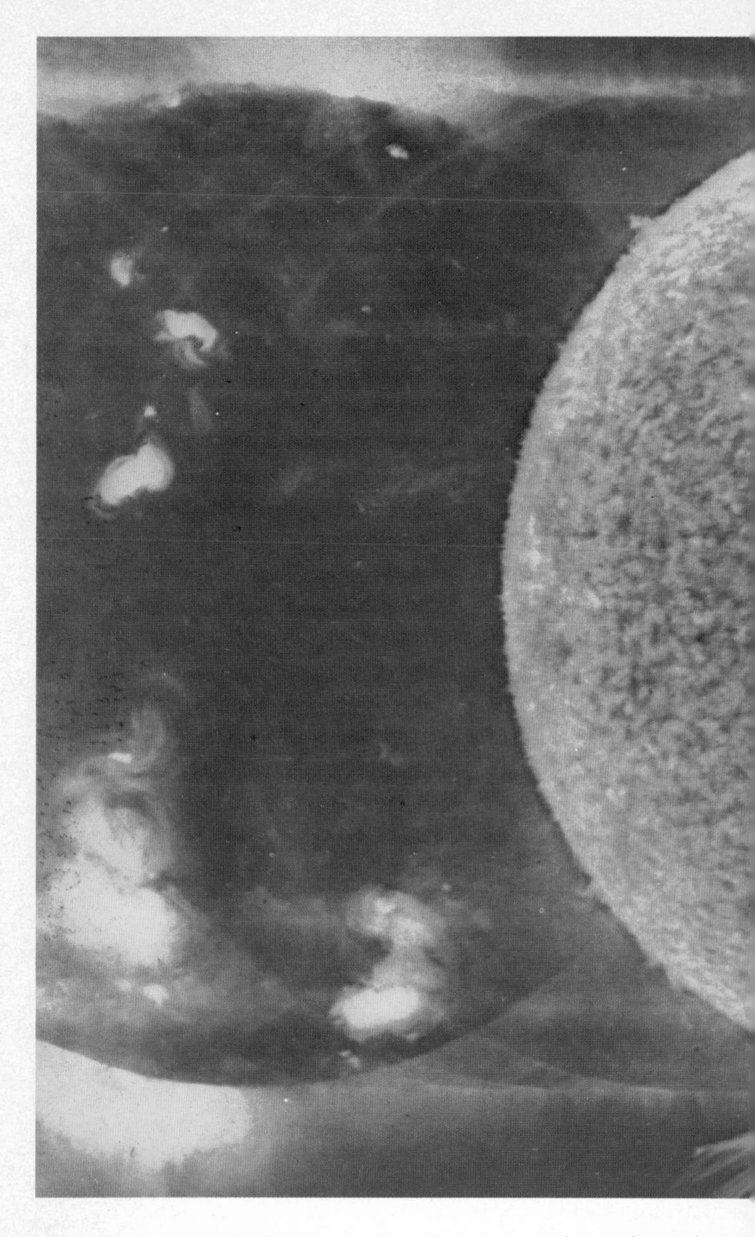

Fig. 47 Skylab view of the Sun showing the cellular structure of the surface and associated phenomena which are not visible to the naked eye.

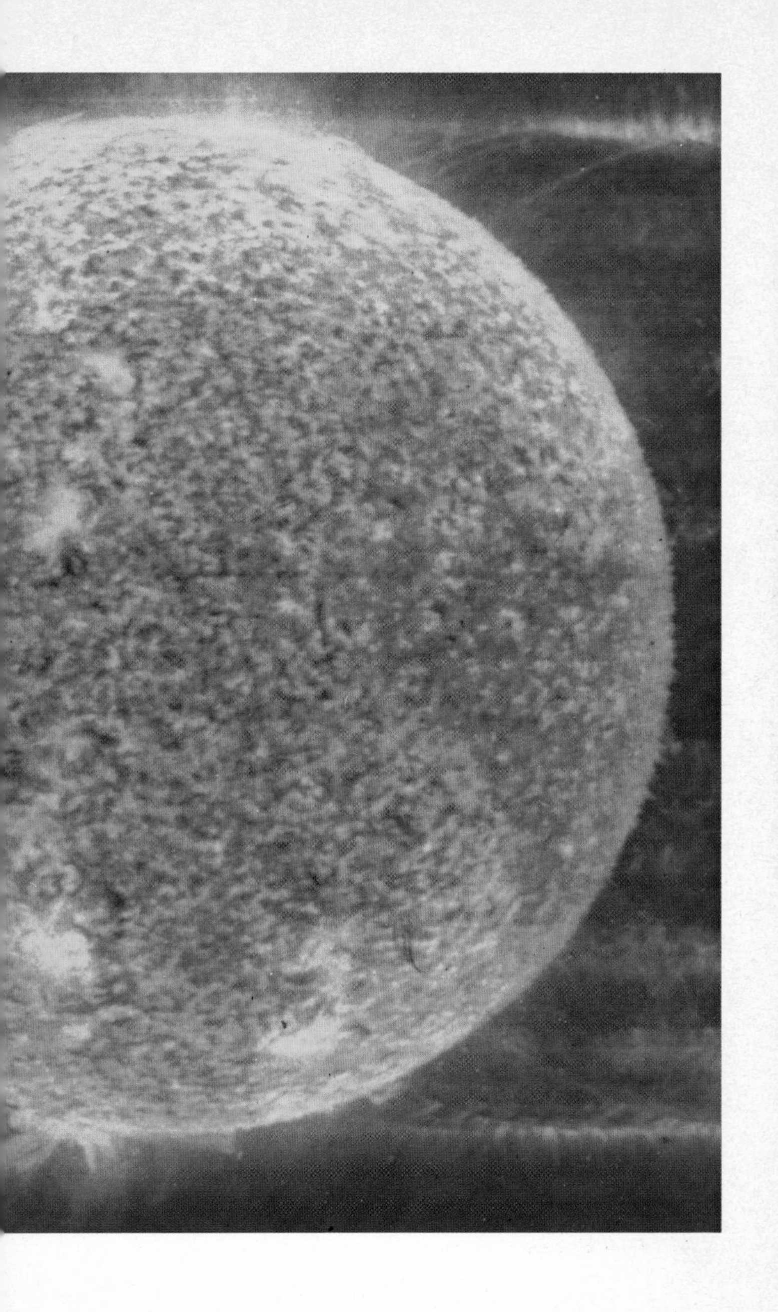

Fig. 48 Horsehead Nebula. A region in Orion where many new stars are born.

The Constellations

The celestial sphere

Throughout this guide emphasis will be placed on the description of celestial objects as they exist in reality, and planets, comets, stars, novae, supernovae and nebulae will be explained in the context of modern astronomical science. When considering the coordinates of celestial objects, by which their positions and motions may be measured, it is necessary to adopt the older (or rather, the ancient) interpretation of the heavens as a sphere surrounding and centred on the Earth, but this certainly does not reflect the current view of the universe.

It will, for instance, be necessary to assume that the Sun revolves round the Earth and that celestial objects move across the surface of a sphere. This is not as confusing as it may at first seem but it is very important to appreciate the artificial nature of the selected coordinate systems.

The angle of inclination of the rotational axis of the Earth (or any other body) does not alter as it revolves in its orbit about its primary **(Fig. 49)**. This angle, measured from a line perpendicular to the plane of the body's orbit, is about 23·5° for the Earth. Although there are variations over long periods of time, in the short term the orientation of the axis can be considered to be fixed in space.

It is precisely this mechanism that provides the Earth with its seasons, and points of solstice and equinox are adopted to describe the effect this has on the spinning planet. Equinox positions mark the days, in spring and autumn, when day and night are of equal duration. At such times the Earth's polar axis is at right angles to the Earth-Sun line. Solstice positions, in summer and winter, mark the points at which the Earth's tilt causes the Sun to appear at the greatest distance from the plane of the equator.

In **Fig. 50** the Earth's equatorial plane is projected onto a sphere, becoming the celestial equator, and the plane of the Sun's path around the Earth is also drawn on the sphere to become the plane of the ecliptic. The celestial equator has now been given a more usual orientation and the apparent path of the Sun, the ecliptic, is tilted at 23·5°. The points at which the two planes cross, the nodes, establish a geographical coordinate using the line from the centre of the two

Fig. 49 (above right) As Earth orbits the Sun it moves through positions of equinox (where day and night are equal in length) and solstice, marking the points at which the Earth's tilt displays the greatest angle with the plane of the orbit. **Fig. 50 (right)** The Earth's equatorial plane is here projected onto a sphere making it the celestial equator, and the plane of the Sun's path around the Earth is included to become the plane of the ecliptic. The points where the two intersect are called 'nodes' and at present these point to the First Point in Aries.

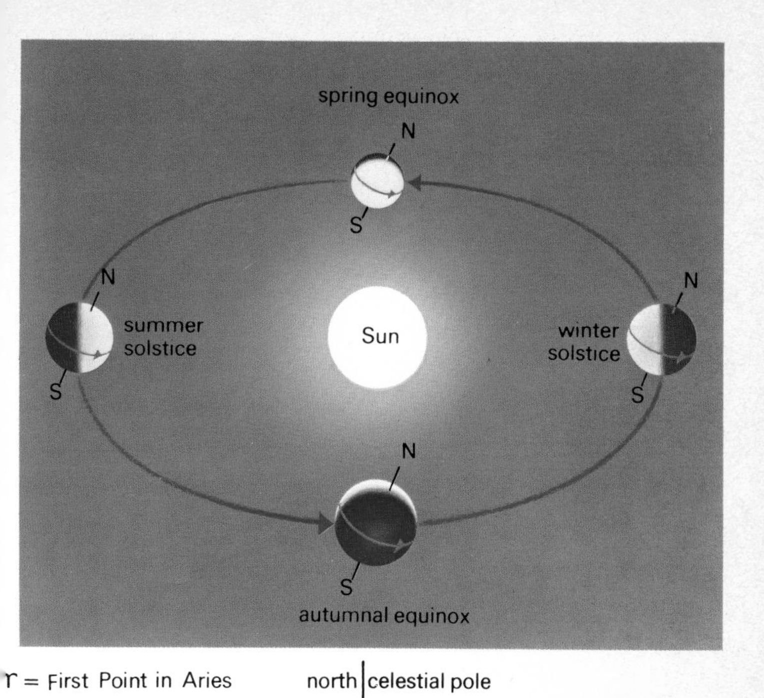
spring equinox
N
S
N
summer solstice
S
Sun
N
winter solstice
S
N
S
autumnal equinox

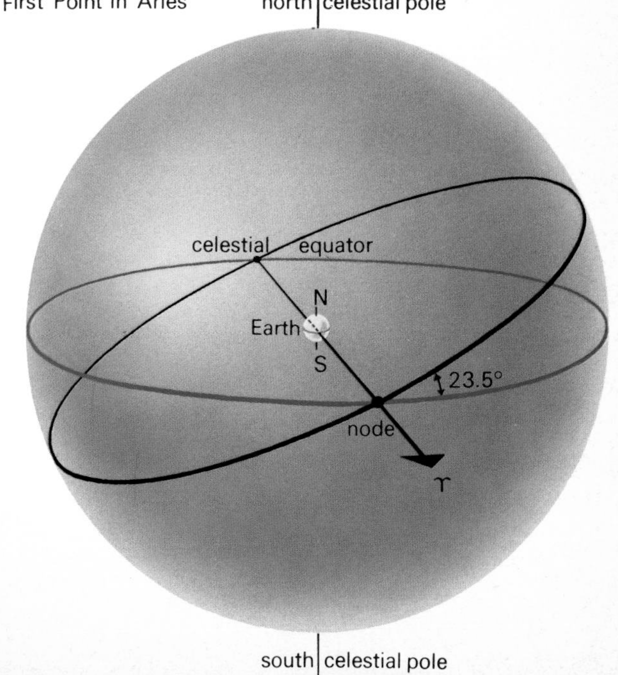
ϒ = First Point in Aries
north celestial pole
celestial equator
N
Earth
S
23.5°
node
ϒ
south celestial pole

planes (the Earth) through the spring (vernal) equinox to set up a system of longitude.

This point is known as the First Point of Aries. The north celestial pole is the point on the celestial sphere directly above the Earth's geographic pole. The lines of celestial longitude are marked eastwards 360° around the equator past the autumnal equinox position and back to the Aries line as in **Fig. 51**. For purposes of clarity the plane of the ecliptic has been deleted. Unlike the conventional method adopted for the geographic coordinate system, the celestial sphere uses a system of hours, minutes and seconds to mark off longitudes, rather than degrees, minutes and seconds of arc.

Moving eastward from the First Point of Aries, or vernal equinox, the 360° circle is divided first into 24 equal hours of longitude (corresponding to 15° of longitude in the geographic system) with each 'hour' divided into 60 equal minutes and each minute further divided into 60 equal seconds; 1 min of celestial arc is 0·25° across and 1 sec of celestial arc is 0·004° across. This system of celestial longitude is known universally as Right Ascension (RA).

Having established longitude, it is necessary to construct a system for measuring latitude or elevation from the celestial equator now subdivided into RA. In **Fig. 52**, the geographic system is adopted whereby a circle equals 360° and, instead of expressing lines of latitude in degrees north or south, the symbols + and − are used for positions of celestial latitudes north and south of the celestial equator respectively.

To obtain a coordinate for any celestial object, first, the RA is read in hours, minutes and seconds from the zero meridian, the arc running from the north pole through the vernal equinox to the south pole. The latitude, or Declination (Dec), is then found lying between 0° and 90° + or −. This coordinate system is applied to all objects in the universe and although it purports to show celestial phenomena projected onto a sphere, the sophistication of the human mind will accept that the RA and Dec values merely point the eye, or telescope, along a specified line of sight.

Any two objects which appear close when viewed from Earth may, in reality, be themselves separated by a very considerable distance. To map the distribution of the local group of galaxies would require first, a knowledge of the coordinate positions on the celestial sphere and secondly, a measured value of the respective distances from the Earth, thereby establishing the third dimension. Without an appreciation of the radial distance any number of objects close together can give the impression of being part of a single cluster.

Fig. 51 (above right) Lines of celestial longitude are marked in hours, minutes and seconds rotating east around the celestial equator, starting and finishing at the First Point in Aries. This system is known as RA. **Fig. 52 (right)** Lines of celestial latitude from the plane of RA are marked in degrees, + for latitudes north of the plane and − for latitudes south of the plane. Latitude values are referred to as Dec.

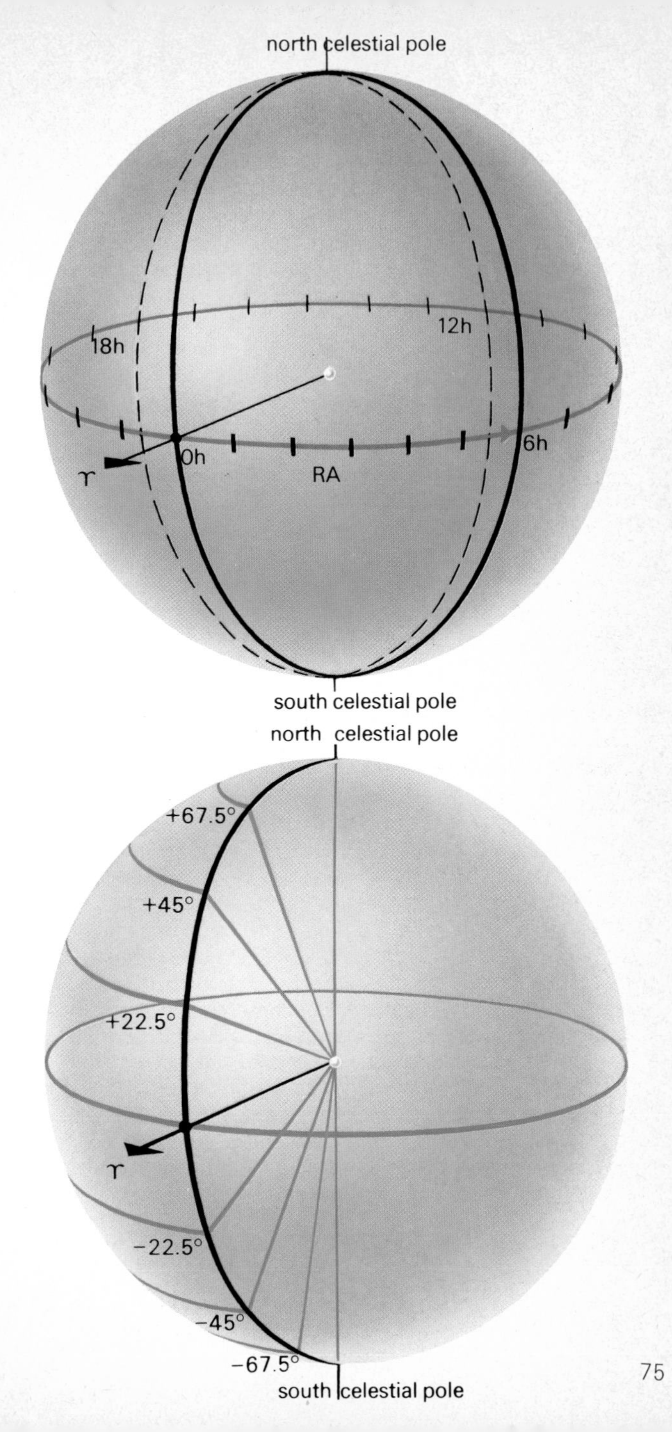
north celestial pole
12h
18h
0h
6h
RA
south celestial pole
north celestial pole
+67.5°
+45°
+22.5°
−22.5°
−45°
−67.5°
south celestial pole

Reference to **Figs 49 and 50** will show how the Sun moves round the plane of the ecliptic in just over 365 days, and how this causes the day–night cycle to migrate through the constellations. Therefore from a given geographic latitude, observation of the entire visible sky will take one full year. Although it has been necessary to construct a fixed geometrical relationship between the ecliptic and the celestial equator this is only the current situation and the gradual precession of the polar tilt **(Fig. 53a)** causes a precession of the equinox and a continually changing position for the celestial north pole **(Fig. 53b)**. The unique nature of the celestial sphere provides a coordinate system centred on a minor planet around a moderate Sun located towards the edge of an undistinguished galaxy among billions in a boundless universe. Since the mapping system is geocentric, it has limitless potential for accommodating objects at successively greater distances.

Sometimes it is convenient to establish a coordinate system based on the galactic equator rather than the ecliptic and in this case **(Fig. 54)** the zero meridian is set up along a line passing from the Earth through the centre of the galaxy; the galactic equator is tilted 62·5° to the celestial equator and positions are given in terms of galactic longitude (in degrees) and galactic latitude.

The constellations

The following 62 pages of maps and notes are provided so that the reader can develop an understanding of the configuration of respective constellations. The first four maps **(Figs 55–58)** show northern and southern skies for orientation with the celestial sphere. The grouped clusters of stars were first arranged in the constellations according to supposed similarity with mythological figures or heroes. More than 1800 years ago. Ptolemy, drew up a list of 48 constellations which formed the basis for subsequent additions.

In order that the constellations may be readily recognized, the most significant stars and objects have been included on the charts, some of which therefore show fainter stars than others.

The names of the constellations are presented in Latin followed by the English derivation. The allocation of Greek letters to the stars within each constellation follows the 17th century tradition whereby, in the main, the alphabet proceeds through a sequence of decreasing apparent magnitude (mag). The Greek alphabet is presented in the appendix to assist with interpretation. For definition of magnitudes and values of the stellar classes see p. 32. Constellations below 50° south are shown with south at top.

Fig. 53 (above right) a Precession of the Earth's polar tilt causes a precession of the equinox. This takes place gradually due to the changing relationship of the ecliptic to the celestial equator. **b** Precession of the polar axis also changes the position of the celestial north pole. **Fig. 54 (right)** A coordinate system based on the galactic equator. Galactic and celestial equators are inclined 62·5° to each other.

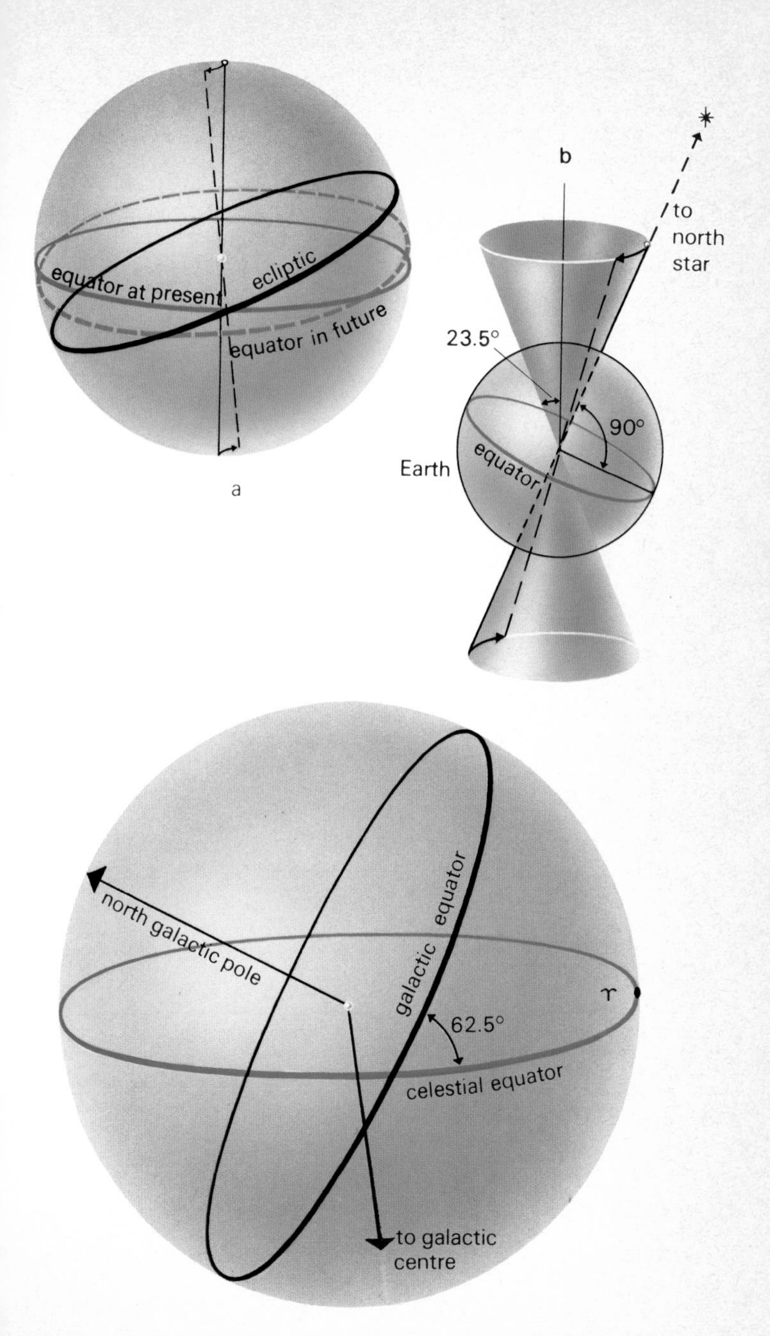

equator at present
ecliptic
equator in future
a
b
to
north
star
23.5°
90°
Earth
equator
north galactic pole
galactic equator
♈
62.5°
celestial equator
to galactic
centre

Fig. 55 Northern sky constellations 1.

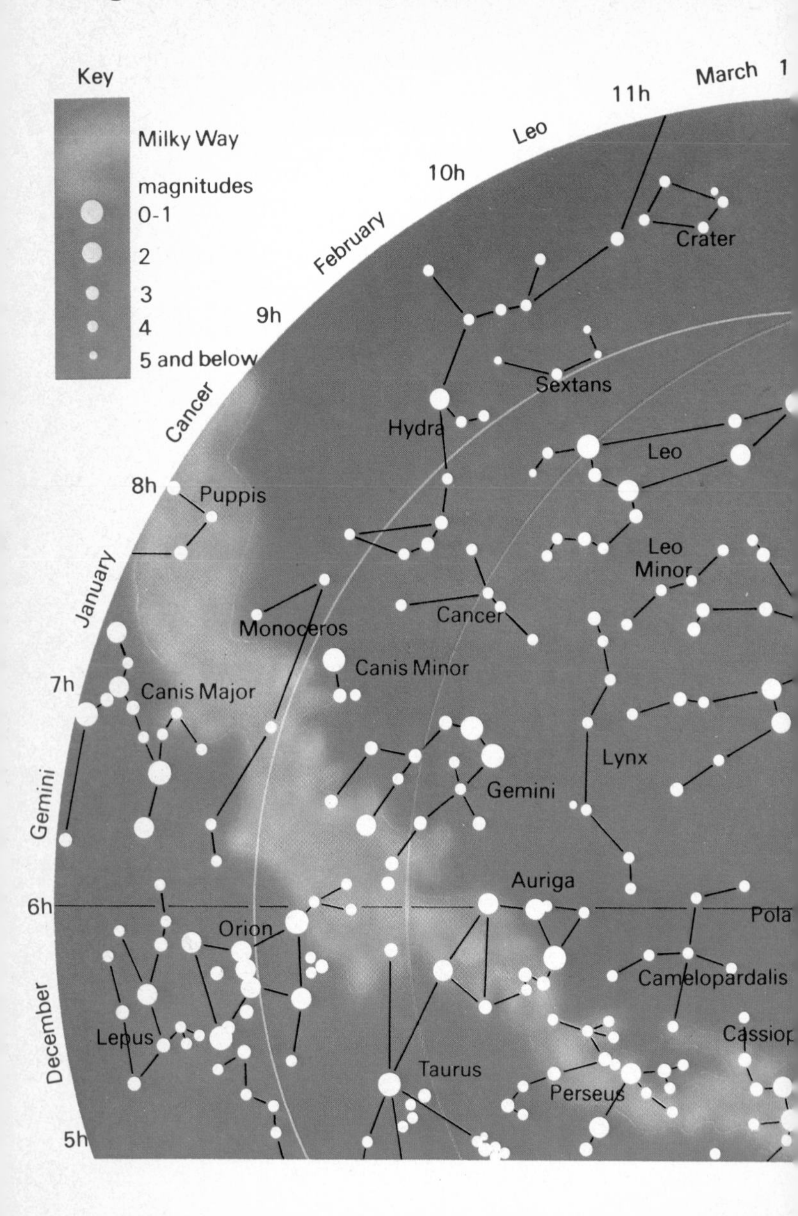

LMC = Larger Magellanic Cloud

SMC = Smaller Magellanic Cloud

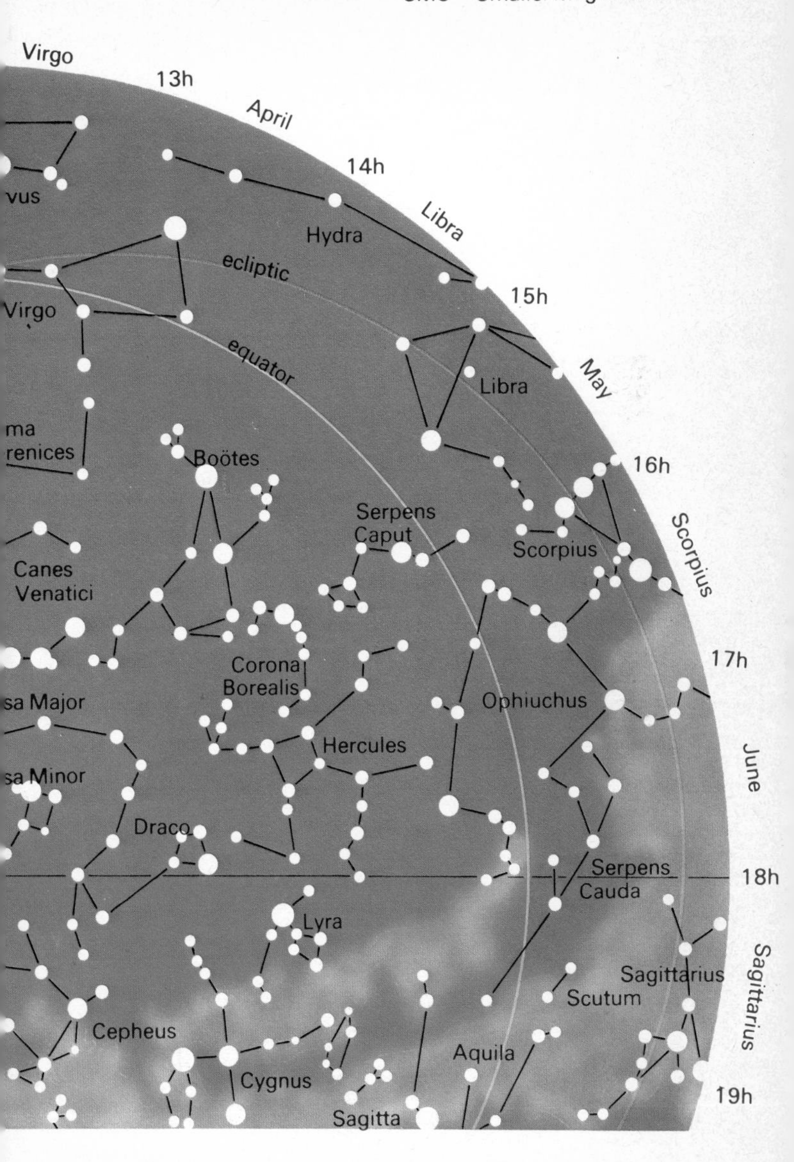

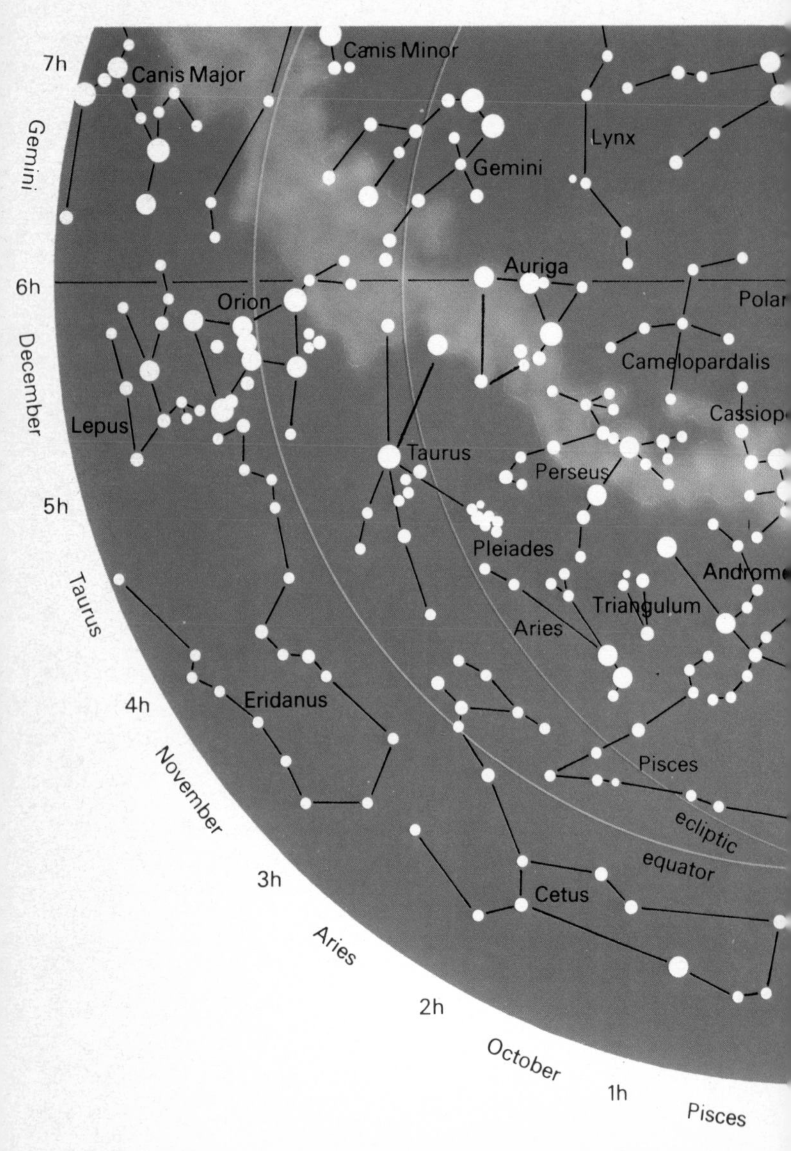

Fig. 56 Northern sky constellations 2.

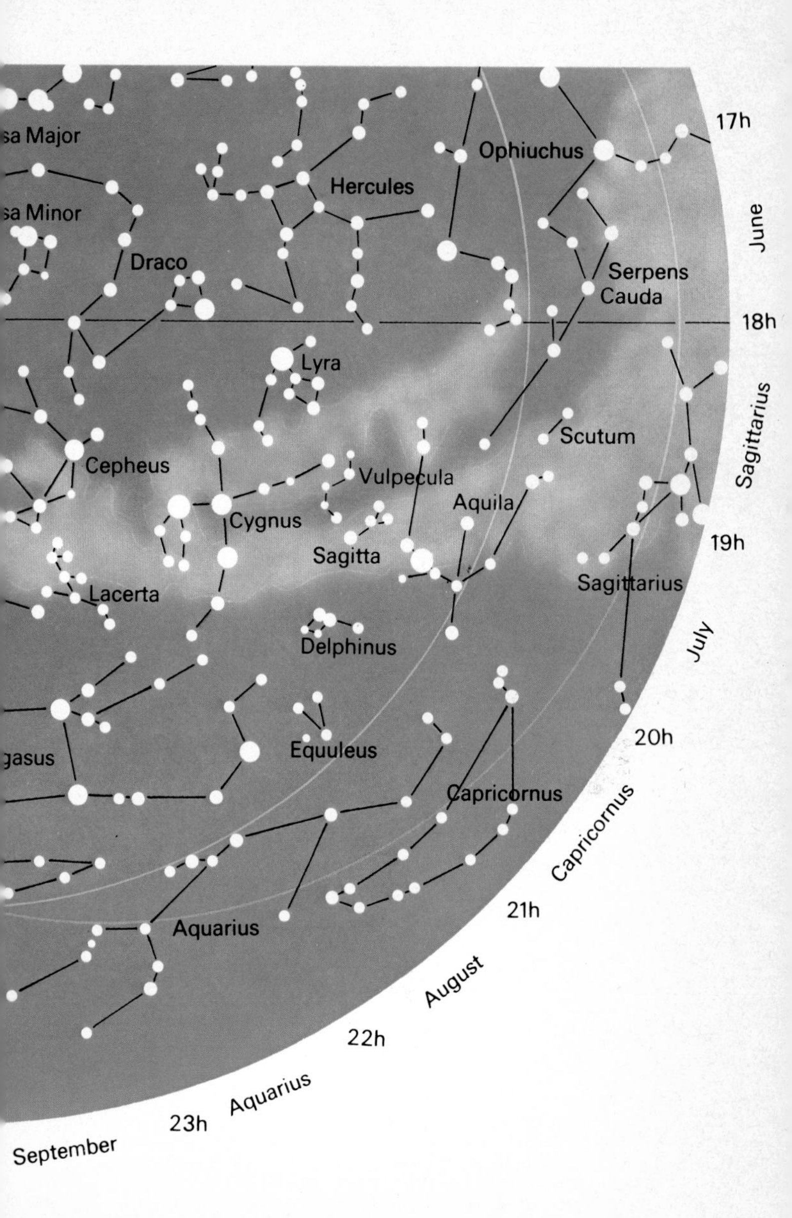
sa Major
sa Minor
Draco
Hercules
Ophiuchus
Serpens
Cauda
Lyra
Cepheus
Cygnus
Vulpecula
Aquila
Scutum
Sagitta
Sagittarius
Lacerta
Delphinus
Equuleus
gasus
Capricornus
Aquarius
17h
June
18h
Sagittarius
19h
July
20h
Capricornus
21h
August
22h
Aquarius
23h
September

Fig. 57 Southern sky constellations 1.

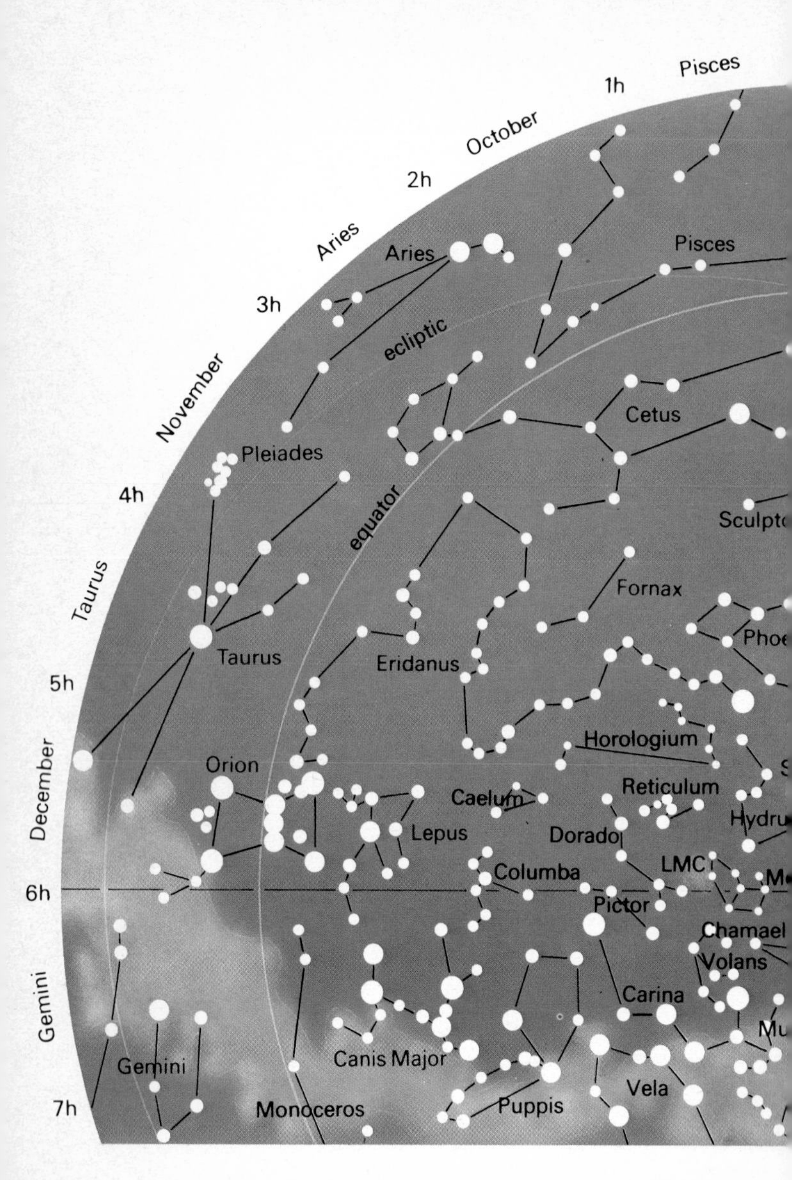

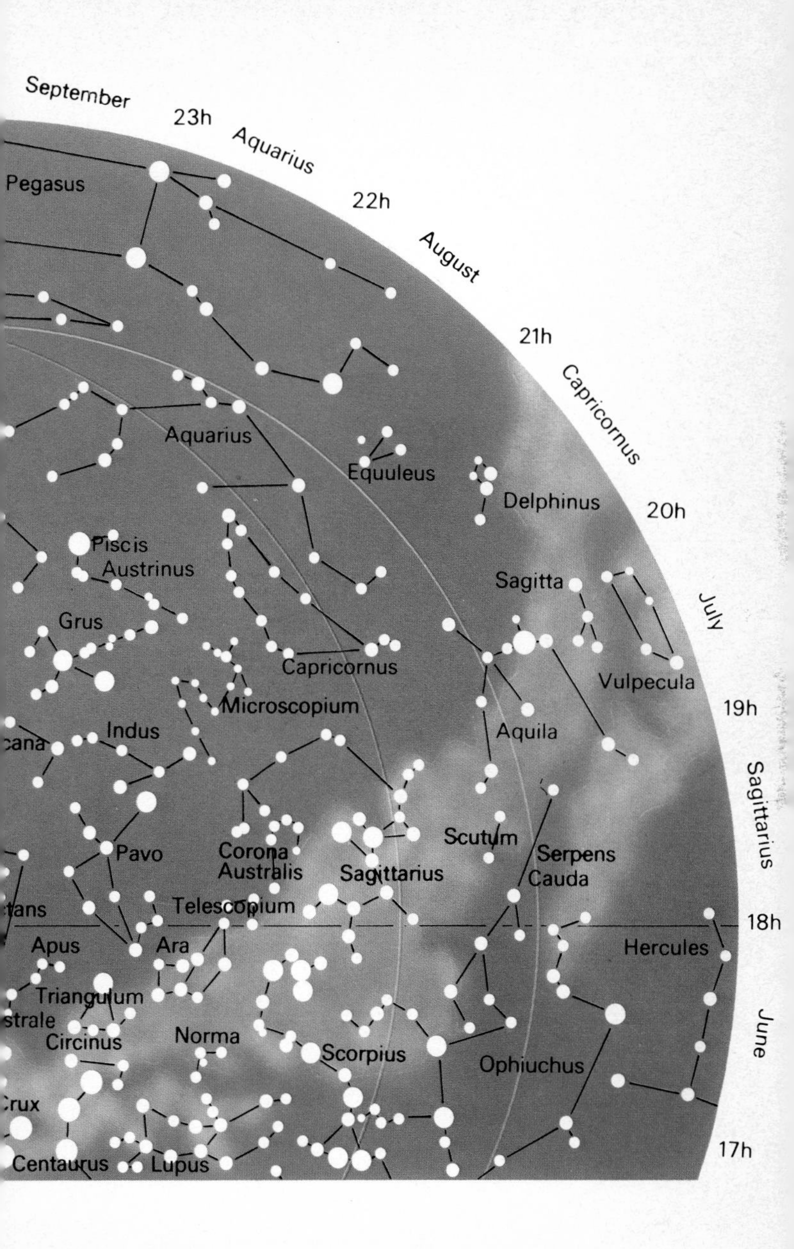

September
23h
Aquarius
22h
August
21h
Capricornus
20h
July
19h
Sagittarius
18h
June
17h
Pegasus
Aquarius
Equuleus
Delphinus
Piscis
Austrinus
Sagitta
Grus
Capricornus
Vulpecula
Microscopium
Indus
Aquila
Pavo
Corona
Australis
Sagittarius
Scutum
Serpens
Cauda
Telescopium
Apus
Ara
Hercules
Triangulum
Circinus
Norma
Scorpius
Ophiuchus
Centaurus
Lupus

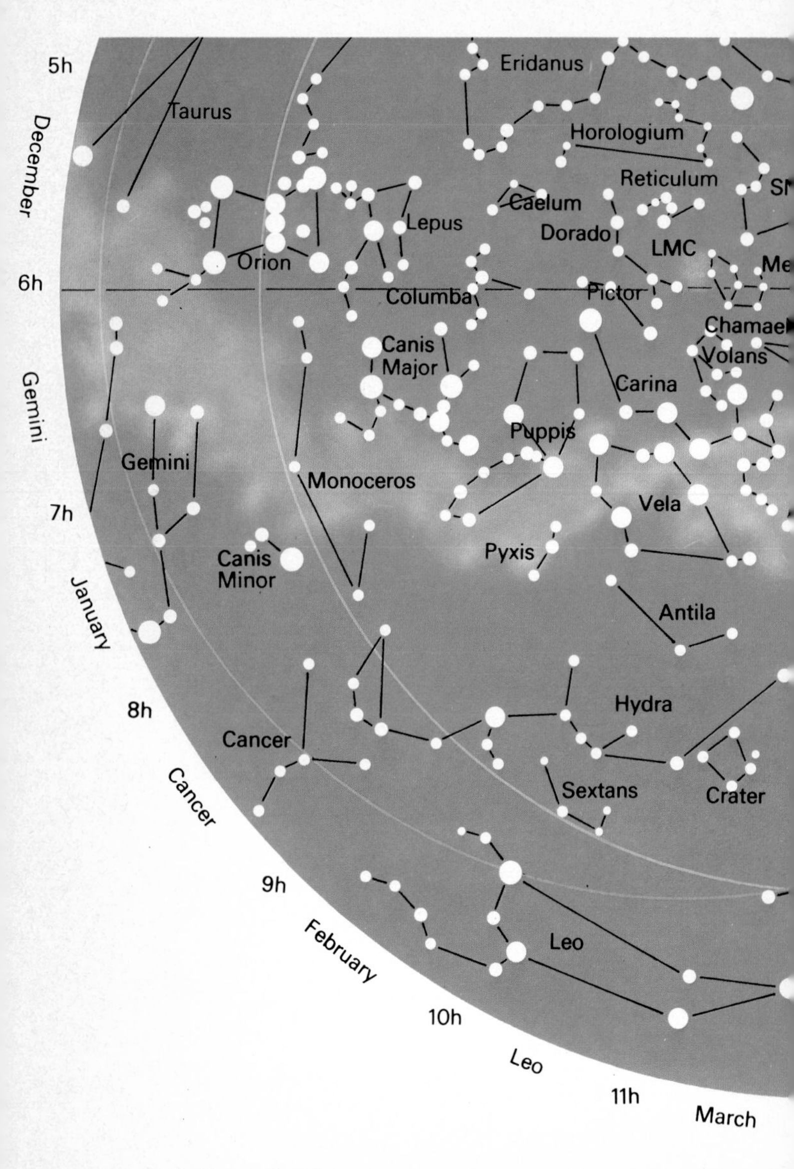

Fig. 58 Southern sky constellations 2.

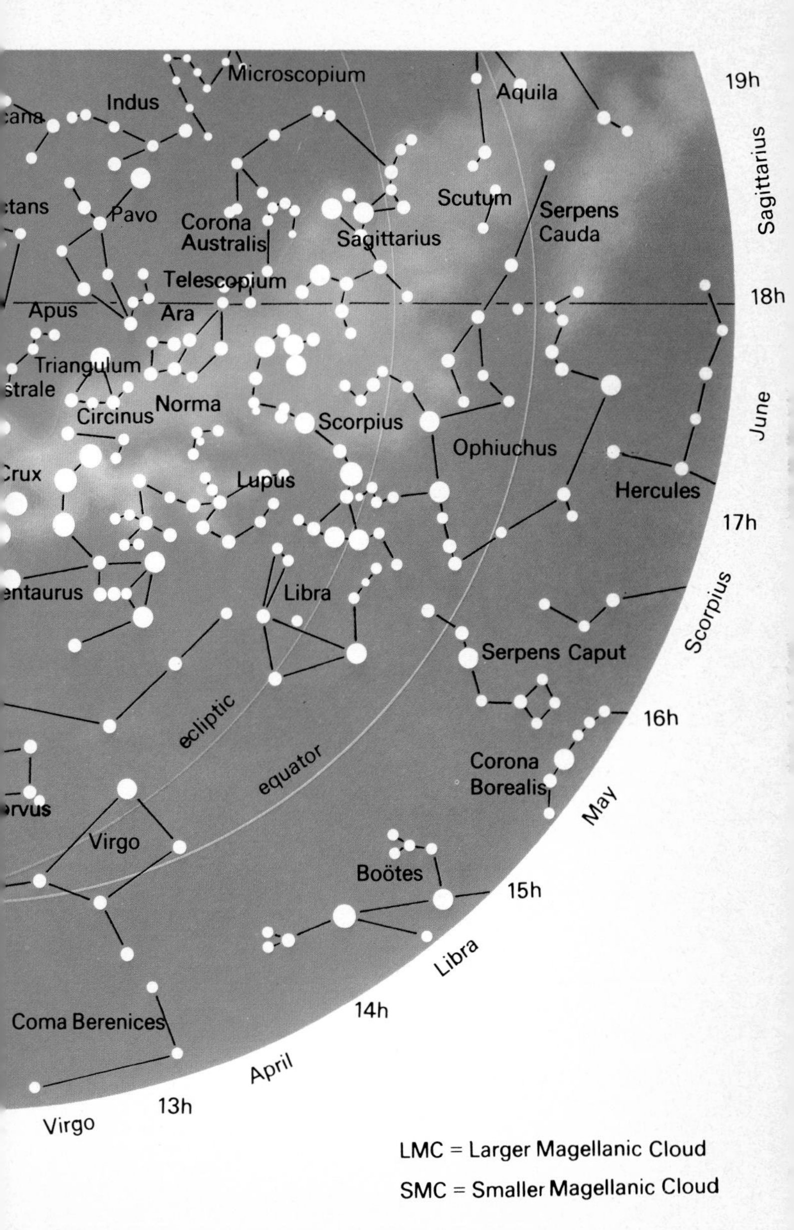

LMC = Larger Magellanic Cloud

SMC = Smaller Magellanic Cloud

The constellations: a description

Andromeda (Andromeda) RA: 01 hr, Dec: 35

This is one of the oldest named constellations and formed part of Ptolemy's catalogue of the second century AD. Mythological legend surrounds the story of the naming of Andromeda and it is one of the best-known groups in the sky, not so much perhaps for the stars it contains but because it includes M31, a giant spiral galaxy very close to our own. These two are the largest in the Local Group of galaxies which contains about 20 members.

The constellation can be found in an area from RA 22 hr 55 min–2 hr 35 min and between Dec 21—53, bordered by Cassiopeia, Lacerta, Pegasus, Pisces, Triangulum and Perseus. The three most prominent stars are laid out along the Dec 30—40 and virtually span the constellation. Star β, Mirach, is a mag 2·02 MO type with absolute mag 0·2, lying at a distance of 76 light years. Originally in the constellation Pegasus, and indeed right on that border, can be found α, Alpheratz, a mag 2·06 B9 star with absolute mag −0·1 lying 90 light years from the solar system. In the opposite direction, and across towards Perseus, γ, Almaak, is seen as a mag 2·14 K3 type with absolute mag −2·4 some 260 light years distant. It is actually a multiple star with companions of mag 5·4 and 6·2.

Heading almost directly north from star δ, the keen observer will see with the naked eye a faint hazy patch lying on a similar declination to Almaak. This, when resolved by a telescope, is seen as the M31 galaxy with a total system mag of 5·0. The galaxy is about $2{\cdot}2 \times 10^6$ light years away.

Antlia (Pump) RA: 10 hr, Dec: −30

This is a southern sky constellation flanked by Centaurus, Hydra, Pyxis and Vela, which is inconspicuous and comparatively unimportant for object content or direction seeking. It lies between RA 9 hr 25 min–11 hr 5 min and Dec −24 — −40.

Apus (Bird of paradise) RA: 15 hr, Dec: −75

Apus is another southern sky constellation which was added by Bayer in the 17th century. It was originally joined by the constellation Avis Indica, a cluster now eliminated and absorbed in adjacent groups. Apus is flanked by the constellations Musca, Chamaeleon, Octans, Pavo, Ara, Triangulum Australe and Circinus. The constellation spans a region of space from RA 13 hr 45 min–18 hr 10 min and between Dec −68 — −82. No interesting stars are contained in this group and the most prominent is the mag 3·8 α. θ is an irregular variable of class M and varies between mag 5·0 and 6·6.

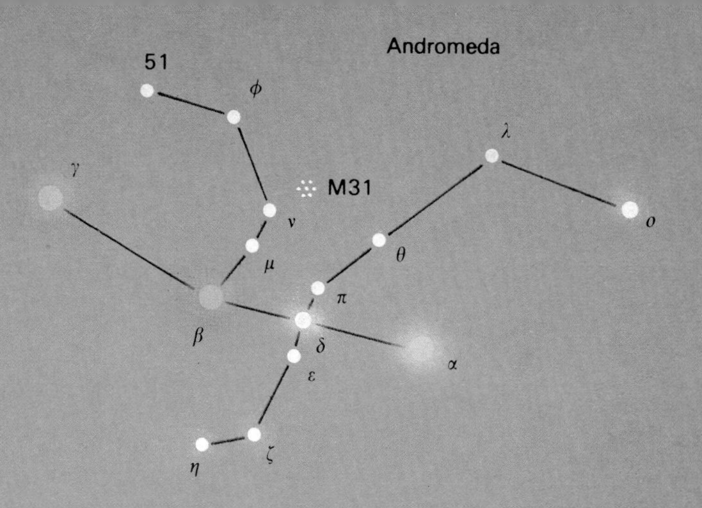

Antila

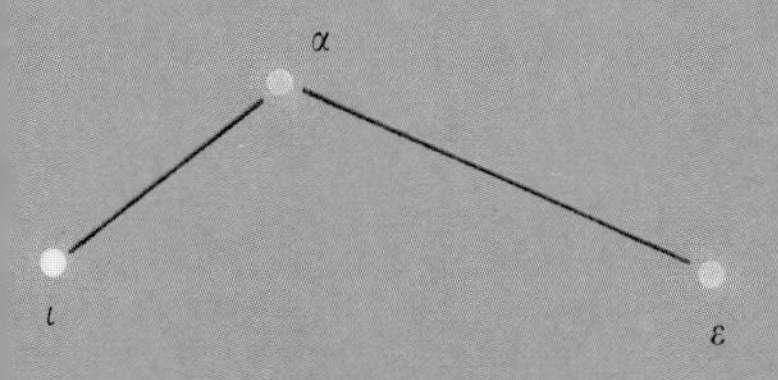

Apus

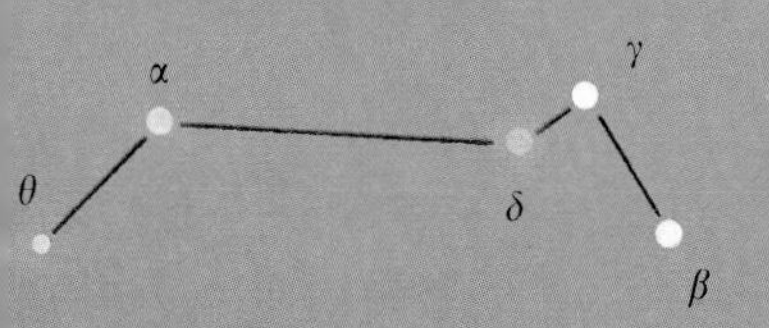

Magnitudes

- 1 and over
- 2
- 3
- 4
- 5
- Nebulae and Clusters

Spectral Classes

- O – B
- A0 – A9
- F – G
- K – M

Aquarius (Water-bearer) RA: 22 hr, Dec: −15

Aquarius is one of the 12 constellations of the zodiac, and one of the oldest named groups. The mythological representation of a water-bearer influenced the Egyptians to believe that its ascendant appearance over the horizon with the Sun brought fertility to the land.

The constellation is flanked by Pegasus, Equuleus, Delphinus, Aquila, Capricornus, Piscis Austrinus, Sculptor, Cetus and Pisces. It extends from RA 20 hr 35 min–23 hr 55 min and from Dec 3 −25, a sprawling expanse of the ecliptic plane. The brightest star in the group is β, Sadal Suud, of visual mag 2·86 and absolute mag −4·6, which is very similar to the Sun in class but a supergiant by type. Star α, called Sadal Melik, is of mag 2·96 and from here a triangle of stars can be seen representing the jug carried by the mythological Water-bearer. The constellation also contains the Saturn Nebula of mag 8, so named because of its similarity to the ringed planet.

Aquila (Eagle) RA: 20 hr, Dec: 05

This is a prominent summer constellation named after the mythological eagle sent to carry Ganymedes to Olympus. Aquila has consistently been associated with birds and the triangular outline is seen to represent a bird with outstretched wings. Aquila occupies a region of the sky from RA 18 hr 40 min–20 hr 35 min and Dec from 8 — −12, flanked by Aquarius, Delphinus, Sagitta, Hercules, Ophiuchus, Serpens, Scutum, Sagittarius and Capricornus.

Generally in the direction of the Milky Way, Aquila is populated with numerous stellar features and an interesting nebula. Star α, Altair, is a very bright white star to the east of the apex, only 16 light years from Earth, and is an A7 class source of mag 0·77 (absolute mag 2·2). It is flanked by γ Aquilae, 340 light years away (mag 2·7) and Alshain β (mag 3·9). Star η Aquilae is a prominent Cepheid variable (mag 3·7–4·4) with a period of seven days.

Ara (Altar) RA: 17 hr, Dec: −55

Ara has precessed considerably from the position it held when named. It lies far to the south but was visible from the Mediterranean in 1000 BC and was so named because of its apparent similarity to an altar. Ara occupies a region of the sky from RA 16 hr 30 min–18 hr 10 min and from Dec −45 — −67.

Flanked by Telescopium, Pavo, Apus, Triangulum Australe, Scorpius and Norma, it is a mediocre collection of stars with members β and α of mag 2·9 and spectral types K3 and B2 respectively (absolute mags are −4·3 and −2·4 due to the 1030 and 390 light year distance).

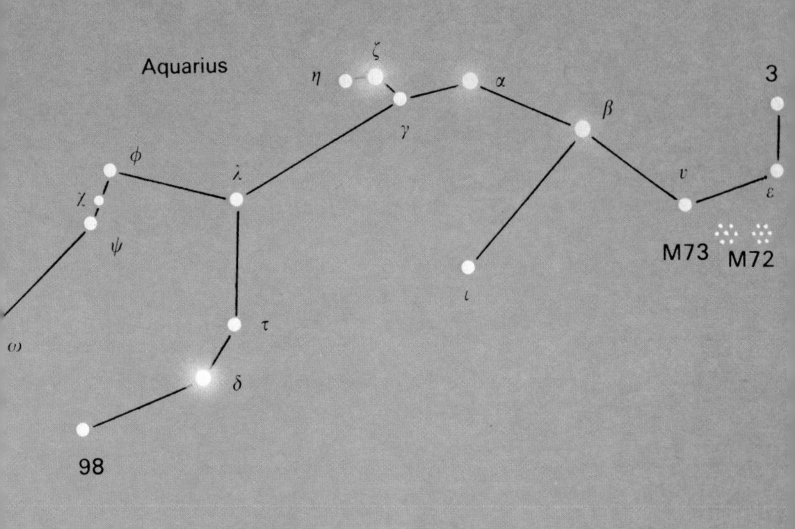

Aquarius
ζ
η
α
3
β
γ
φ
λ
υ
χ
ε
ψ
M73
M72
ι
τ
ω
δ
98

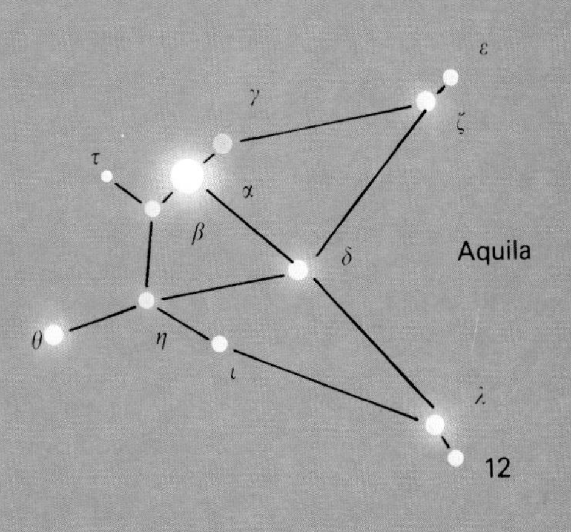

ε
γ
ζ
τ
α
β
Aquila
δ
θ
η
ι
λ
12

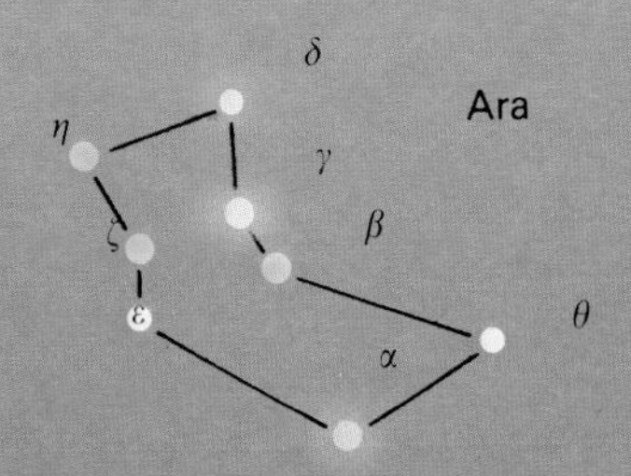

δ
Ara
η
γ
β
ζ
ε
θ
α

Aries (Ram) RA: 02 hr, Dec: 20
Aries, named by the Greeks after the ram with the Golden Fleece, bequeathed its name to the zero point of the system of RA, because the vernal equinox was once located in the constellation. Precession has shifted the equinox into Pisces which, together with Triangulum, Perseus, Taurus and Cetus, borders the constellation. Aries has only two stars above magnitude 4.

These are α, Hamal, with mag 2·0 (76 light years distant) and β, Sheratan, of mag 2·7 (52 light years distant) of spectral types K2 and A5 respectively. The third named star, γ, Mesartim, is double with mag's 4·2 and 4·4. Its name may come from the Arabic word for 'The Sign' and refers to the alignment with the vernal equinox in early history.

Auriga (Charioteer) RA: 05 hr, Dec: 40
Although it is accepted that Auriga represents a bearded man carrying a goat, the Assyrian's viewed this as a chariot and the Greeks saw it as a lame man riding a horse. Auriga achieves fame today from the star ε Aurigae, an eclipsing binary (with 27-year period) with a mag 3 component, orbiting the largest star yet observed – an infrared body some $3{\cdot}0 \times 10^9$ km across, both of which lie 3400 light years from Earth.

The brighter component can be seen through the tenuous envelope of the larger star, and therefore the system appears to be a single variable. ζ Aurigae is an eclipsing binary with a period of 3 years, and the star α, Capella, is interesting since it is a bright (mag 0·05) spectroscopic binary 45 light years away. It has components of 4·3 and 3·3 solar masses and a period of 104 days.

Other prominent stars are β, Menkalina, an A2 type of mag 1·86, ι of K3 type and mag 2·64, and θ a B9 star of mag 2·65.

Boötes (Herdsman) RA: 15 hr, Dec: 30
Boötes is one of the oldest constellations and is mentioned in *The Odyssey*. It is flanked by Serpens Caput, Corona Borealis, Hercules, Draco, Ursa Major, Canes Venatici, Coma Berenices and Virgo and occupies a large portion of the sky from RA 13 hr 35 min–15 hr 50 min and Dec from 8—55.

The most important star is α, Arcturus, of mag −0·06, a red giant K2 type 40 light years away and 30 times the diameter of the Sun. Arcturus was one of the stars first measured by Halley to have motion relative to the Sun. ε, Izar, of mag 2·37, is a double star (companion mag 5·9) often referred to as one of the most beautiful stars in the sky.

Other visible elements of the constellation are η, Saak, of mag 2·69, and γ, Seginus, of mag 3·05, 32 and 118 light years away respectively.

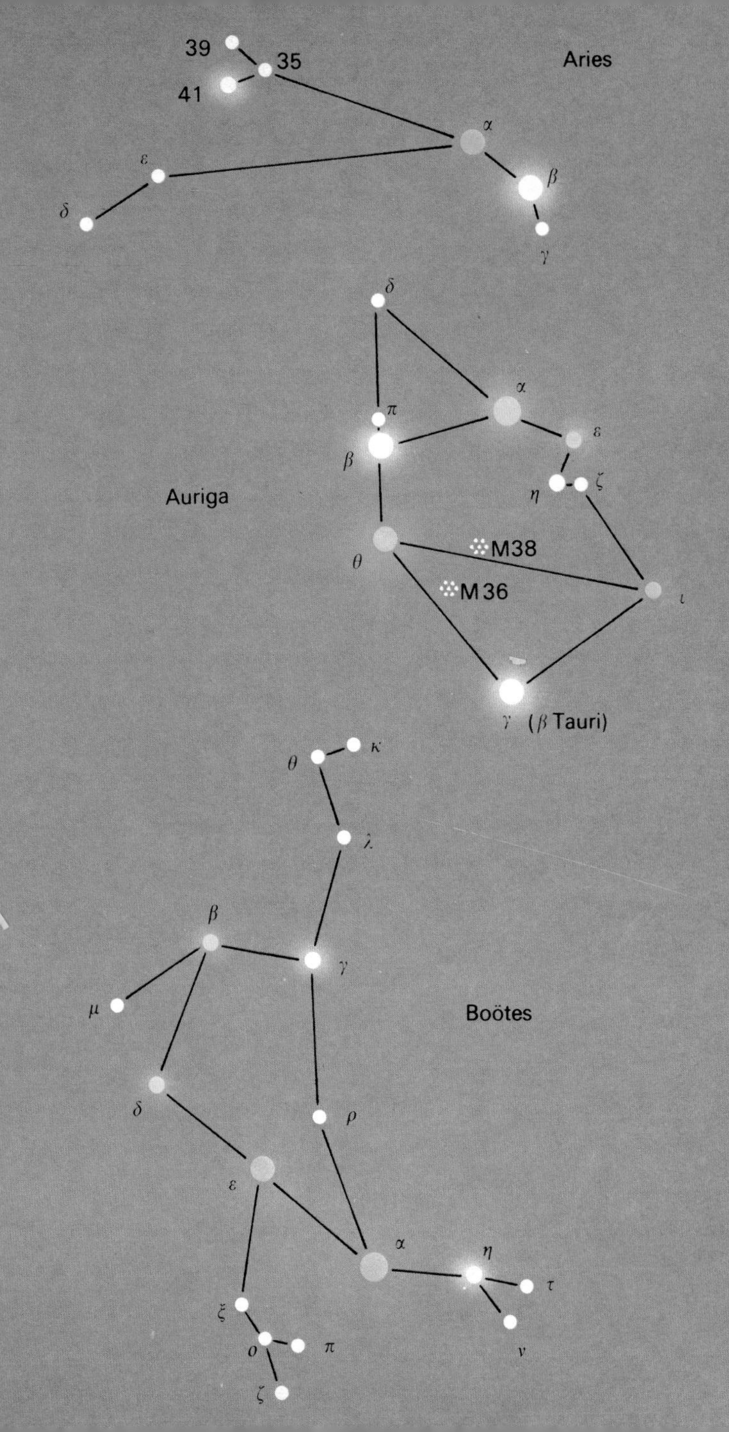

39
35
41
Aries
α
ε
β
δ
γ
δ
α
π
ε
β
ζ
η
Auriga
θ
M38
θ
M 36
ι
γ (β Tauri)
κ
θ
λ
β
γ
μ
Boötes
δ
ρ
ε
α
η
τ
ξ
ο
π
υ
ζ

Caelum (Chisel) RA: 05 hr, Dec: −40

This inconspicuous southern constellation is best dedicated to the memory of the little known astronomer Lacaille than to any serious observational activity. Lacaille studied at the Paris Observatory and made major contributions to establishing an accurate measure of the arc of the meridian. Later, from 1751–3, he derived the positions of some 10 000 stars during which time he set up the constellation Caelum. No other astronomer has made a greater contribution to the mapping of southern constellations.

Caelum spreads from RA 4 hr 20 min–5 hr 5 min and from Dec −27 — −49 and is a winter group hardly visible from 50° north with no interesting components.

Camelopardalis (Giraffe) RA: 06 hr, Dec: 70

This is an inconspicuous constellation occupying a region between Cassiopeia, Perseus, Auriga, Lynx, Ursa Major, Draco, and Ursa Minor from RA 3 hr 10 min–14 hr 30 min and Dec 53—86. Its name, first used by Bartschius in 1614 but believed to have been derived earlier, has been variously given as Camelopardus and Camelopardis.

The constellation occupies a relatively barren region of space and the seven brightest stars are all between mag 4—5. This constellation contains little of interest among the bright objects, although some variable stars are visible with binoculars. Its boundaries are very irregular, especially the western one, which meanders towards Ursa Minor with numerous changes of direction.

Cancer (Crab) RA: 09 hr, Dec: 20

Several thousand years ago the constellation framed the background of the Sun when the latter reached the point of summer solstice and achieved maximum elevation above the celestial equator (23·5°). The Sun was directly overhead along latitude 23° N and this line around the Earth was known as the Tropic of Cancer. Precession has now displaced this constellation from that position.

Cancer is one of the old constellations and comprises one of the 12 zodiacal groups. It occupies a region of the sky from RA 7 hr 55 min–9 hr 20 min and from Dec 7—33, flanked by Lynx, Gemini, Canis Minor, Hydra and Leo. Cancer contains no stars greater than 4th mag and the only interesting objects are two open clusters, M44 and M67. M44 is displaced from a line joining stars γ and δ centred on Dec 20 and contains more than 300 stars between mags 6 and 12, a group otherwise known as Praesepe. The other cluster, M67, lies slightly west of the star α towards β.

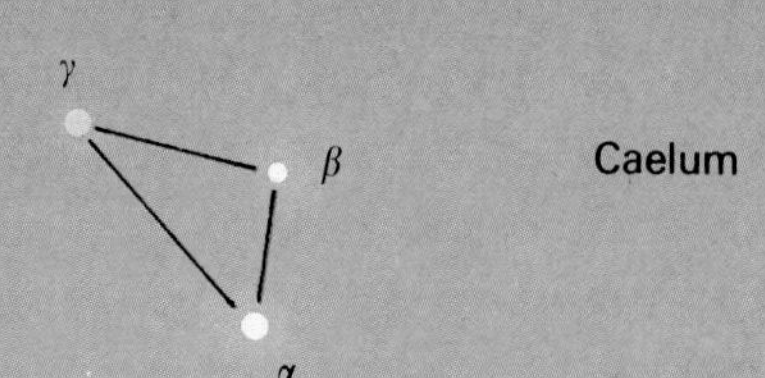
γ
β
Caelum
α

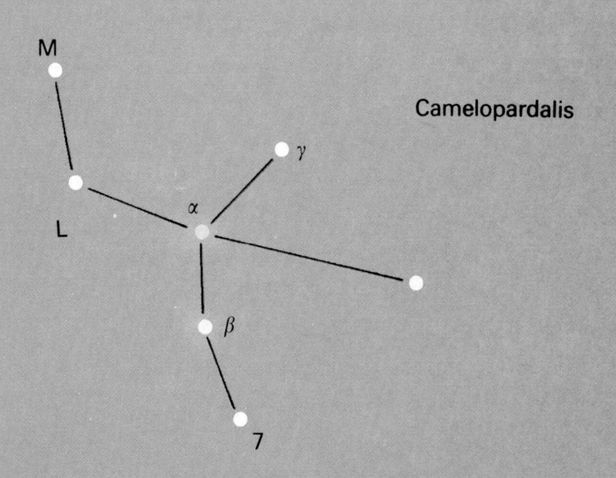
M
Camelopardalis
γ
α
L
β
7

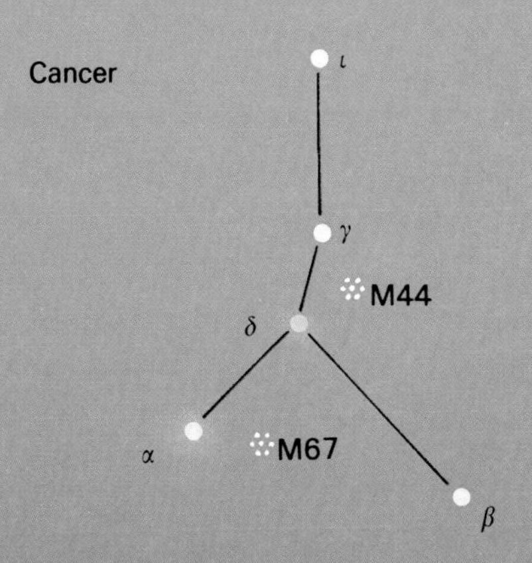
Cancer
ι
γ
M44
δ
α
M67
β

Canes Venatici (Hunting dogs) RA: 13 hr, Dec: 40
Canes Venatici was set up in the late 17th century to fill gaps in Ptolemy's original 48 constellations. It is flanked by Ursa Major, Coma Berenices and Boötes, and spreads from RA 12 hr 5 min–14 hr 5 min and from Dec 28—52. The only useful object for amateurs is the star α, Cor Caroli, a mag 2·9 A0 type 91 light years from Earth. An optical double, Cor Caroli lies visually close to a mag 5·6 star at a remote distance.

The constellation contains four interesting objects other than Cor Caroli. A globular cluster, M3, is located on the extreme southern boundary of the constellation containing over 10^5 stars in a sphere 65 light years across, 6×10^4 light years away. M51, close to the north-west boundary, is the famous Whirlpool Nebula, a spiral galaxy. M63, north-east of Cor Caroli, is another spiral galaxy (mag 9·6) as is NGC 4258 (mag 9·2) north-east of star β. All three galaxies are between 6×10^6 and 9×10^6 light years away.

Canis Major (Big dog) RA: 07 hr, Dec: −25
This constellation lies between RA 6 hr 10 min–7 hr 25 min and Dec −11 to −33, and is framed by Monoceros, Lepus, Columba and Puppis. The most important star is α Canis Majoris, Sirius the Dog star, so named because it is the brightest component of Canis Major which, to the Egyptians, represented Anubis, the jackal-headed god. Sirius rose just before the Sun when the Nile was about to begin its yearly flood, and therefore was of great importance in the Egyptian calendar.

Sirius is a mag −1·43 A1 type less than 9 light years away and is the brightest star in the sky, accompanied by a faint companion orbiting $2{\cdot}9 \times 10^9$ km away in 49·9 years with mag 9·1. Sirius B was the first white dwarf to be discovered. Canis Major contains four other stars brighter than mag 2·5 and also, M41, an open cluster 1300 light years away. Many of the attendants to Sirius are in fact intrinsically brighter but at very great distance, rendering them visually subservient to the prime star.

Canis Minor (Small dog) RA: 07 hr, Dec: 05
Canis Minor is bounded on two sides by Monoceros and contains two stars of interest: Procyon, a mag 0·37 F5 star 11·5 light years distant, and β, Gomeisa, of mag 2·9 but of luminosity −1·1 due to its 210 light year distance. Procyon is six times solar luminosity, twice the size of the Sun but only 1·1 times the mass. Canis Minor lies between RA 7 hr 5 min–8 hr 10 min and Dec 0—13. Procyon and Gomeisa are the only two prominent objects in the constellation.

M51

Canes Venatici

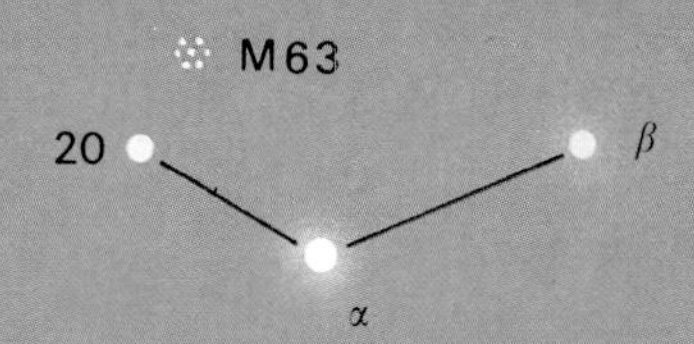

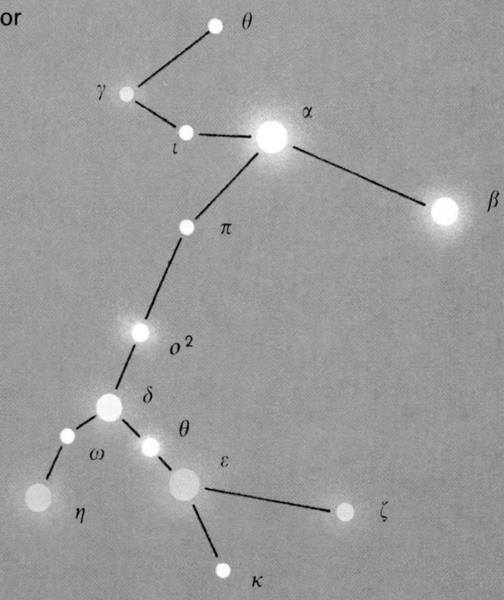

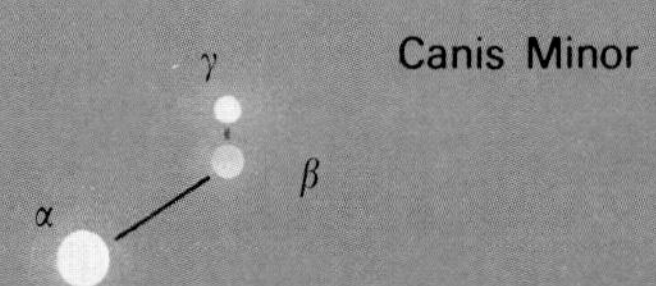

Capricornus (Sea goat) RA: 21 hr, Dec: −15
This constellation was associated with the low point reached by the Sun as it dipped below the celestial equator (23·5°) at winter solstice and therefore gave its name to the Tropic of Capricorn for latitude 23° S. It covers an area from RA 20 hr 5 min–21 hr 55 min and from Dec −9 — −28 and is flanked by Aquila, Sagittarius, Microscopium, Piscis Austrinus and Aquarius.

The main star, δ, Deneb Algedi, is a mag 2·95—2·88 variable A6 type about 50 light years distant. Two naked eye doubles are of interest: α, Prima Giedi, is of mag 3·2 with a binary companion of mag 9, and Secunda Giedi, is mag 3·8 with a binary companion of mag 11. The second pair, β, Dabih, is very close with mags 3·3 and 6. A globular cluster, M30, lies below Deneb Algedi.

Carina (Ship's keel) RA: 09 hr, Dec: −60
Once part of the great sprawling Argo Navis, Carina is now separated (as are Puppis and Vela) and occupies a portion of the sky from RA 6 hr 5 min–21 hr 55 min and from Dec −51 — −75. The constellation is flanked by Volans, Chamaeleon, Musca, Centaurus, Vela, Puppis, and Pictor, and contains the second brightest star in the sky – Canopus, used by planetary spacecraft as a point of automated navigation.

Canopus is mag −0·73, and an F0 type star with an absolute mag of +1·4. Other companions are β, Miaplacidus, a mag 1·67 A0 type 86 light years away, ε Carinae a mag 1·97 K0 type at a distance of 340 light years and ι, Tureis, a 2·25 mag F0 star 750 light years distant. A rich globular cluster, NGC 2808, lies due east of υ, a mag 2·97 A7 star. An interesting variable, η Carinae, grew from mag 4 to rival Sirius by 1843 and then, 10 years later, dimmed to its current mag 8.

Cassiopeia (Cassiopeia) RA: 01 hr, Dec: 60
Cassiopeia extends from RA 23 hr–3 hr 35 min and Dec 47–77. α, Schedir, and γ, Tsih, are variable stars (about mag 2·16 and 1·6–2·9, respectively) and the latter has a mag 8·18 companion. Of the remaining three stars making up the famous 'W', β, Chaph, is a mag 2·26 F2 type, δ, Ruchbah, is a mag 2·67 A5 (probably an eclipsing variable with a 759-day period) and ε Cassiopeia is a mag 3·3.

All but the last are less than 150 light years distant but ε is 500 light years away with an absolute mag of −2·7. The southern half of the constellation is flooded by the Milky Way and many open clusters populate the region including M52 and M103. Cassiopeia lies opposite Ursa Major across the celestial north pole.

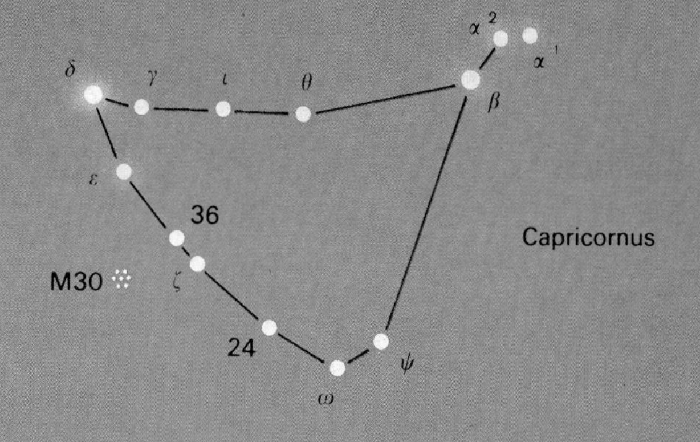
Capricornus
δ
γ
ι
θ
α 2
α 1
β
ε
36
M30
ζ
24
ψ
ω

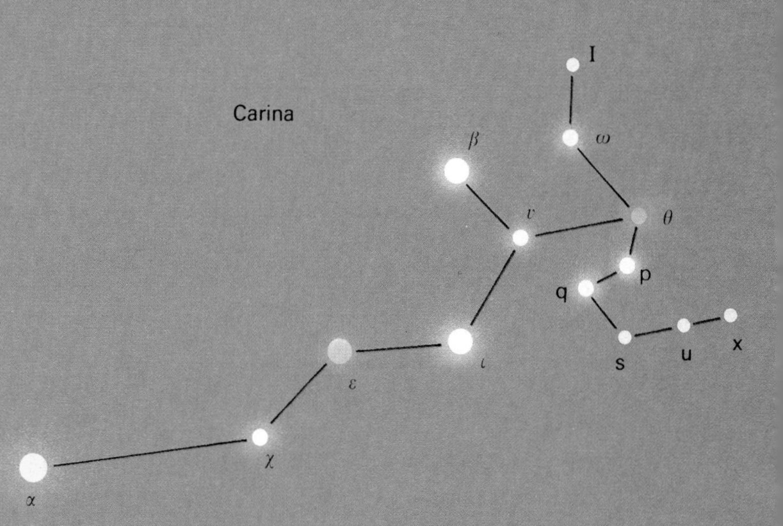
Carina
I
β
ω
υ
θ
p
q
s
u
x
ι
ε
χ
α

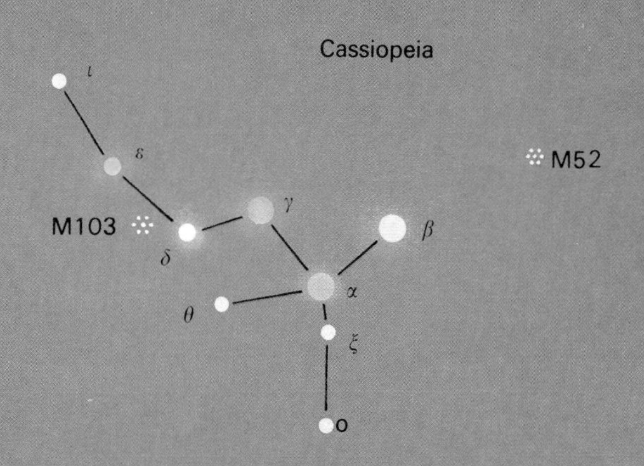
Cassiopeia
ι
ε
M52
M103
γ
δ
β
α
θ
ξ
o

Centaurus (Centaur) RA: 13 hr, Dec: −50
Centaurus is a large, irregular form, lying between RA 11 hr–15 hr and Dec −30 — −65. It contains the α Centauri system only 4·3 light years away (the closest star except the Sun) which is actually a binary with components of mag 0·01 and 1·3. A mag 11·1 companion, called Proxima Centauri, appears to orbit this binary and actually comes between the α system and the Sun making it the solar system's closest companion. β Centauri (Hadar) is 390 light years away with an absolute mag of −5·2.

Cepheus (Cepheus) RA: 21 hr, Dec: 55
This constellation lies between RA 20 hr–8 hr 20 min and Dec 53—88, and contains δ Cephei, the prototype Cepheid Variable with a mag of 3·51 —4·42 in a period of 5·4 days. α, Alderamin, is an A7 2·44 mag star 52 light years away and β, Alfirk, is a variable (mag 3·14—3·19) binary. The only other stars of note are γ, Er Rai, a mag 3·2 object, and ζ Cephei of mag 3·31.

Cetus (Whale) RA: 02 hr, Dec: −5
Cetus is located between RA 23 hr 55 min–3 hr 20 min and Dec −25—10, and contains a very striking variable star. Known as Mira the variable star *o* Ceti changes between mag 2—10 in a period of 332 days, taking seven months to wane and three months to wax. The star is a supergiant M6 type about 424 × 10^6 km across attended by a B-type binary companion in a 14-year orbit. The B star appears to interact with matter ejected from the pulsating supergiant.

Mira radiates 3·5 times as much energy at maximum brightness as it does at minimum and the star is the prototype for the class of long-period (or Mira) variables. Other interesting stars are β, Diphda, a K1 type with mag 2·02, and α, Menkar, a M2 with mag 2·54.

Chamaeleon (Chameleon) RA: 11 hrs, Dec: −80
Surrounded by Octans, Apus, Musca, Carina, Volans and Mensa, this constellation lies between RA 7 hr 40 min–13 hr 35 min and Dec −75 — −82. The brightest stars in the group are of mag 4, and δ and ε Chamaeleontis are visual binaries.

The constellation is best found by rotating 45° around the celestial south pole from Apus or by fixing Crux and then the south pole so that Chamaeleon is seen to lie between the two. The constellation was one of a series named by Bayer very early in the 17th century.

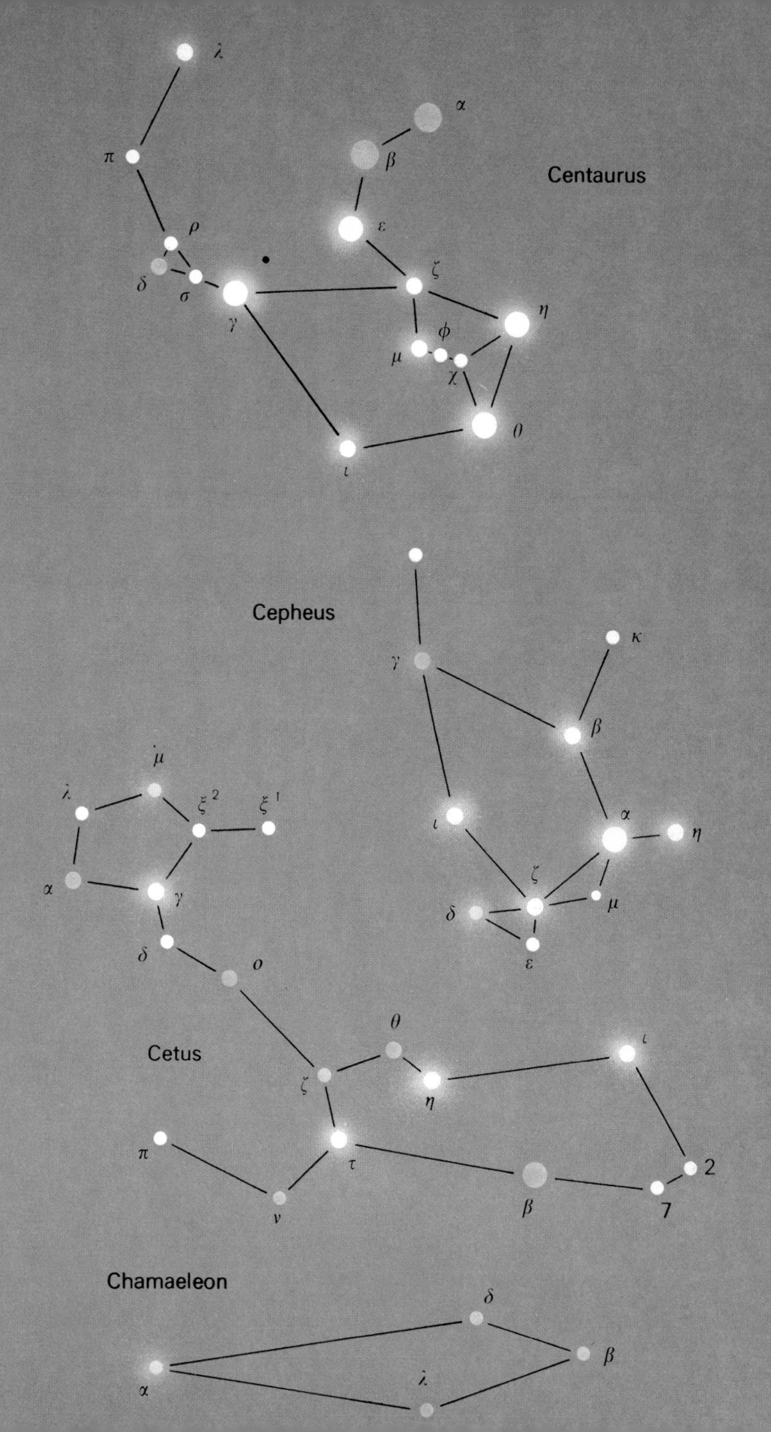
Centaurus
λ
α
π
β
ε
ρ
δ
σ
γ
ζ
η
φ
μ
χ
θ
ι
Cepheus
κ
γ
β
ι
α
η
ζ
δ
μ
ε
μ
λ
ξ 2
ξ 1
α
γ
δ
ο
θ
Cetus
ι
ζ
η
π
τ
2
ν
β
7
Chamaeleon
δ
β
α
λ

Circinus (Compass) RA: 16 hr, Dec −65
This constellation occupies a small region of the southern sky between 13 hr 35 min–15 hr 25 min RA and Dec −55 — −70. It appears to form a broad, inverted 'L' and is flanked by Triangulum Australe, Norma, Lupus, Centaurus, Musca and Apus. Were it not for the separation of this constellation by Lacaille in 1763 (when he contributed 14 constellations to the charts) it would more properly be seen as a part of the constellation Centaurus branching, perhaps, from α Centauri the closest star system to the Sun. In any event, positive location of the Centaurus components helps with the identification of Circinus. The only really interesting star in this group is α Circini, a double (mags 3·4 and 8·8) of yellow and reddish appearances respectively.

Columba (Dove) RA: 06 hr, Dec: −35
This southern constellation is flanked by Lepus, Caelum, Pictor, Puppis and Canis Major and extends across a region of space between RA 5 hr–6 hr 40 min and Dec −27 — −43. The constellation's name is a genuine attempt to immortalize biblical events and was originally Columba Noae, which by literal interpretation means The Dove of Noah. The abbreviated expression 'Columba' is almost always used but the original is still the official title.

The only two stars of interest are α, Phakt, and β, Wezn. Phakt is a B8 star with an apparent mag of 2·64 and an absolute mag of −0·06 and is about 140 light years away. Wezn is of mag 3·2 and lies at the centre of the irregular 'T' formed by the constellation.

Coma Berenices (Berenice's hair) RA: 13 hr, Dec: 20
This constellation forms one of Brahe's list from the 17th century and is flanked by Ursa Major, Leo, Virgo, Boötes and Canis Venatici, lying between RA 11 hr 55 min–13 hr 35 min and Dec 14—34. The origin of the name lies in Ptolemaic Egypt when a Pharaoh's sister, Berenice, promised to offer her severed hair to Venus if her husband returned safe from the Syrian wars. He did, but the locks were lost from the temple of Venus and the story was developed that Jupiter had removed them to form the constellation. Among others, Eratosthenes mentions the constellation.

There are no stars greater than mag 4·5 in Coma Berenices but the north galactic pole lies close to a line connecting β and γ and the region is rich in extragalactic nebulae. Since it lies far above the galactic plane many external galaxies, some in clusters of hundreds, are to be found here.

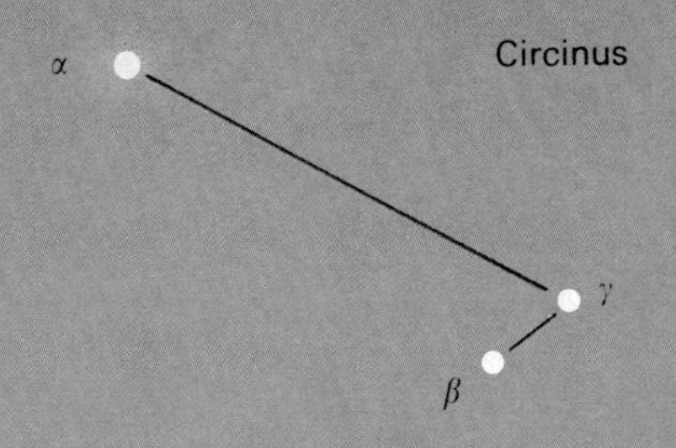

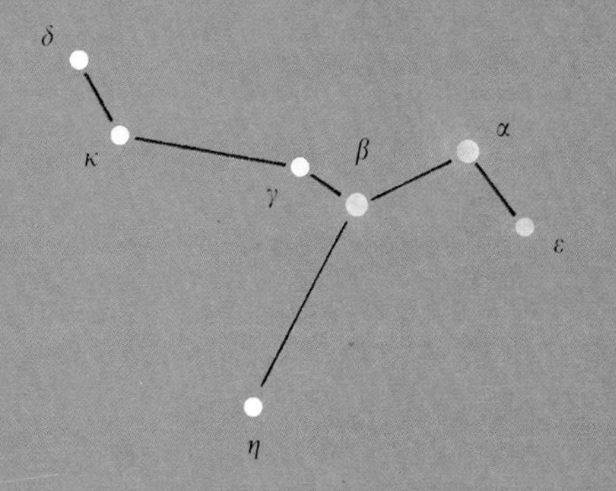

Columba

Coma Berenices

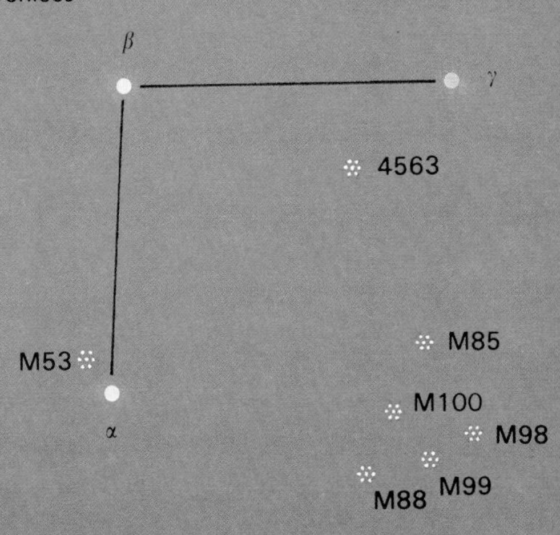

Corona Australis (Southern crown) RA: 19 hr, Dec: –40
The constellation was one of Ptolemy's original group of 48 drawn up in the second century AD. It lies between RA 17 hr 55 m–19 hr 15 min, Dec −37 — −46, bordered by Sagittarius on the east, Telescopium and Ara on the southern flank and Scorpius to the west. Corona Australis forms an arc apparently lying within the embrace of Sagittarius.

Corona Australis can be found by first locating the stars α, β and ε in Sagittarius which form a triangle. The constellation is then found to lie within this triangle and has been so named because of its apparent similarity to the Corona Borealis which lies in the northern celestial hemisphere. No prominent objects lie within the constellation and the brightest stars are of mag 4 making it a relatively inconspicuous and comparatively unimportant constellation.

Corona Borealis (Northern crown) RA: 16 hr, Dec: 30
This group looks very much like a more luminous, northerly duplicate of the Corona Australis. It lies between 15 hr 15 min–16 hr 25 min in RA and 26—40 in Dec, flanked by Boötes and Hercules with Serpens Caput to the south.

The most interesting star in the group is α, Alphecca, an eclipsing variable (Δ mag 0·1) with a mean mag of 2·23 and a period of 17·4 days. The A0 type star is 76 light years away with an absolute mag of 0·4. Star R Coronae Borealis is the prototype of irregular variables observed to fade randomly from its usual steady brightness, a process caused by absorption by carbon particles in the stellar atmosphere, and not due to an eclipsing companion.

Corvus (Crow) RA: 12 hr, Dec: −20
Shaped rather like a kite this southern constellation is flanked by Crater, Hydra and Virgo, seeming to form the south-western corner of the latter, and lies between RA 11 hr 55 min–12 hr 55 min and Dec −11 — −25. The constellation is one of the original 48 groups.

The four brightest stars, γ, β, δ and ε, are all brighter than mag 3·1 although the fifth star in order of apparent mag is α, an alpha designation usually applied to the brightest star in a given group. Star γ, Gienah, is a mag 2·59 B8 type at a distance of 450 light years and absolute mag of −3·1. β is a mag 2·66 G5 type (very similar to the Sun) at 108 light years. Star δ is a double (mags 2·7 and 8·26) with the secondary called The Raven. ε is a mag 3·04 K3 object at 140 light years distance.

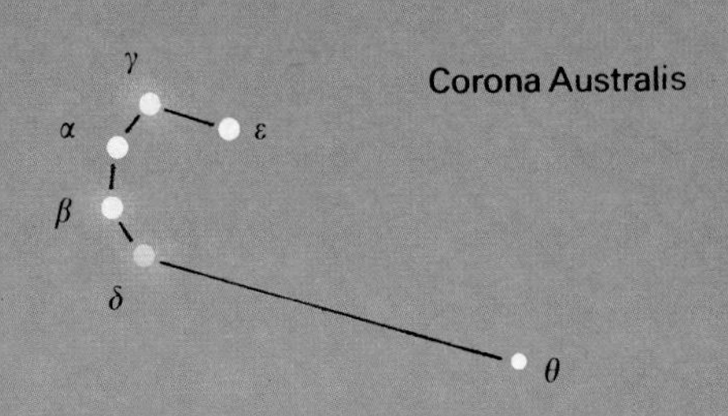
Corona Australis
γ
α
ε
β
δ
θ

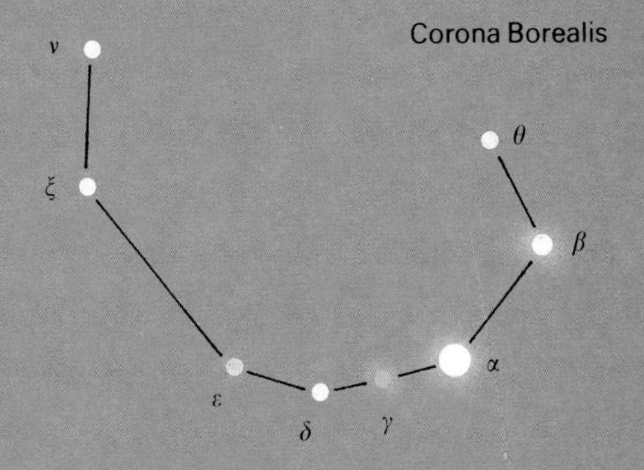
Corona Borealis
ν
θ
ξ
β
α
ε
δ
γ

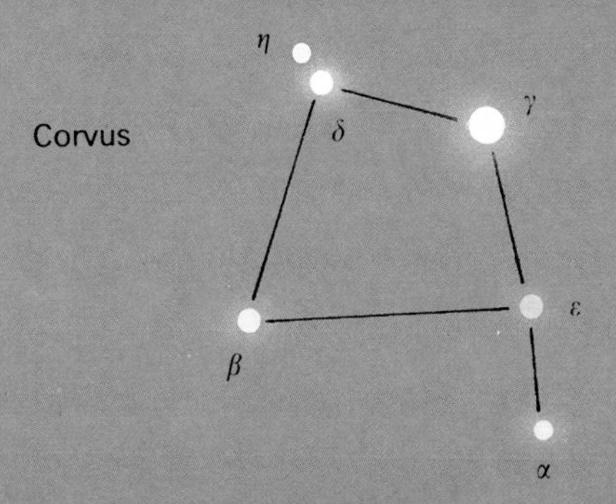
Corvus
η
γ
δ
ε
β
α

Crater (Cup) RA: 11 hr, Dec −15
This is a southern sky constellation flanked by Leo, Virgo, Sextans, Hydra and Corvus, lying between RA 10 hr 50 min–11 hr 55 min and Dec −6 — −25. It is very inconspicuous but, nevertheless, one of Ptolemy's original 48 constellations. All the stars are of mag 4 or less.

Crux (Southern cross) RA: 12 hr, Dec: −60
Crux is a comparatively small constellation between RA 11 hr 55 min–12 hr 55 min and Dec −55 — −64 which seems to lie within the folds of Centaurus and is surrounded on three sides by this constellation with Musca to the south. The Southern cross was added to the list of constellations in the 17th century, and is most famous for its almost exact axial alignment with the south celestial pole.

The prime star is α, Acrux, actually a triple system with components of mag 1·39, 1·86 and 4·9. The brighter pair is seen as a single light source at mag 0·87 at a distance of 370 light years with absolute mags of −3·9 and −3·4. Both are class B stars. β, Mimosa, is even further away, at 490 light years, and being a B0 star it has an absolute mag of −4·6. γ is mag 1·68 at a distance of 220 light years and is a M3 star of absolute mag −2·5. δ, Crucis, is a variable star with a mean mag of 2·81 (2·78–2·84) at a distance of 570 light years.

Star ε is of mag 3 and seems ill-placed in this symmetrical system. In the area between α and β lies the Coalsack, a famous non-luminous nebula imposing in aspect and filled with inert dust and gas.

Cygnus (Swan) RA: 21 hr, Dec: 40
For an obvious reason Cygnus is sometimes referred to as the northern cross, and the constellation occupies a region between 19 hr 5 min–22 hr RA and 28—60 Dec. It is notable for containing one of the first sources of detected radio emission and also for including the famous Cygnus X-1 X-ray source now thought to be the product of matter spiralling into a black hole. α, Deneb, is one of the brightest stars in the sky and, with a mag of 1·26 and a distance of 1600 light years, the star is seen to be an A2 type with an absolute mag of −7·1, 30 000 times the luminosity of the Sun.

One of the most rewarding sights in the sky is the optical double β, Albireo, of mag 3·07, a rich blue K type star 410 light years away accompanied by a golden partner of mag 5. Most of the stars in this constellation are intrinsically bright and the constellation notes (see p. 144) contain additional information.

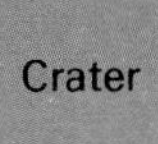
Crater

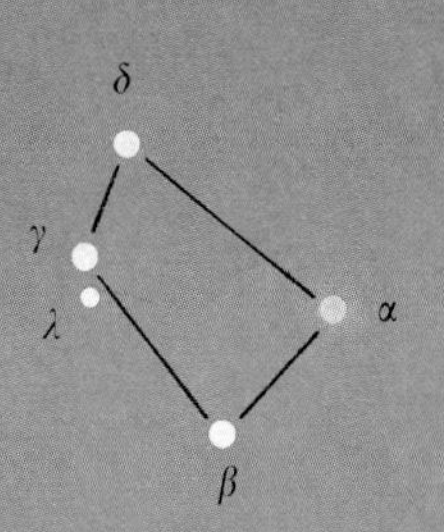
δ
γ
λ
α
β

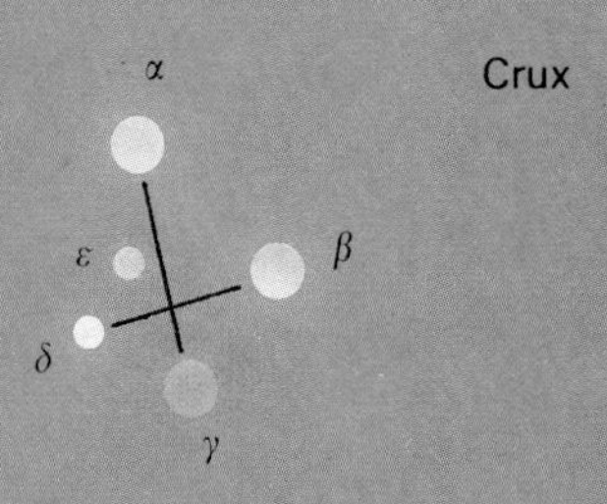
α
Crux
ε
β
δ
γ

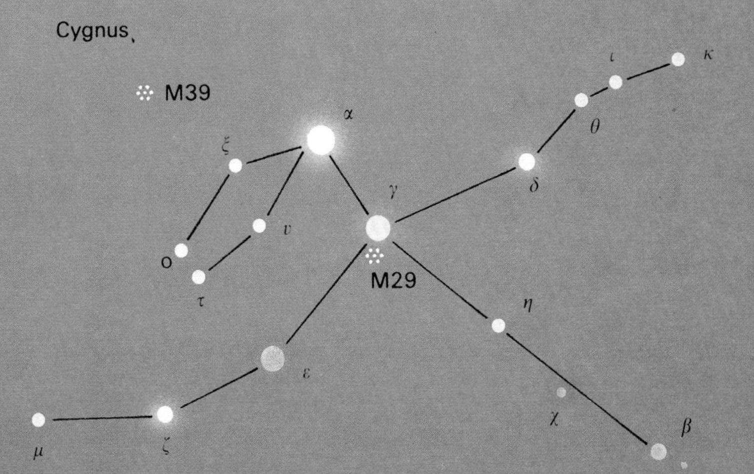
Cygnus
ι
κ
M39
α
θ
ξ
δ
γ
υ
ο
M29
τ
η
ε
χ
β
μ
ζ

Delphinus (Dolphin) RA: 21 hr, Dec: 15

Delphinus lies in a part of the sky populated by several constellations of marine affiliation (Aquarius, Capricornus, and so on) and extends from RA 20 hr 10 min–21 hr 10 min and from Dec 2—21 flanked by Vulpecula, Sagitta, Aquila, Aquarius, Equuleus and Pegasus. Looking somewhat like a petard it contains doubles α, Sualocin, with components of mag 4·0 and 9·5, and β, Rotanev. These names perpetuate the memory of Nicolo Cacciatore, assistant to Piazzi the astronomer who discovered the first asteroid. Cacciatore means 'hunter' and the reverse spelling in Latin is Rotanev. The Latin for Nicolo is Nicolaus and spelt backwards this reads Sualocin.

Dorado (Swordfish) RA: 05 hr, Dec: −60

This constellation is one of the 'modern' designated star groups named by Bayer and lies between RA 3 hr 50 min–6 hr 35 min and Dec −48 — −70. It is flanked by Mensa, Volans, Pictor, Caelum, Horologium, Reticulum and Hydrus. There are no particularly outstanding stars and the only object of note in the group is α Doradus of mag 3·5. However, the constellation does contain the Large Magellanic Cloud (LMC) in its southern sector which appears as a faint patch of incandescence more than 1 hr of longitude in width and centred on Dec −70.

The LMC is so called because it was discovered by the circumglobal Magellan expedition early in the 16th century. Visible to the keen, naked eye the LMC is an irregular galaxy lying some $1{\cdot}7 \times 10^5$ light years from the Sun with about 10 per cent of the volume of the Milky Way galaxy. An extremely bright double star, S Doradus, displays visual mag 8·2—9·4 and at the distance of the LMC has an absolute mag greater than −8 making it 6×10^5 times brighter than the Sun.

Draco (Dragon) RA: 18 hr, Dec: 60

This is a circumpolar constellation occupying a sinuous spread of sky from RA 9 hr 20 min–20 hr 40 min and Dec 47—86. The brightest star in Draco is γ, Etamin, a mag 2·21 K5 object more than 100 light years away. In 3000 BC α, Thuban, a spectroscopic binary of mag 3·6, was the pole star but subsequent precession of the equinoxes has now moved Polaris to this esteemed position. β, Rastaban, of mag 2·77, is actually a double star with component mags of 2·7 and 11·5. Draco contains the north pole of the ecliptic lying roughly between stars δ and ζ on the 18 hr RA arc. It is one of the oldest constellations on record, known to the Arabs, Egyptians, Chinese and Greeks under different designated names.

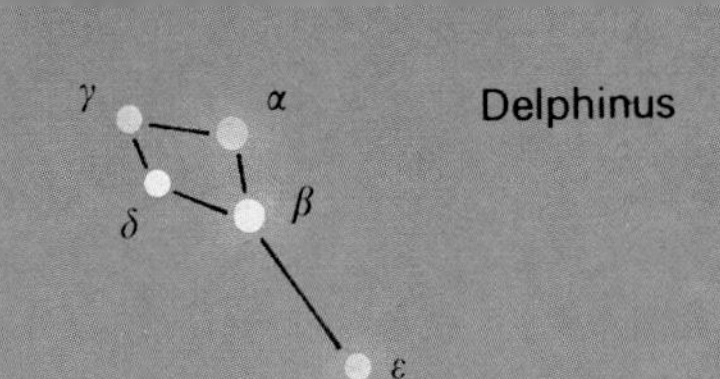
Delphinus
γ
α
δ
β
ε

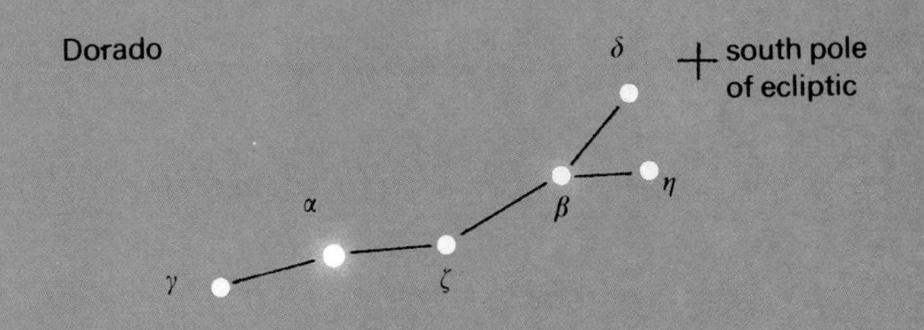
Dorado
δ
south pole
of ecliptic
η
α
β
γ
ζ

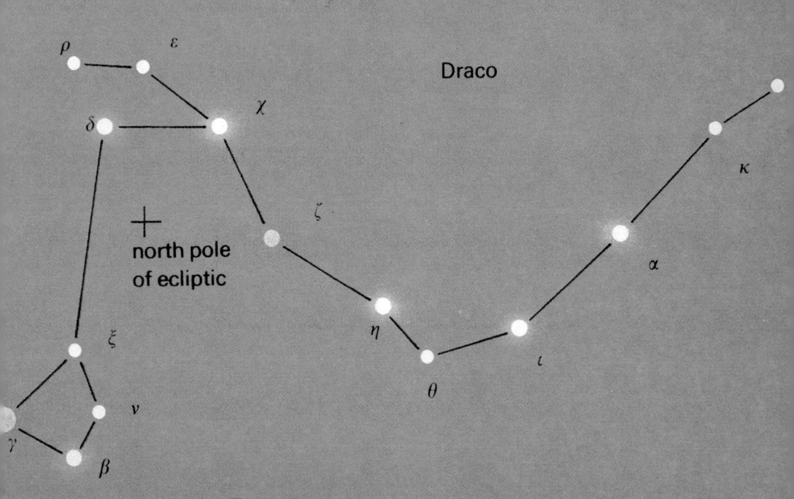
ρ
ε
Draco
δ
χ
κ
ζ
north pole
of ecliptic
α
ξ
η
ι
θ
ν
γ
β

Equuleus (Small horse) RA: 21 hr, Dec: 10

Equuleus is a small constellation flanked by Pegasus, Delphinus and Aquarius lying between RA 20 hr 55 min–21 hr 25 min and Dec 2–13. Although it is a very inconspicuous group, the constellation is a member of the old list known to the Babylonians. It is best found by following a line established by the stars λ, γ and α in the constellation of Aquarius and is displaced from α by as great a distance as that separating α from λ in Aquarius.

Only three stars of mag 4 are contained in Equuleus and star β is a mag 3 object north-west of α. Nothing of greater interest is seen to lie within the boundaries of this constellation but it presents a good location between Aquarius and Delphinus for the observer.

Eridanus (River Eridanus) RA: 03 hr, Dec: −25

Eridanus is supposedly a representation of the celestial equivalent of the Nile to the Egyptians and the Euphrates to the Babylonians as it snakes its exceedingly sinuous path flanked by Taurus, Cetus, Fornax, Phoenix, Hydrus, Horologium, Caelum, Lepus and Orion. In the extremes of its sinuosity, which includes many celestial bays, the constellation covers an area from 1 hr 25 min–5 hr 10 min RA and from 0 — −58 Dec.

The brightest star in the group, α, Achernar, is not visible to observers in the northern latitudes because of the extreme celestial latitude covered by the constellation. The star has a mag of 0·53 and it lies at a distance of 120 light years with an absolute mag of −2·3. β, Kursa, is a mag 2·8 star of A3 class with an absolute mag of 0·9 and it lies at a distance of 80 light years. Star θ Eridani is a triple system with the brightest member visible as a yellow dwarf to the naked eye and two extremely faint companions, one of which is a red dwarf and the other a white dwarf.

Fornax (Furnace) RA: 03 hr, Dec: −30

This constellation lies surrounded by Cetus, Phoenix and Eridanus, between RA 1 hr 40 min–3 hr 50 min and from Dec −24 — −40, and is a boxed region presenting almost a perfect rectangle. The name Fornax (Latin for furnace) was given to the constellation by the roving astronomer Lacaille in the mid-18th century and after naming a nearby group 'The Sculptor' he proceeded to assign this craftsman a furnace with which to accomplish his assumed tasks.

The three stars making up the main configuration take on the appearance of a flattened 'V' and lie in the centre of the rectangular area covered by this constellation. No stars brighter than mag 4 are found in this region.

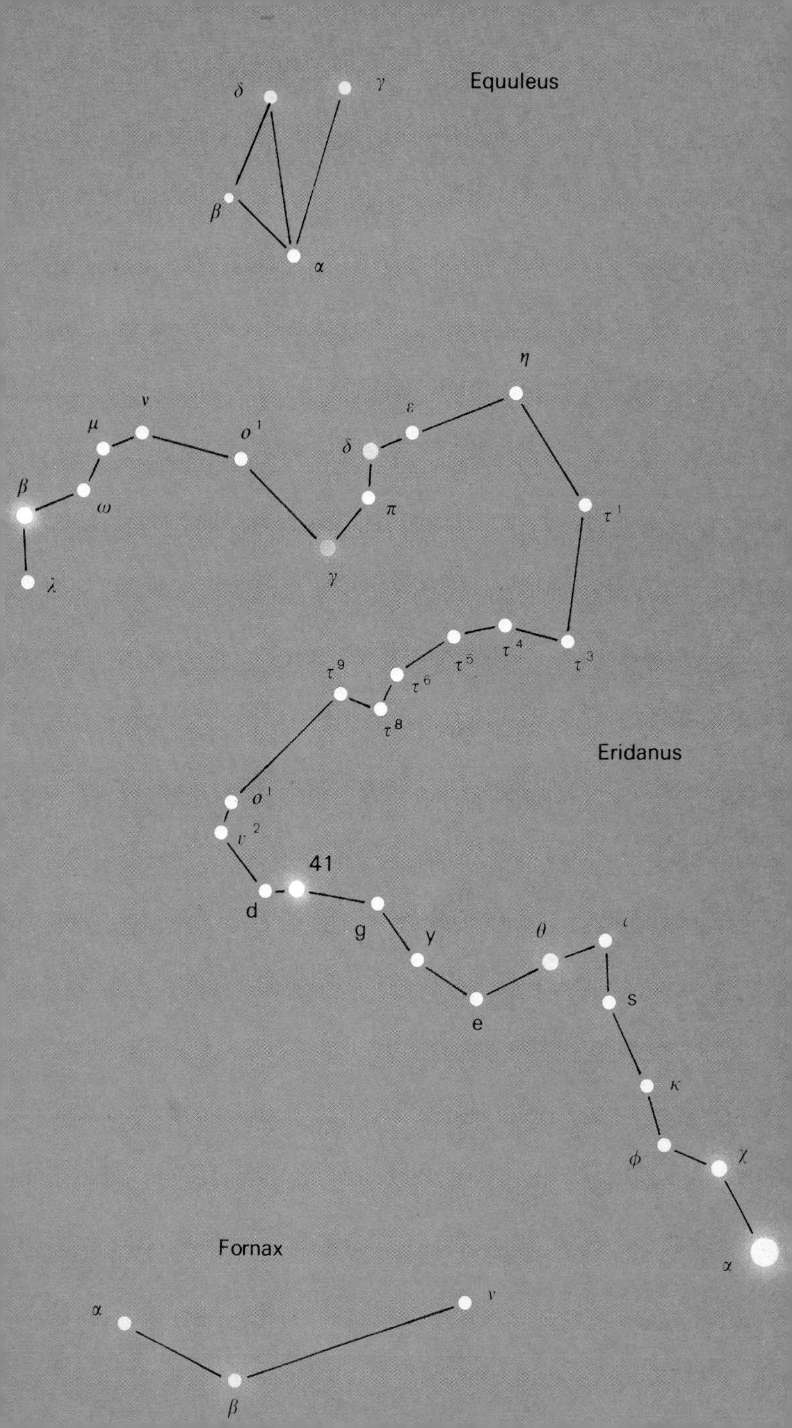
Equuleus
δ
γ
β
α
η
ε
ν
μ
o^1
δ
β
ω
π
τ^1
λ
γ
τ^4
τ^3
τ^5
τ^9
τ^6
τ^8
Eridanus
o^1
v^2
41
d
g
y
θ
ι
e
s
κ
φ
χ
α
Fornax
α
ν
β

Gemini (Twins) RA: 07 hr, Dec: 25

Gemini is one of the constellations of the zodiac, and is surrounded by Auriga, Taurus, Orion, Monoceros, Canis Major, Cancer and Lynx. It lies between RA 5 hr 55 min–8 hr 5 min and Dec 10—35.

Gemini is most famous for the stars α, Castor, and β, Pollux, which have come to signify the entire group. Star α is of mag 1·62 and consists of three spectroscopic binaries of stars (mags 1·97, 2·95 and 9·08) at a distance of 45 light years. Each of these has a coupled motion and the six stars pursue interlinked orbits about each other. Pollux is the brightest member of the constellation, and a mag 1·16 K0 star 35 light years distant but intrinsically dimmer than the Castor system.

Alhena is an A star of mag 1·93 at a distance of 105 light years, μ, Tejat, is a variable with a mean mag of 2·92 at 160 light years and ε, Mebsuta, is a mag 3 G8 star more than 1000 light years from the Sun. The constellation accommodates a planetary nebula and open star clusters, and Neptune and Pluto were discovered when passing through Gemini.

Grus (Crane) RA: 22 hr, Dec: −45

Grus is a 17th-century constellation added by Bayer which lies between RA 21 hr 25 min–23 hr 25 min and from Dec −37 — −56 flanked by Piscis Austrinus, Microscopium, Indus, Tucana, Phoenix and Sculptor.

The most prominent stars are α, Alnair, a mag 1·76 B5 star 65 light years distant, and β, a slightly variable mean mag 2·17 M3 star nearly 300 light years away. Star γ Al Dhanab is a B8 object of mag 3·03 and is the brightest of all three, intrinsically at mag −3·1 but reduced in apparent mag by its 550 light year distance. A 4th mag star, δ, is a naked eye double with two components designated $\delta 1$ and $\delta 2$.

Hercules (Hercules) RA: 17 hr, Dec: 30

This is one of the early constellations, and is surrounded by Draco, Boötes, Corona Borealis, Serpens Caput, Ophiuchus, Aquila and Lyra from RA 15 hr 45 min–18 hr 55 min and from Dec 4—51.

The most important star, α, Ras Algethi, is a cool red M type super-giant variable of mean mag 3·5 with a mag 5·4 G type companion. Ras Algethi has been estimated to be up to $4{\cdot}5 \times 10^{10}$ km in radius which, if true, makes it the largest known star. β, Kornephoros, is a G type 2·8 mag star 100 light years distant, while ζ and μ Herculis are multiple systems.

An interesting globular cluster, M13, can be found between η and ζ. It is believed to contain more than 10^5 stars in a group 100 light years across. The cluster is 34 000 light years away.

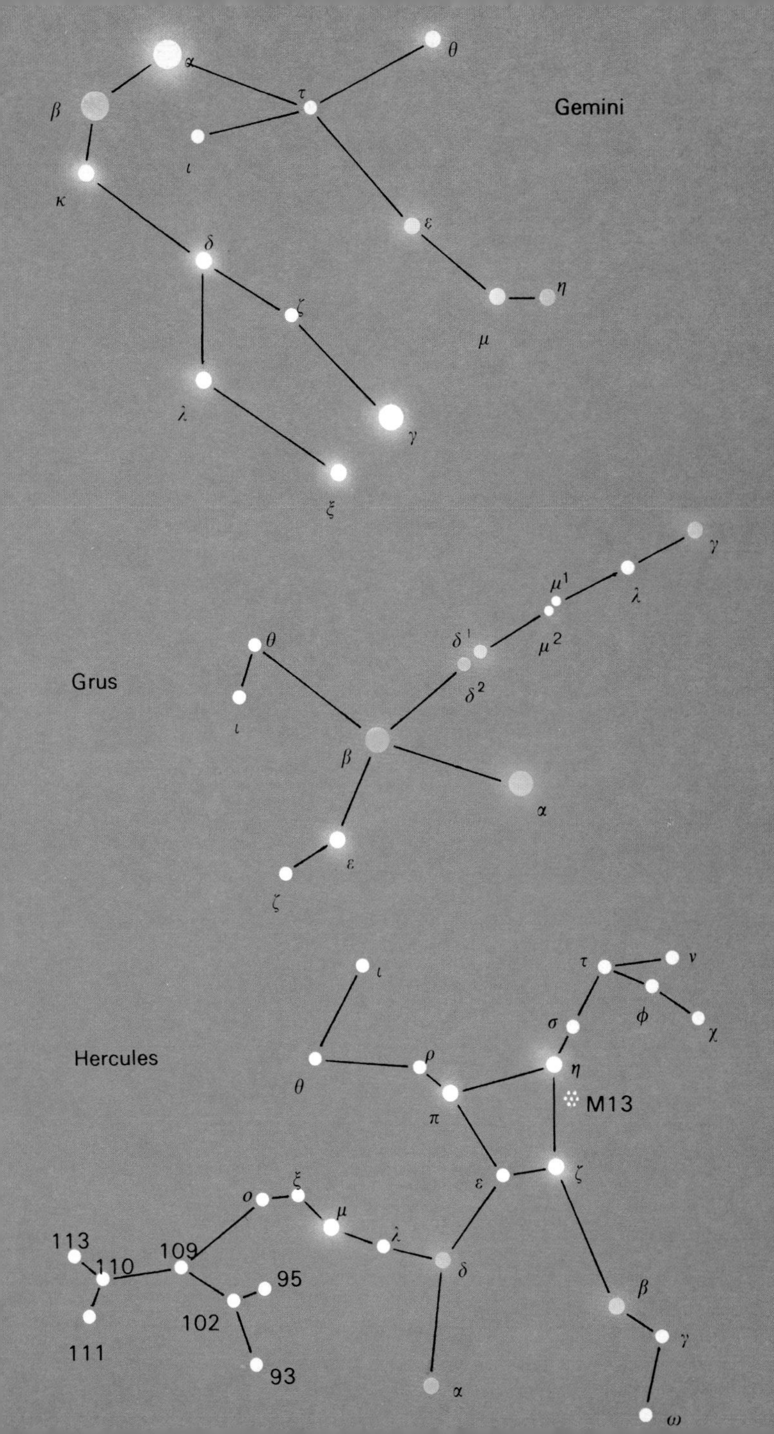
Gemini
α
β
θ
τ
ι
κ
ε
δ
ζ
η
μ
λ
γ
ξ
Grus
γ
μ1
λ
θ
δ1
μ2
δ2
ι
β
α
ε
ζ
Hercules
ι
τ
ν
φ
σ
χ
ρ
η
θ
M13
π
ε
ζ
ξ
ο
μ
λ
113
109
110
δ
95
102
β
γ
111
93
α
ω

Horologium (Clock) RA: 03 hr, Dec −55

Horologium which was added to the list of constellations by Lacaille in the mid-18th century is flanked by Eridanus, Hydrus, Reticulum, Dorado and Caelum. It lies between RA 2 hr 15 min–4 hr 20 min and Dec −40—67. It seems to occupy a series of steps between Eridanus and Hydrus and can more properly be thought of as an adjunct to Eridanus than as a constellation in its own right. In fact its brightest components lie between Eridanus and Caelum.

The most prominent star in Horologium is of mag 3·8 and there is nothing of significance throughout the rest of the sky occupied by this constellation. It is said that Lacaille named it as the Latin word for clock so that the Argo Navis (now divided into constellations Carina, Puppis and Vela) would have a chronometer for navigation!

Hydra (Water monster) RA: 10 hr, Dec: −15

Hydra, the largest of the constellations, snakes a long path between RA 8 hr 10 min–15 hr and Dec 7—35 flanked by Sextans, Leo, Cancer, Pyxis, Antlia, Centaurus, Libra, Virgo, Corvus and Crater. The head of the Water monster contains six stars with only two greater than mag 4.

α, Alphard, is a red K4 type mag 1·98 object with an absolute mag of −0·3 located 94 light years from the Sun and found south-east of the head. ε is one of the two prominent stars in the head and is actually four stars of mags 3·7, 5·2, 6·8 and 12·1. On the borders of Hydra and Centaurus, just below star γ is the M83 nebula, a galaxy catalogued as NGC 5253 while M68, a globular cluster, can be found between stars γ and ξ. Observers will search in vain for a nebula charted as M48, shown to lie in the general area of the head. It does not exist!

Hydrus (Water snake) RA: 01 hr, Dec: −70

This constellation is surrounded by Octans, Mensa, Dorado, Reticulum, Horologium, Eridanus and Tucana and spreads through an extreme from RA 0 hr–4 hr 35 min and from Dec −58 — −82. It was included in the Bayer catalogue of 1603. Technically, its only claim to fame is that it accommodates part of the Small Magellanic Cloud (SMC) but 95 per cent of this nebula lies in the neighbouring Tucana. The nebula is a companion galaxy to the LMC, both of which are irregular and lie some $1{\cdot}7 \times 10^5$ light years away.

β Hydri is a mag 2·78 star of G1 type at a distance of just over 20 light years. Star α, although not the brightest, is a mag 2·84 F0 type more than 30 light years distant.

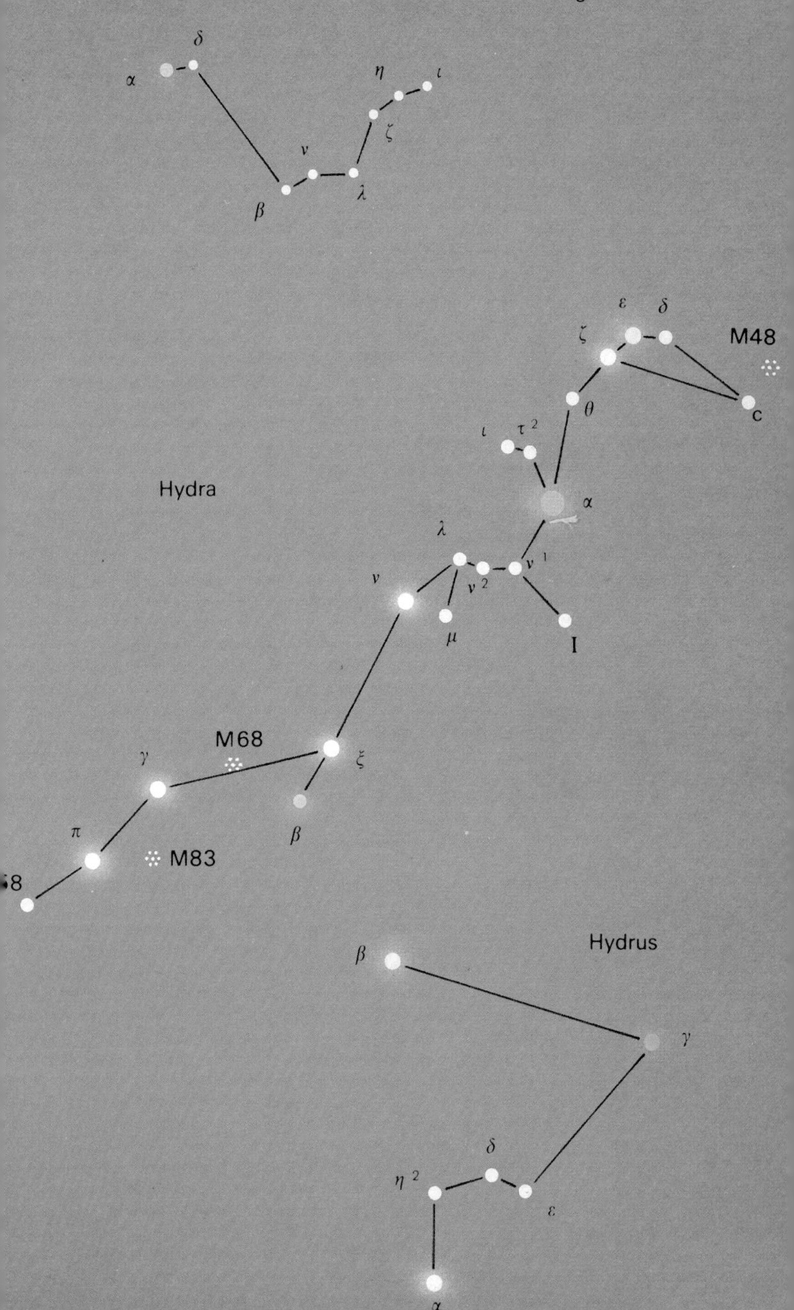

Horologium
α
δ
β
ν
λ
ζ
η
ι
ε
δ
ζ
M48
θ
c
ι
τ 2
Hydra
α
λ
ν 1
ν
ν 2
μ
I
M68
γ
ξ
β
π
M83
Hydrus
β
γ
δ
η 2
ε
α

Indus (Indian) RA: 20 hr, Dec: −50

Indus is an inconspicuous constellation in the southern sky. It was named by Bayer in the early 17th century to perpetuate the Indian nations about whom he had the gravest fears regarding the continuation of their way of life. The constellation is surrounded by Octans, Pavo, Tucana, Grus, Microscopium and Telescopium. It lies between RA 2 hr 25 min–23 hr 25 min and Dec −45 — −75.

It would be very easy for the casual observer to mistake the star α Pavonis in the constellation Pavo for a component of Indus. Star α Indi is a mag 3·2 object, the brightest in the constellation. β Indi is another 3rd mag star and the other three prime components are all of 4th mag. ε Indi is one of the nearest stars to the Sun, about 11·4 light years distant, with a mag of 4·7 and absolute mag of 7·0.

Lacerta (Lizard) RA: 22 hr, Dec: 40

Named by Hevelius in 1690, Lacerta is flanked by Andromeda, Cassiopeia and Cygnus and bordered to the north by Cepheus and to the south by Pegasus. Situated between RA 21 hr 55 min–22 hr 55 min and Dec 35 —56 this constellation contains only eight prime members, all of mag 4, with several variable stars. No other interesting objects are contained within the boundaries of this northern constellation.

Leo (Lion) RA: 10 hr, Dec: 20

Leo is one of the oldest constellations recognized and was close to the rising and setting positions of the Sun at summer solstice; the Egyptians respected the audacity of the Lion. Babylonians venerated the lion and depicted it in the sky and Greeks wove a mythological story around the constellation. It lies between 9 hr 20 min–11 hr 55 min and Dec −6–33, and is flanked by Leo Minor, Cancer, Hydra, Sextans, Crater, Virgo, Coma Berenices and Ursa Major.

The most interesting objects in the constellation are α, Regulus, a magnificent double (mags 1·36 and 10·8) of B7 spectrum, absolute mag −0·7, and lying at a distance of 84 light years. γ, Algeiba, is a worthy K0 type of mag 1·99 at a distance of nearly 200 light years, and β, Denebola, is a mag 2·14 A3 type just 43 light years away.

One of the brightest, intrinsically, in the constellation is ε, Asad Australis, a mag 2·99 G0 type star with an absolute mag of −2·1 which is 340 light years away. As would be expected from its close proximity to Coma Berenices and Virgo, Leo contains many external galaxies visible above the galactic plane.

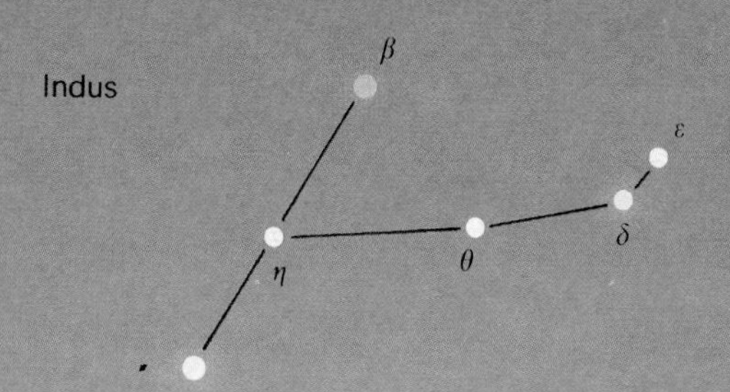
Indus
β
ε
δ
θ
η
α

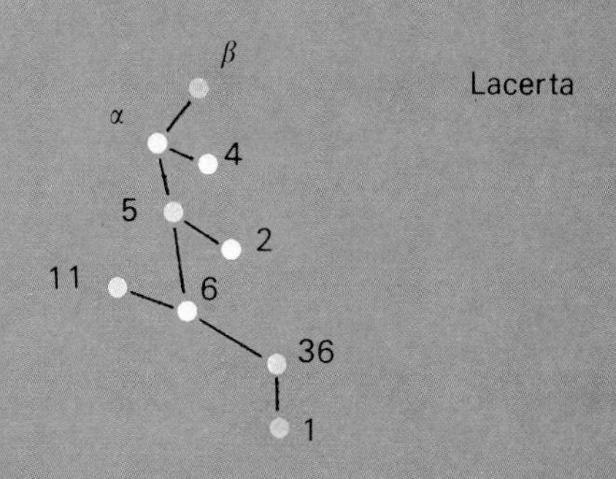
β
Lacerta
α
4
5
2
11
6
36
1

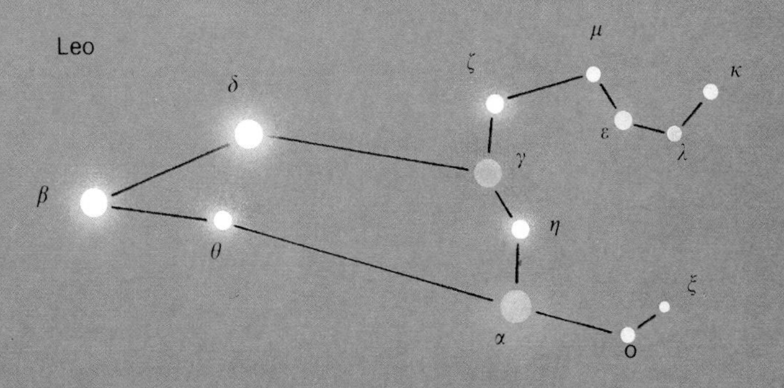
Leo
μ
ζ
κ
δ
ε
λ
γ
β
η
θ
ξ
α
o

Leo Minor (Small lion) RA: 10 hr, Dec: 35
This constellation was added to the list by Hevelius late in the 17th century and it can be found providing the northern cap to Leo. It lies between RA 9 hr 20 min–11 hr 5 min and Dec 23—42, flanked by Ursa Major, Lynx and Leo.

The prime star, carrying the unusual designation β is a mag 4 member as are the remaining three objects in the prime configuration of this constellation. Leo Minor is useful only for its application to stellar orientation and contains nothing of interest to the amateur.

Lepus (Hare) RA: 05 hr, Dec: −20
Lepus is one of the 48 constellations set up by Ptolemy and was mythologically placed in the sky to represent a hare although the origin is lost to antiquity. It lies between RA 4 hr 55 min–6 hr 10 min and Dec −11 — −27, flanked by Orion, Eridanus, Columba and Canis Major.

The prime star, α, Arneb, is a mag 2·58 F0 type with an absolute mag of −4·6 and a distance of 900 light years. Intrinsically less bright, β, Nihal, is a mag 2·81 G5 star (absolute mag 0·1) lying just 113 light years distant. The star is a double with a mag 9·4 companion.

The star ε Leporis is a mag 3·2 object while μ Leporis is of mag 3·3. A 430-day-period variable, R Leporis, is located in the direction of Eridanus and being of mag 6 at its brightest it is just visible to the naked eye but totally invisible when it dims to mag 10·4.

The M79 nebula is found south-west of Nihal; the distance from that star is equal to the distance separating α and β and on a line extended beyond these two stars.

Libra (Balance) RA: 15 hr, Dec: −15
Libra is one of the old constellations, and is flanked by Serpens Caput, Virgo, Hydra, Lupus, Scorpius and Ophiuchus. It occupies a region of the sky between RA 14 hr 20 min–16 hr in longitude and 0 — −30 in latitude, straddling the ecliptic and therefore a zodiacal constellation. The brightest star in the group is the mag 2·61 β, Zubenelschamali, of B8 type, absolute mag −0·6 located 140 light years from the solar system.

Star α, Zubenelgenubi, actually a double, is an A3 class object with an absolute mag of 1·2. Its companion is said by some to be a vivid green even to the naked eye and it has an apparent mag of 5·2. Several noted astronomers, amateur and professional, have refuted this and the coloration seems to be very much a reflection of personal visual interpretation.

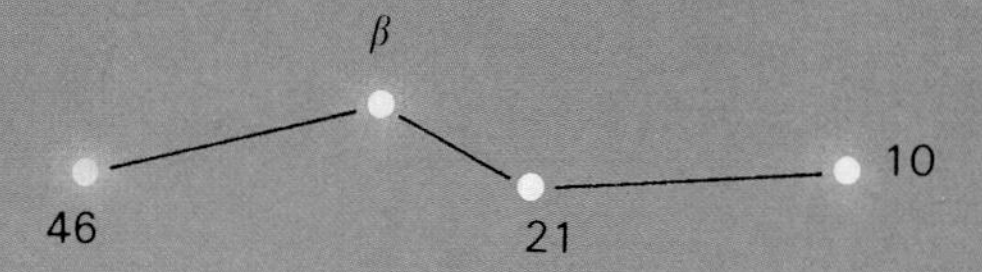
Leo Minor
β
10
46
21

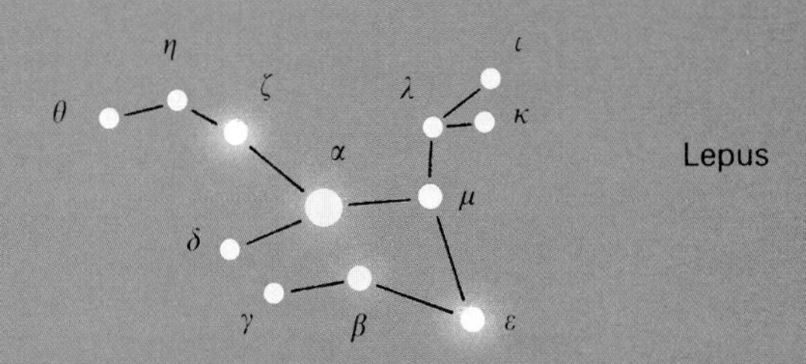
η
ι
θ
ζ
λ
κ
α
Lepus
μ
δ
γ
β
ε

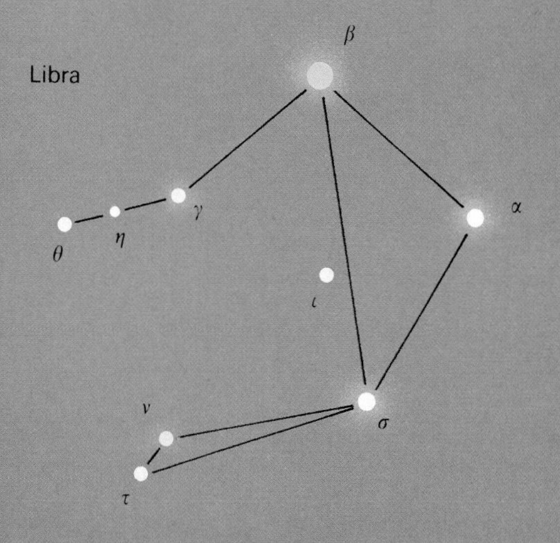
β
Libra
α
γ
θ
η
ι
ν
σ
τ

Lupus (Wolf) RA: 15 hr, Dec: −40

Lupus is an original constellation and is seen to be flanked by Centaurus, Libra, Scorpius, Norma and Circinus. It lies between RA 14 hr 15 min–16 hr 5 min and Dec −30 — −55.

The principle objects in this constellation are five stars above mag 3·5 forming the main geometric frame of the group. The three brightest stars, α, β and γ are all brighter than mag 2·8 and all three exhibit B1 or B2 class spectra. Star α is a mag 2·3 object lying at a distance of 430 light years, and star β is of mag 2·69 at a distance of 540 light years. These have absolute mags of −3·3 and −3·4 respectively, intrinsically very bright sources. Star γ has a visual mag of 2·8, an absolute mag of −2·7, and is found to lie 570 light years from the Sun. This latter star is a double with individual mags of 3·5 and 3·7.

Star ζ Lupi is another double with mag 3·4 and 3·8 components, and star η is a third double with components exhibiting mags 3·5 and 7·7.

Lynx (Lynx) RA: 09 hr, Dec: 40

Towards the end of the 17th century, Hevelius reputedly named this constellation, Lynx, because of the remarkable eyesight demanded of any observer studying this apparently barren area. Lynx is bound by constellations Camelopardalis, Auriga, Gemini, Cancer, Leo Minor and Ursa Major. It is found in an area between RA 6 hr 10 min–9 hr 40 min and Dec 33—62.

Star α Lyncis is a mag 3·2 object 180 light years away. The rest are less than mag 4 in brightness.

Lyra (Harp) RA: 19 hr, Dec: 35

Lyra was one of the first constellations to be described. It is surrounded by Draco, Hercules, Vulpecula and Cygnus and lies between RA 18 hr 10 min–19 hr 30 min and Dec 25—47.

Star α, Vega, third brightest from Britain, is a class A0 object of mag 0·04 at a distance of 26 light years (absolute mag 0·5) with a mag 10 companion. Vega and the Sun will pass close by each other in about $0{\cdot}3 \times 10^6$ years from now.

Star β, Shelyak, is an eclipsing variable (mags 3·4—4·1) with a mag 7·8 companion. These stars, each between 16×10^6–24×10^6 km in radius orbit each other with a separation of less than 5×10^6 km. Star ε is actually two stars, one of mag 4·6 with a mag 6·3 companion and one of mag 4·9 with a mag 5·2 companion. Each pair is a binary and the entire group is bound gravitationally. It is not unusual to find a multiple system of this nature and complex gravitational interactions can theoretically involve up to a dozen stars; the lone stance of our own star, the Sun, seems all the more remarkable in the light of this likely possibility.

Yet another double, ζ Lyrae, is found midway between δ and α with components of mag 4·2 and 5·5. Lyra incorporates a reported nova from 1919 due south of the midpoint between stars β and γ and

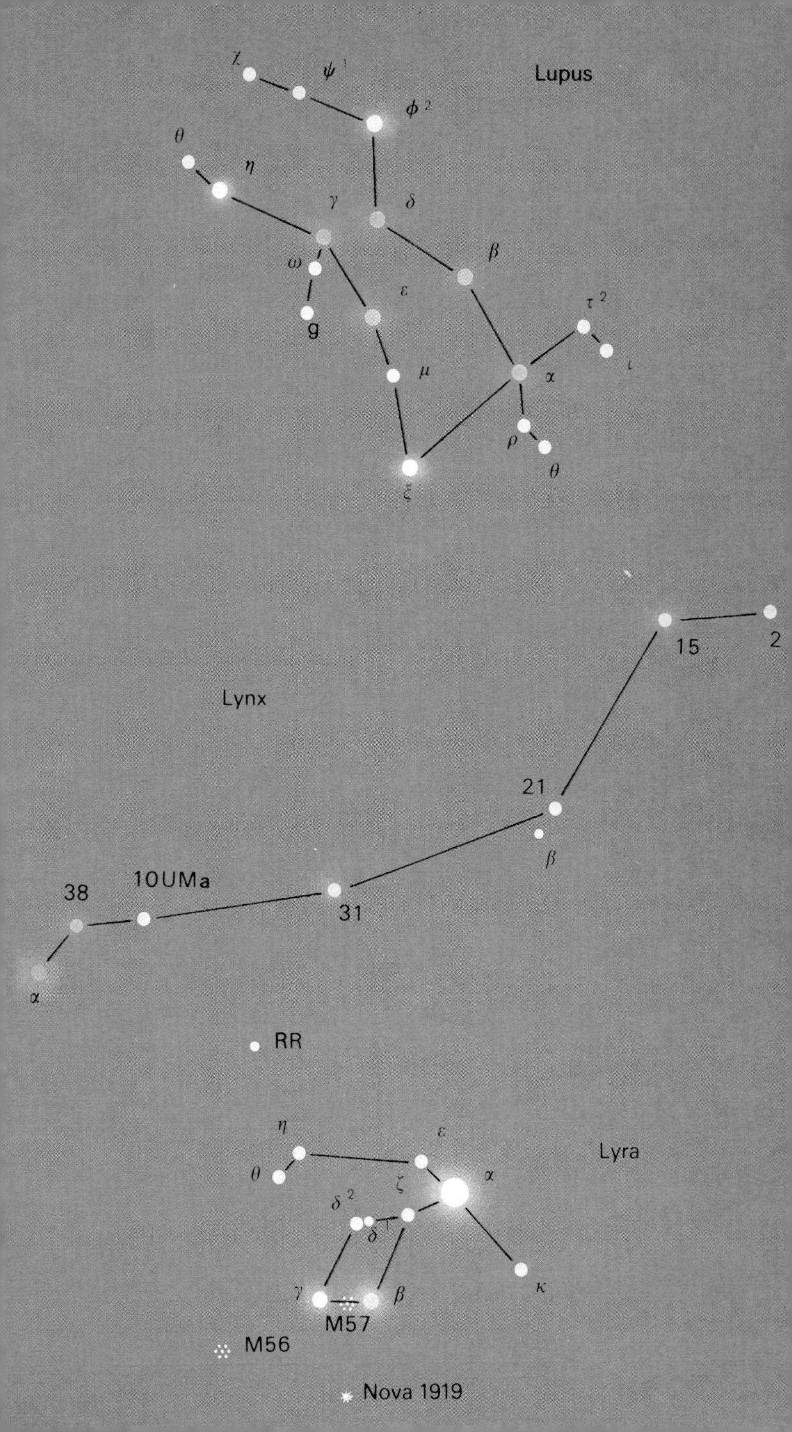
Lupus
χ
ψ 1
φ 2
θ
η
γ
δ
β
ω
ε
g
τ 2
ι
μ
α
ρ
θ
ξ
Lynx
15
2
21
β
10UMa
38
31
α
RR
Lyra
η
ε
θ
α
ζ
δ 2
δ 1
γ
β
κ
M57
M56
Nova 1919

a globular cluster, catalogued as M56 or NGC 6779, south-west of star γ and in almost a direct line drawn from star α through γ.

Far to the north-east and very close to the Cygnus border lies RR Lyrae, the prototype of a class of star used for distance measurement. It can be found by following a line drawn from star γ through star η. The famous Ring Nebula, catalogued as M57 (NGC 6720) is located almost half way along a line drawn from γ to β. This is a classic planetary nebula and was caused by the central star of mag 15 shedding a shell of material.

Mensa (Table) RA: 06 hr, Dec: −75

Originally named Mons Mensae by Lacaille in the mid-18th century, this constellation is bordered by Octans, Chamaeleon, Volans, Dorado and Hydrus. It occupies a portion of the sky between 3 hr 25 min–7 hr 40 min RA and −70 — −85 in Dec and was named to perpetuate a celestial nostalgia to Table Mountain in South Africa.

The constellation contains about 20 mag 5 stars but none of these is particularly interesting. The most notable claim to fame for this constellation comes from its partial occupation by a section of the LMC, itself more properly sited in Dorado.

Microscopium (Microscope) RA: 21 hr, Dec: −35

Microscopium is one of the few square-section constellations totally devoid of steps, angular divergence or irregularity, and is an insignificant area named by Lacaille and surrounded by Capricornus, Sagittarius, Indus, Grus and Piscis Austrinus. It lies between 20 hr 25 min–21 hr 25 min RA and Dec −28— −45, but there is little of observational interest in the entire constellation.

Monoceros (Unicorn) RA: 07 hr, Dec: 00

Firmly ensconsed on the celestial equator, this constellation is bordered by Gemini, Orion, Lepus, Canis Major, Puppis, Hydra and Canis Minor. Monoceros is located between RA 5 hr 55 min–8 hr 10 min and Dec −11 —12 and contains very few interesting objects for amateurs, among them NGC 2506, 2244 and 2323. The central portion of the constellation straddles the Milky Way.

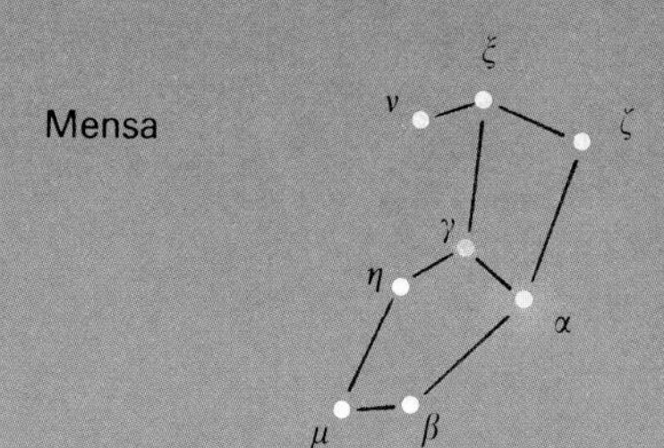
Mensa
ξ
ν
ζ
γ
η
α
μ
β

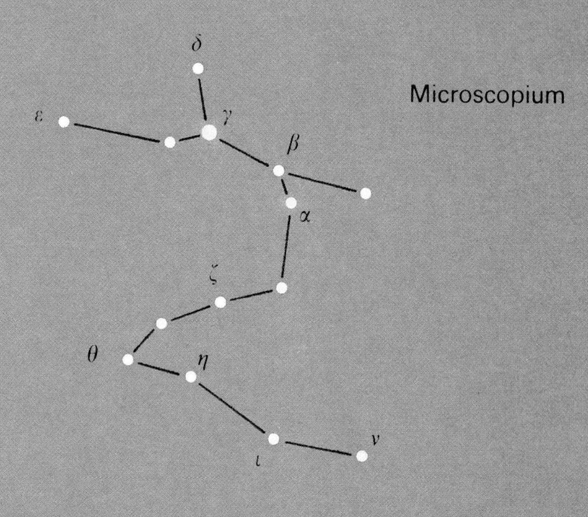
δ
Microscopium
ε
γ
β
α
ζ
θ
η
ν
ι

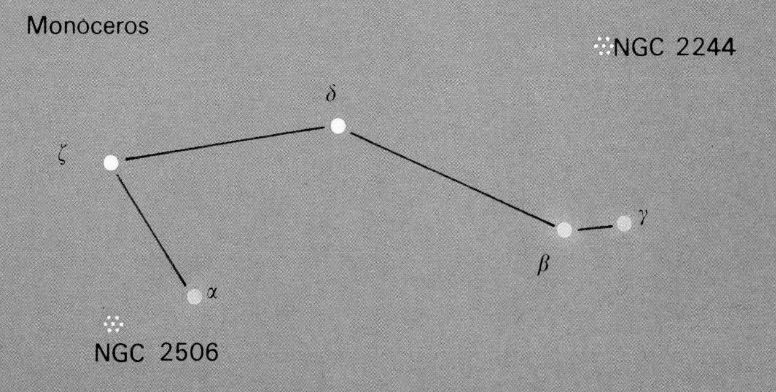
Monoceros
NGC 2244
δ
ζ
γ
β
α
NGC 2506

Musca (Fly) RA: 13 hr, Dec: −70

This constellation is sometimes known as Musca Australis (Southern fly) and is one of those entered in the 1603 Bayer catalogue. It is located in an area of the southern sky between 11 hr 20 min–13 hr 45 min RA and between −64 — −75 Dec, surrounded by Chamaeleon, Apus, Circinus, Centaurus, Crux and Carina.

The star α Muscae is a mag 2·7 variable (mag 2·66–2·73) of B3 spectral type (absolute mag −2·9) situated 430 light years from the solar system. Star β Muscae, mag 3·06, is a B3 double with component mags of 3·7 and 4·1 and the system lies at a distance of about 470 light years.

Norma (Square) RA: 16 hrs, Dec: −55

This southern constellation was named by Lacaille and the uninteresting region is surrounded by Triangulum Australe, Ara, Scorpius, Lupus and Circinus. It is seen to occupy a portion of the celestial sphere lying between 15 hr 10 min–16 hr 30 min RA and between −42 — −60 in Dec.

There are no particularly interesting stars in the constellation and all are dimmer than mag 5. An interesting open cluster – NGC 6067 – is located on a line projected from ε through γ and as far from γ as ε is from γ.

Octans (Octant) RA: 22 hr, Dec: −85

Octans is the south circumpolar constellation also discovered by Lacaille. It is surround by Hydrus, Mensa, Chamaeleon, Apus, Pavo, Indus and Tucana but is irregular in circumpolar outline and meets Pavo, Indus and Tucana a full 15° from the celestial south pole.

The brightest star, υ, is mag 3·7 and the closest star to the actual pole is σ, a dim object of nearly mag 6·0. Apart from the important relationship with the celestial sphere, the constellation contains little of observational interest.

Ophiuchus (Snake bearer) RA: 17 hr, Dec: 00

Ophiuchus is located across the celestial equator but is not a zodiacal constellation although its angular spread in hours of RA would ordinarily qualify it for inclusion. Ophiuchus spans an area of sky from RA 16 hr–18 hr 45 min and from Dec −30—14.

The constellation is flanked by Hercules, Serpens Caput, Scorpius, Sagittarius and Serpens Cauda. It is located between Serpens Caput and Serpens Cauda – effectively dividing them from each other.

The most prominent star in the constellation is Ras Alhague, north of the celestial equator and just south of Hercules. The star is a mag 2·09 A5 type located 58 light years from the solar system with an absolute mag of 0·8. Star η, Sabik, is a mag 2·46 object, actually a double with components of mag 3·0 and 3·4, with an A3 spectrum and an absolute mag of 1·4 at a distance of 70 light years.

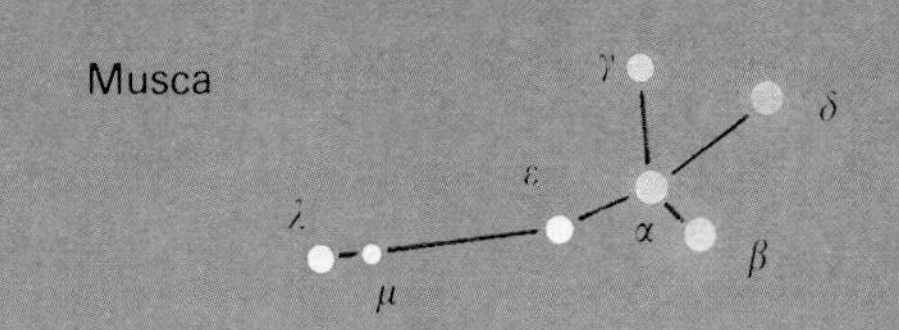
Musca
γ
δ
ε
λ
α
β
μ

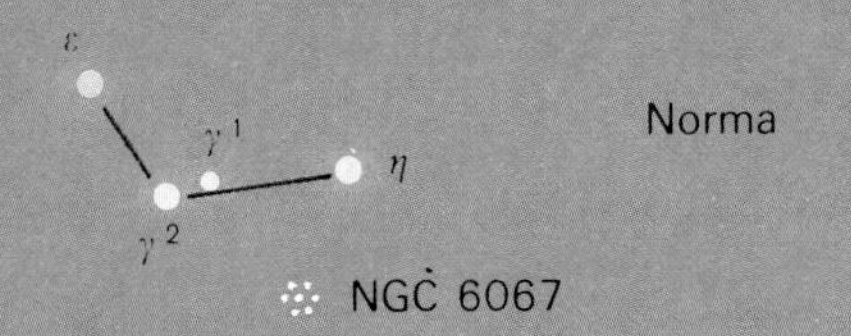
ε
Norma
γ1
η
γ2
NGC 6067

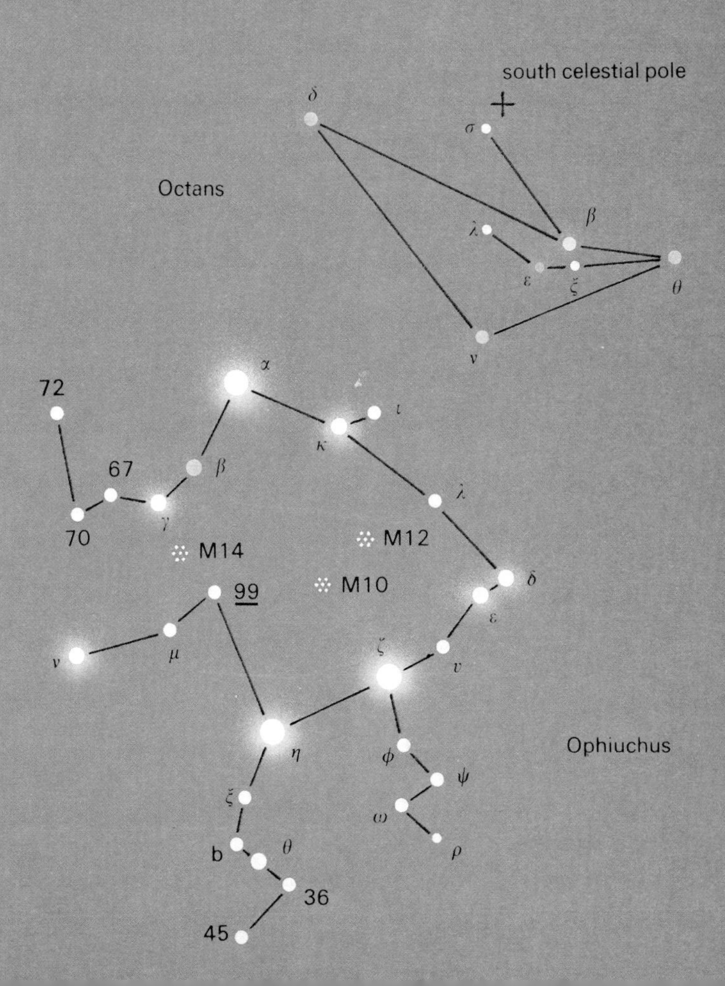
south celestial pole
δ
σ
Octans
β
λ
ε
ξ
θ
ν
α
72
ι
κ
67
β
λ
70
γ
M12
M14
M10
δ
99
ε
ν
μ
ζ
υ
η
Ophiuchus
φ
ψ
ξ
ω
b
θ
ρ
36
45

Star ζ, Han, found almost due north-east from Sabik, is a mag 2·57 O9 type star with an absolute mag of −4·3. It lies at a distance of more than 500 light years. At the other end of the spectral scale the observer can find δ, Yed Prior, on the boundary with the constellation Serpens Caput. This is an M1 type object with mag 2·72, absolute mag −0·5, and a separation distance of 140 light years.

Star β, Cheleb, another interesting object, is a K2 type mag 2·77 object with absolute mag of −0·1 at a distance of 124 light years. Three faint globular clusters can be seen located within the area of a triangle set up by straight lines joining Yed Prior and stars γ and μ. These clusters, M10, M12 and M14, are all at extremely remote distances. Globular cluster M9 is adjacent to a line joining stars η and ξ. M19 is south-west of θ.

Orion (Orion) RA: 05 hr, Dec: 00

This interesting constellation lies between RA 4 hr 40 min–6 hr 20 min and Dec −11—23, flanked by Taurus, Eridanus, Lepus, Monoceros and Gemini. The three most interesting stars in this group form the famous belt of Orion: ζ, Alnitak, ε, Alnilam, and δ, Mintaka. Alnitak is a remote double with component mags of 1·9 and 4·05 (system mag of 1·79) and absolute mag of −6·6; Alnilam is a super-giant star of type B0 at a distance of 1600 light years (as is Alnitak) with an absolute mag of −6·8; and Mintaka is an eclipsing variable (mag 2·2—2·35) with a mag 6·47 companion and a period slightly less than six days.

The upper part of the constellation is marked by α, Betelgeuse, (mag 0·5–1·1 variable, class M2) and γ, Bellatrix, (mag 1·64, B2 type). To the south lie κ, Saiph (mag 2·06, B0 type), and β, Rigel (mag 0·08, B8 type of absolute mag −7·1, distance 900 light years). See notes on p. 144 for further details.

Pavo (Peacock) RA: 20 hr, Dec: −60

This constellation lies between RA 17 hr 35 min–21 hr 25 min and Dec −57 — −75 and is bordered by Octans, Indus, Telescopium, Ara and Apus. α Pavonis, a mag 1·95 B3 star (absolute mag −2·9) is 310 light years away and κ is a mag 4·0—5·5 Cepheid variable with a period of just over nine days.

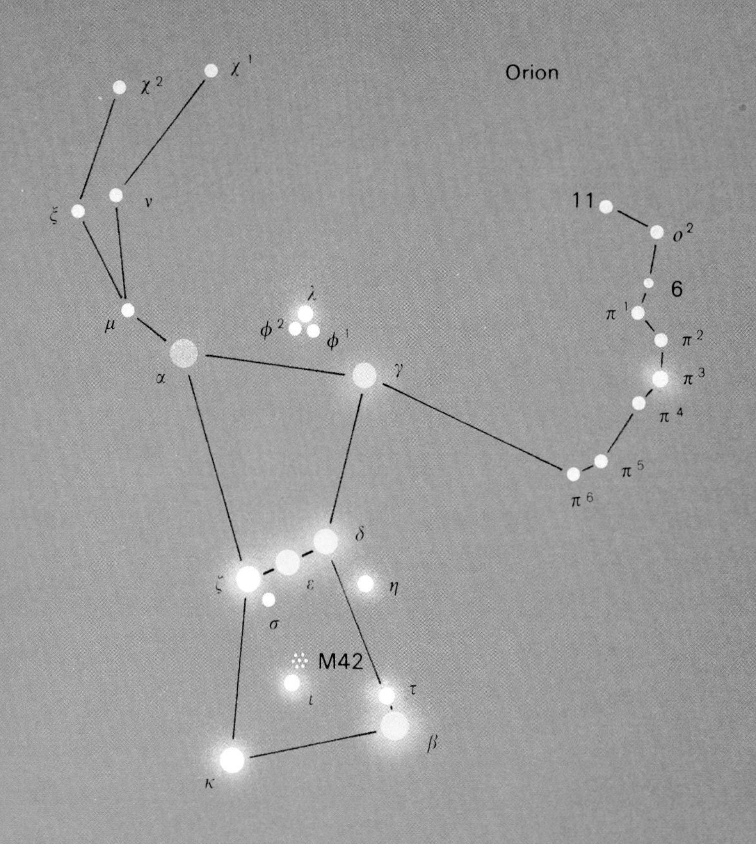
Orion
χ 2
χ 1
ξ
ν
μ
α
λ
φ 2
φ 1
γ
11
o 2
6
π 1
π 2
π 3
π 4
π 5
π 6
δ
ζ
ε
η
σ
M42
ι
τ
β
κ

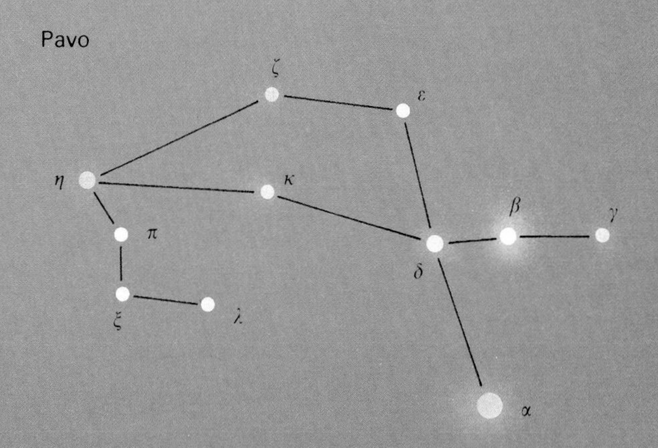
Pavo
ζ
ε
η
κ
π
β
γ
δ
ξ
λ
α

Pegasus (Pegasus) RA: 22 hr, Dec: 20

Pegasus is another very interesting constellation and lies between RA 21 hr 5 min–0 hr 15 min and Dec 2—36. It is surrounded by neighbouring constellations Lacerta, Cygnus, Vulpecula, Delphinus, Equuleus, Aquarius, Pisces and Andromeda. Pegasus can be recognized by the square format of three member stars (β, α and γ) and α, Alpheratz, in the neighbouring constellation of Andromeda. Alpheratz lies on the very border of the two designated field areas.

Star ε, Enif, in Pegasus is a mag 2·3 K2 type star with a mag 9 companion. Enif is nearly 800 light years away and has an absolute mag of −4·6. The three corners of the square of Pegasus that lie in this constellation are all above mag 3 and good observational objects.

Star β, Scheat, is a large M2 variable red giant (absolute mag −1·5) at a distance of 210 light years and may range between 2·4 and 2·7. Star α, Markab, is a white B9 of mag 2·5 and γ is a mag 2·84 B2 type (absolute mag −3·4) at distances of 110 and 570 light years respectively. North-west of ρ lies η, Matar, a mag 2·95 G8 star.

Perseus (Perseus) RA: 03 hr, Dec: 40

Perseus, one of the earliest named constellations, is flanked by Cassiopeia, Andromeda, Triangulum, Aries, Taurus, Auriga and Camelopardus. It occupies a region of the sky between RA 1 hr 25 min–4 hr 45 min and Dec 31—59.

The most interesting object in Perseus is β, Algol, an eclipsing binary and a prototype of this class of astronomical phenomena. Algol, a mag 2·06—3·28 B8 type star, lies at a distance of 105 light years and is accompanied by a companion of similar size just $1{\cdot}6 \times 10^7$ km away. The period of the components is just under three days but a third, much smaller, star orbits the binary system in 23 months.

The brightest star in Perseus is α, Mirfak, a giant F5 of mag 1·8 (absolute mag −4·4) surrounded by several much fainter stars. ζ Persei is a double with mag 3 and 9 components; ε is another double with mags 3 and 8 and ρ Persei is a mag 3·2—3·8 variable. The open clusters h and χ are magnificent objects.

Phoenix (Phoenix) RA: 01 hr, Dec: −45

This is another Bayer constellation and it is named after the mythological bird that rose centenially from the ashes following repeated sacrificial burning. Phoenix is bordered by Fornax, Sculptor, Grus, Tucana and Eridanus and is found between RA 23 hr 25 min–2 hr 25 min and Dec −40 — −58. The constellation is not very conspicuous and contains only three stars above mag 4. The brightest star in this assemblage is α, Ankaa, a mag 2·39 K0 type lying at a distance of 93 light years and an absolute mag of 0·1. Star β Phoenicis is a double with prime and secondary object mags of 4·1 each.

Star γ is a mag 4 object as is star ζ with a mag 8·4 companion. The constellation is best found by locating Achernar, a mag 0·53 star in Eridanus – one of the brightest in the southern sky.

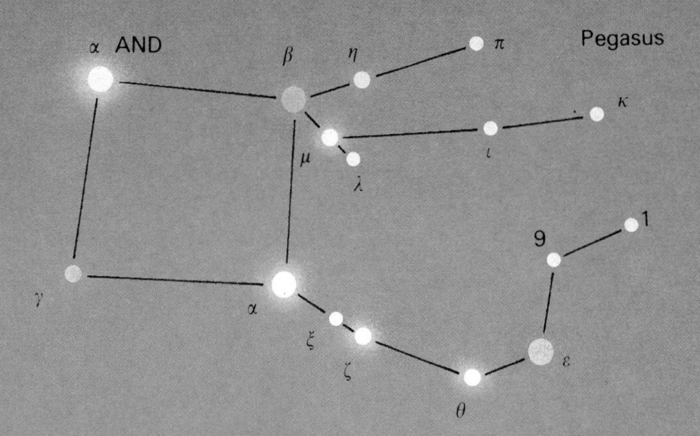
α AND
Pegasus
β
η
π
κ
μ
λ
ι
9
1
γ
α
ξ
ζ
θ
ε

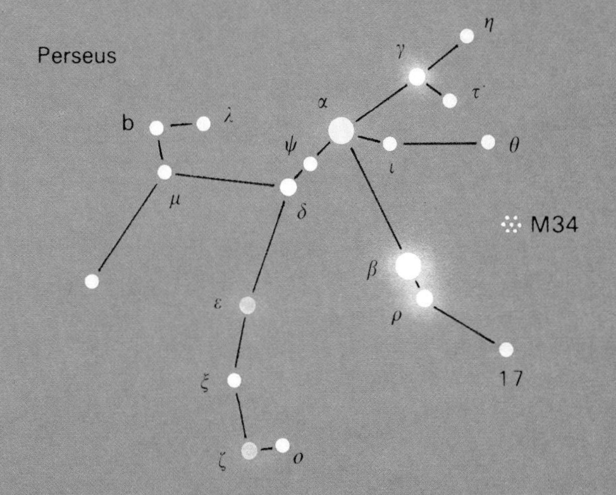
Perseus
η
γ
τ
α
b
λ
ψ
ι
θ
μ
δ
M34
β
ε
ρ
17
ξ
ζ
ο

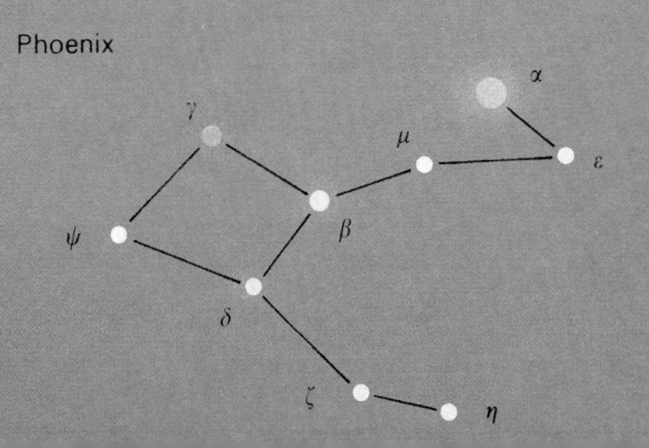
Phoenix
α
γ
μ
ε
β
ψ
δ
ζ
η

Pictor (Painter) RA: 06 hr, Dec: −60

A southern constellation, Pictor was named by Lacaille in 1752 and lies in juxtaposition to Dorado where it seems to occupy a region more properly enclosed by the two constellations. It is flanked to the west, north and north-east by Volans, Carina, Puppis, Columba, Caelum and Dorado. It lies between RA 4 hr 30 min–6 hr 50 min and Dec −43 — −64 and the constellation is found by locating the mag −0·73 star Canopus in the constellation Carina. With the rotation of the heavens Pictor appears to circle Canopus.

The only two stars of note are α and β. Star α Pictoris is a mag 3·27 object and star β Pictoris is of mag 3·9. A nova which flared up in 1925, RR Pictoris, can be found adjacent to star α. It is still visible but a large telescope is required to see anything significant.

Pisces (Fishes) RA: 00 hr, Dec: 10

Although it is unspectacular, Pisces has been recognized as a zodiacal constellation since the days of the earliest civilizations. It lies close to the great square of Pegasus and is flanked by Andromeda, Pegasus, Aquarius, Cetus, Aries and Triangulum. Due to pole precession, Pisces now replaces Aries as the vernal equinox.

The most prominent object in Pisces is η Piscium (mag 3·9) followed by α, Al Rischa, of mag 3·94, actually a double with mag 4·3 and 5·2 components. A Messier recorded galaxy, M74 in the catalogue, is found adjacent to star η.

Piscis Austrinus (Southern fish) RA: 23 hr, Dec: −30

Piscis Austrinus, one of the originally named constellations, is flanked by Aquarius, Microscopium, Grus, Sculptor, Cetus and Capricornus and lies between RA 21 hr 25 min–23 hr 5 min and Dec −25 — −37. The only really interesting object in the entire constellation is the magnificent white star Fomalhaut, an A3 type of mag 1·19 and absolute mag 2·0, lying at the comparatively close distance of 23 light years and 11 times as luminous as the Sun. All the remaining members of this constellation are below mag 3.

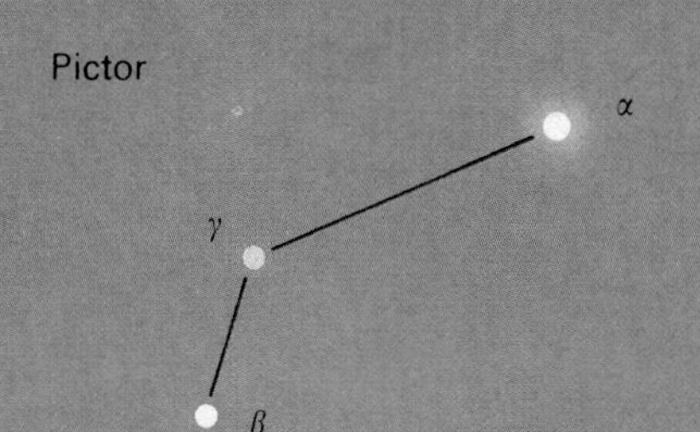
Pictor
α
γ
β

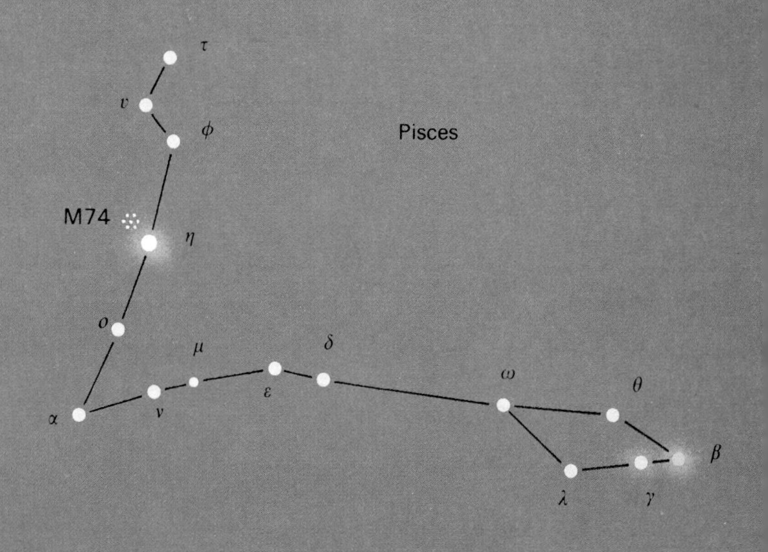
τ
υ
ϕ
Pisces
M74
η
ο
μ
δ
ω
θ
α
ν
ε
β
λ
γ

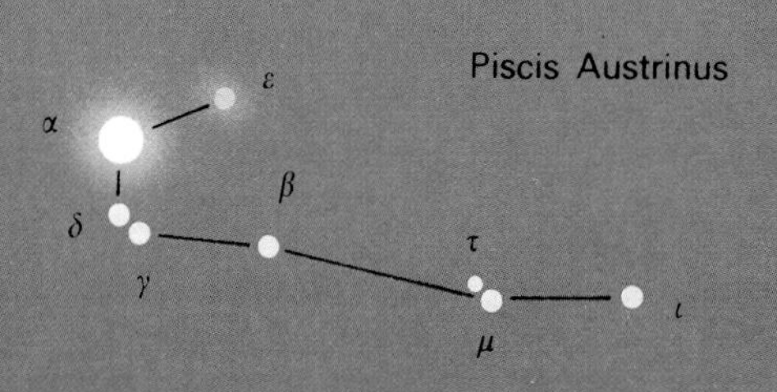
Piscis Austrinus
ε
α
β
δ
τ
γ
ι
μ

Puppis (Ship's stern) RA: 07 hr, Dec: −35

Puppis was one of the three constellations forming the sprawling and unwieldy Argo Navis named as one of the 48 Ptolemaic constellations and now separated into Puppis, Carina and Vela. Puppis can be found best, like Pictor, by first locating Canopus (in the constellation Carina) and seeking the pattern of stars which lies north of this.

Puppis is surrounded by the constellations Canis Major, Columba, Pictor, Carina, Vela, Pyxis, Hydra and Monoceros and can be found to occupy a region of the celestial sphere between RA 6 hr–8 hr 25 min and Dec −11 — −51.

The brightest star in the group is ζ, Suhail Hadar, with a mag of 2·23 at the tremendous distance (comparatively) of 2400 light years giving it an absolute mag of −7·1. It has an O5 spectra. Star ρ, is a variable F6 (mag 2·72—2·87) just 105 light years distant and π Puppis is a K4 mag 2·81 star 140 light years away.

The only other moderately interesting star is τ, a mag 2·97 K0 type with absolute mag 0·1 at a distance of 125 light years. M46 and M93 are two open clusters in the constellation, both lying in the northern part of this region.

Pyxis (Ship's compass) RA: 09 hr, Dec: −35

Probably *the* most inconspicuous of all the constellations, Pyxis was so named by Lacaille and appears insignificant beside the original Argo Navis constellation, now divided into Carina, Puppis and Vela. Pyxis is surrounded by Hydra, Puppis, Vela and Antlia and can be found between RA 8 hr 25 min–9 hrs 25 min and Dec −17 — −37. It contains no stars of note.

Reticulum (Net) RA: 04 hr, Dec: −65

Although included as one of the Lacaille constellations, Reticulum is generally attributed to a German called Habrecht. It is a small constellation in the southern sky flanked by Hydrus, Dorado and Horologium which lies between RA 3 hr 15 min–4 hr 35 min and Dec −53 — −67. Only α Reticuli of mag 3·3, actually a double with component mags of 3·33 and 12, is of interest.

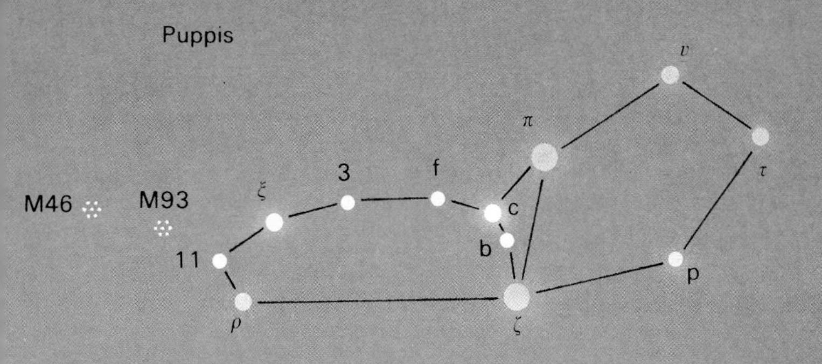

Pyxis

Reticulum

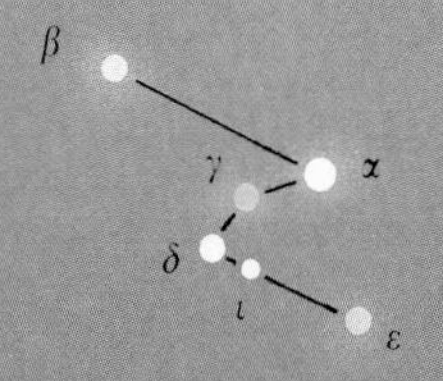

Sagitta (Arrow) RA: 20 hr, Dec: 18
Sagitta was one of the Ptolemaic constellations, and lies between RA 18 hr 55 min–20 hr 20 min and Dec 16—21, surrounded by Vulpecula, Delphinus and Aquila. Sagitta contains only faint, uninteresting stars but it can be seen to possess an open cluster, M71, approximately between stars γ and δ.

Sagittarius (Archer) RA: 18 hr, Dec: –30
Of the 88 constellations, Sagittarius probably contains the most abundant and wide-ranging collection of objects. Within its area lie stars, the galactic nucleus, gaseous nebulae, open and galactic clusters – the list is virtually all embracing.

The constellation is surrounded by Scutum, Serpens Cauda, Ophiuchus, Scorpius, Corona Australis, Telescopium, Microscopium, Capricornus and Aquila. It is found within an area between 17 hr 40 min–20 hr 25 min RA and –12 — –45 in Dec, thereby straddling the ecliptic and joining the list of zodiacal constellations.

The stars in Sagittarius receive Greek designations totally out of order with their measured apparent magnitude and there is little explanation as to why this was allowed. The brightest star is ε, Kaus Australis, a mag 1·81 B9 type at 124 light years. Next is σ, Nunki, mag 2·12, a B2 star at 300 light years, followed by ζ, Ascella, a mag 2·61 of A2 spectra, which is actually a double with mags 3·3 and 3·5. δ, Kaus Meridionalis, is less bright again of mag 2·71 with K2 spectra and an absolute mag of 0·7, at a distance of 85 light years.

Star λ, Kaus Borealis, is another K2, mag 2·8, at 71 light years distance, followed by γ, Al Nasl, of mag 2·97, a K0 type at 124 light years. Star η is actually a double with component mags of 3·17 and 10. A triple system, seen as star π Sagittarii, has mags of 3·7, 3·8 and 6·0. Elsewhere, the constellation displays the Trifid Nebula, M20, a faint and complex gas cloud, the Lagoon Nebula M8 and the Horseshoe Nebula M17. Globular clusters M22, M28, M69, M70, M54, M55 and M75 are to be found as well as open clusters M18, M24, M25, M23 and M21.

Scorpius (Scorpion) RA: 17 hr, Dec: –35
This constellation is bordered by Ophiuchus, Libra, Lupus, Norma, Ara, Crater and Sagittarius lying between 15 hr 45 min–17 hr 55 min RA and Dec –8 — –45. The most interesting object by far is α, Antares, a supergiant variable (mag 0·86—1·02) with a distinctly green mag 5·46 companion. Antares has a diameter of about 563 million km with an M1 spectra and lies at a distance of 520 light years.

Star ε, is a mag 2·28 K2 type at some 66 light years distance and stars δ, Dschubba, β, Graffias, τ, σ, π and μ are all B types within a range of mag 2·34—2·99, situated between 520 and 750 light years in distance. Other objects of interest in Scorpius include two globular clusters, M4 and M80, both of which are to be found fairly close to Antares.

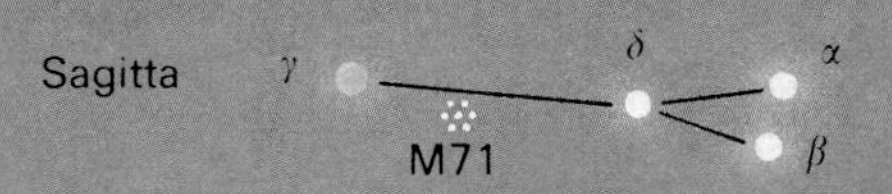

Sagitta
γ
δ
α
β
M71

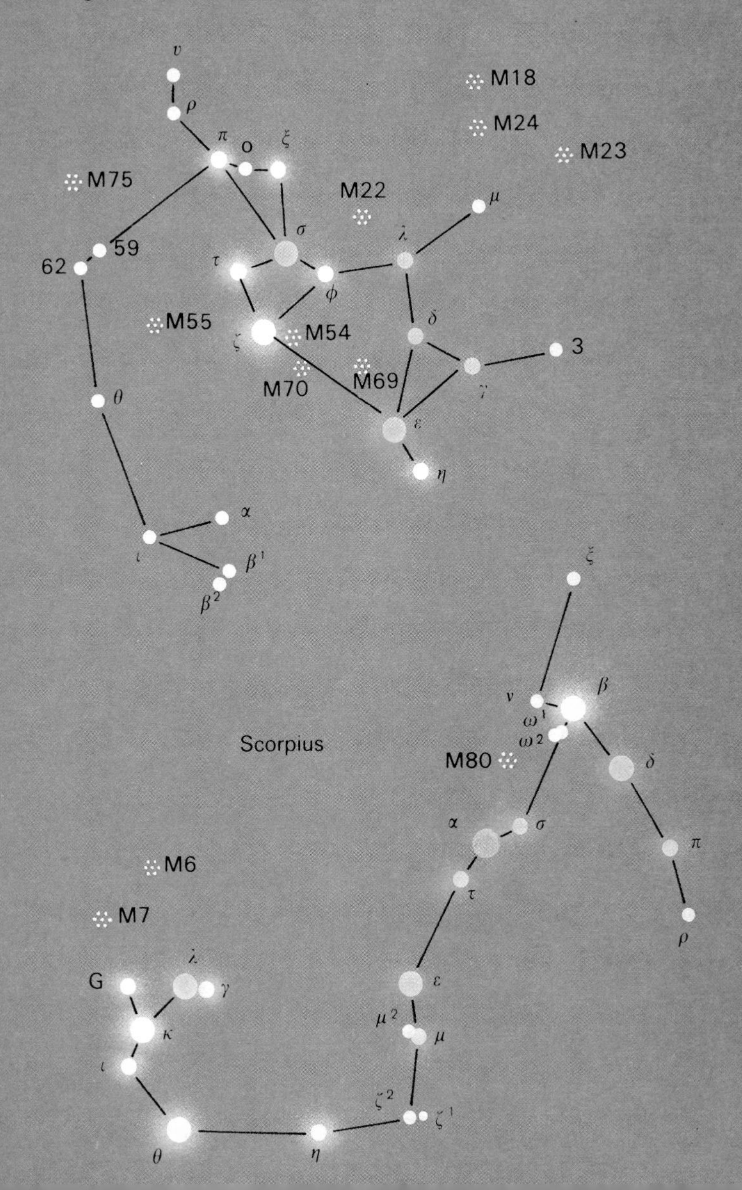

Sagittarius
υ
ρ
π
o
ξ
M18
M24
M23
M75
M22
μ
σ
λ
59
62
τ
φ
M55
M54
δ
ζ
3
M70
M69
θ
γ
ε
η
α
ι
β1
β2
ξ
β
ν
ω1
ω2
Scorpius
M80
δ
α
σ
π
M6
τ
M7
ρ
λ
G
γ
ε
μ2
κ
μ
ι
ζ2
ζ1
θ
η

An open cluster, M6, is located within Scorpius as is M7, a much larger open cluster and one which is nearly lost to the great stellar clouds of the Milky Way. Large areas of the latter are seen across the southern sky and the central regions of this constellation.

Sculptor (Sculptor) RA: 01 hr, Dec: −30

Sculptor was one of the Lacaille constellations, named in 1752, and lies surrounded by Cetus, Aquarius, Piscis Austrinus, Grus, Phoenix and Fornax in an area from RA 23 hr 5 min–1 hr 45 min and Dec −25 — −40. Sculptor contains the south galactic pole and few interesting stars. The most accurate position of the south galactic pole is on a point slightly north of a line drawn from star α to star ι. Nothing of note is contained in this constellation.

Scutum (Shield) RA: 19 hr, Dec: −10

A constellation named by Hevelius in 1690, Scutum is an inconspicuous patch of sky to the naked eye but a glorious stellar forest is visible through even a small telescope. It lies between RA 18 hr 20 min–18 hr 55 min and Dec −4 — −16.

There is absolutely nothing of note here except possibly two stellar clusters designated M11 and M26. M11, the Wild Duck cluster, presents a very striking appearance. It probably contains over 600 stars in a region some 21 light years across, while its distance is about 6000 light years.

Serpens (Snake) RA: 16 hr, Dec: 05

This constellation is divided into two: Serpens Caput and Serpens Cauda. It signifies the snake with which Ophiuchus is struggling and this certainly seems appropriate for the two halves are separated by that constellation. Serpens Caput lies between RA 15 hr 10 min–16 hr 20 min and Dec −3—25.

The only prominent star here is α, Unuk al Hay, a mag 2·65 source. The remainder in Serpens Caput are very faint. Serpens Cauda lies between 17 hr 15 min–18 hr 55 min RA and −16—6 in Dec. This constellation contains a double (mags 4·5) star θ called Alya.

Sextans (Sextant) RA: 10 hr, Dec: 00

This constellation forms boundaries of almost a perfect square and it is flanked by Leo, Hydra and Crater, lying between 9 hr 40 min–10 hr 50 min RA and 7 — −11 in Dec. The constellation was officially named by Hevelius in the late 17th century, but is uninteresting, containing only a few very faint doubles, variables and other objects.

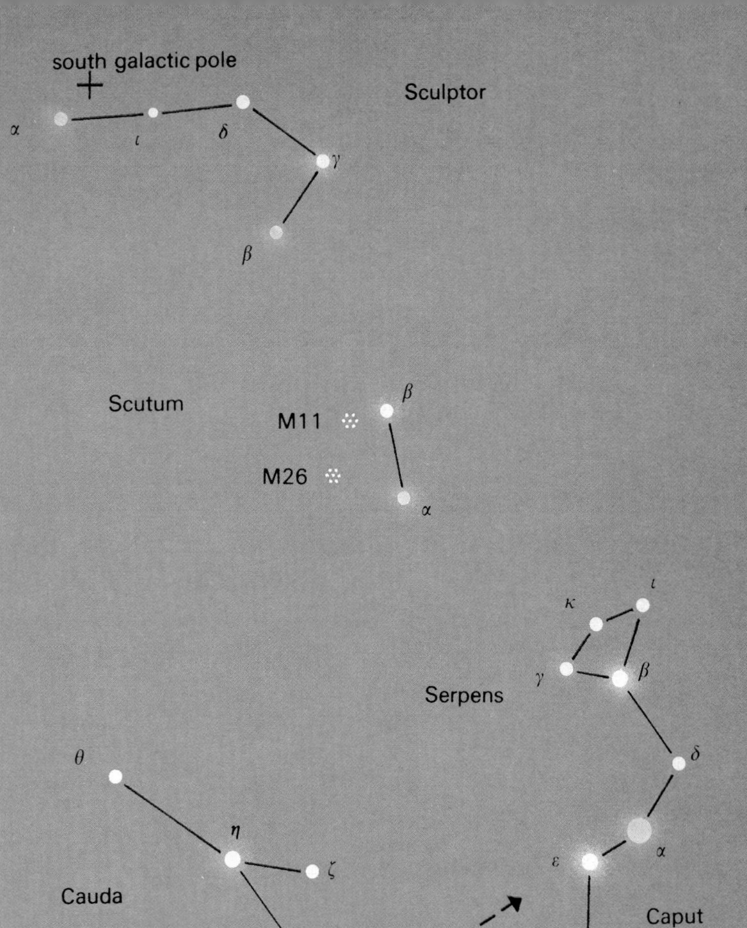
south galactic pole
Sculptor
α
ι
δ
γ
β
Scutum
β
M11
M26
α
ι
κ
γ
β
Serpens
θ
δ
η
α
ε
ζ
Cauda
Caput
ν
M16
μ
ξ

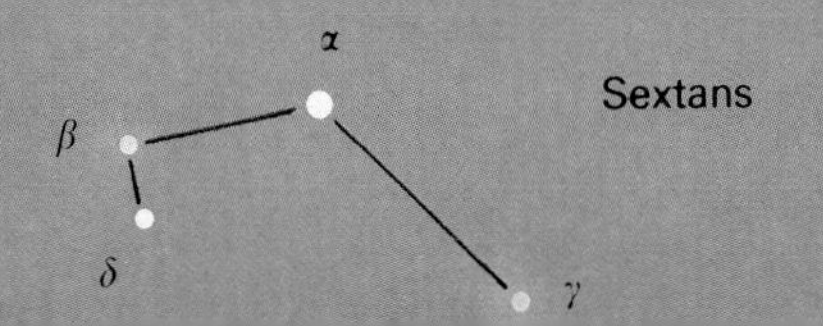
α
Sextans
β
δ
γ

Taurus (Bull) RA: 05 hr, Dec: 20

Taurus the Bull was probably one of the first constellations to be named, and is reminiscent of the oldest domesticated and respected animal. Prior to 3000 BC, Taurus lay across the vernal equinox and so became a stellar system of awe and magnificence. The beast and the constellation have been deified in almost every known civilization in the ancient Middle East. It is a remarkably rich area of the sky, containing some very prominent stars, the Pleiades and Hyades clusters, and the faint but very important Crab Nebula.

The most magnificent star, α, Aldebaran, is a K5 type giant with mag 0·86 and absolute mag −0·7 at a distance of 68 light years. It is a variable star with a diameter of about 50 million km and a luminosity 120 times that of the Sun. Star β, El Nath, is a mag 1·65 B7 type (absolute mag −3·2) at a distance of 300 light years.

Star η, Alcyone, is another B type with a mag of 2·86, as is star ζ with a mag of 3·07. These latter stars are 540 and 940 light years distant respectively. The Pleiades are found around Alcyone for that star is the most prominent member of this 300-star group. When viewed through the telescope, there is hardly a more carnival sight in all the sky, the hot blue stars, all very young, illuminate nebulous regions with a blue incandescent light that swirls and loops from star to star. The system, with all parts moving through space together, is about 500 light years away.

Moving towards Aldebaran the observer will find the Hyades, a cluster of cool red stars almost 130 light years away. Not far from ζ is the Crab Nebula, a supernova more than 900 years old lying at a distance of 3500 light years.

Telescopium (Telescope) RA: 18 hr, Dec: −45

A constellation which should, perhaps, have been more properly included with Corona Australis. It is surrounded by that constellation and Ara, Pavo, Indus and Sagittarius, lying between RA 18 hr 5 min–20 hr 25 min and Dec −46 — −57. The three brightest stars are all nestled on the border with Corona Australis and there is little here of observational interest.

Triangulum (Triangle) RA: 02 hr, Dec: 35

Triangulum is one of the earliest named but inconspicuous constellations lying between RA 1 hr 30 min–2 hr 50 min and Dec 25—37. α Trianguli is only of mag 3·45 with β Trianguli of mag 3, which is again a departure from the correct sequence of decreasing magnitude for successive letter designations. The only object of note is the spiral galaxy M33.

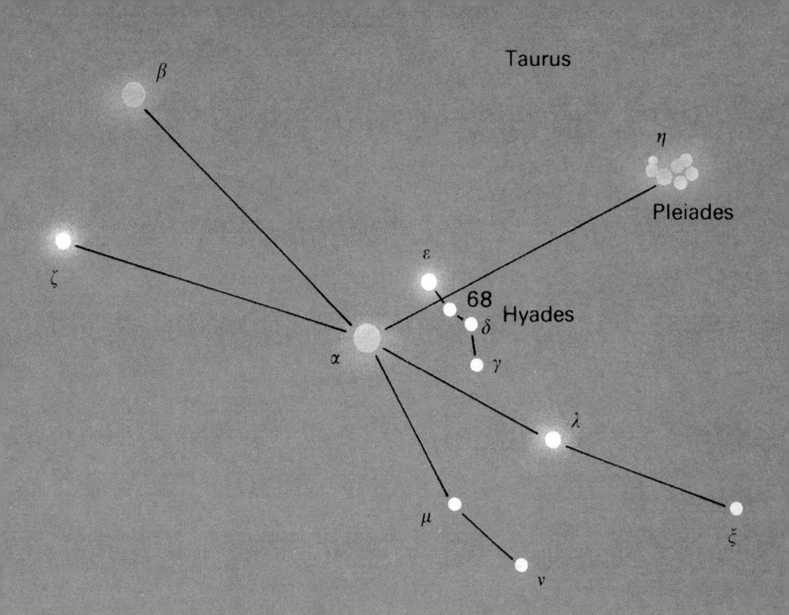

Telescopium

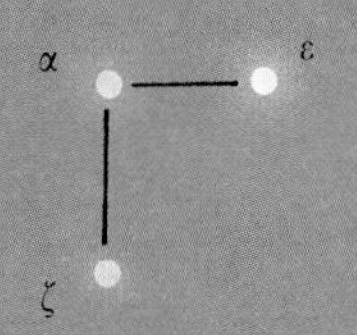

Triangulum

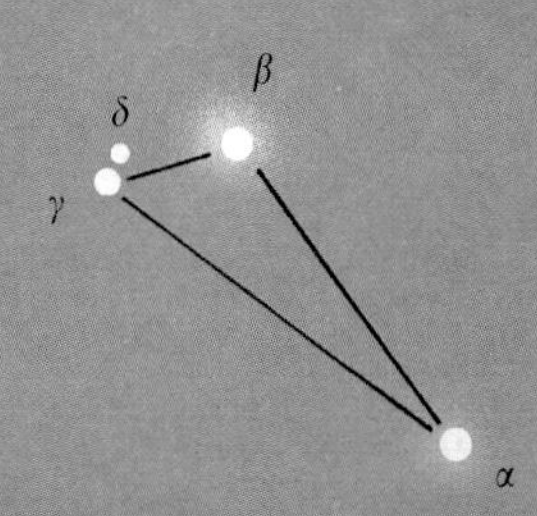

Triangulum Australe (Southern triangle) RA: 16 hr, Dec: –65
Triangulum Australe is a Bayer constellation listed in his early 17th-century catalogue, and is bordered by Ara, Norma, Circinus and Apus between RA 14 hr 50 min–17 hr 10 min and Dec −60 — −70.

Star α is a mag 1·93 K2 type with absolute mag −0·1 at a distance of 82 light years, star β is a F2 type of mag 2·87 at 42 light years distance, and star γ is of mag 2·94 being an A0 type of absolute mag 0·2 at a distance 113 light years. Nothing else of note is found in this constellation.

Tucana (Toucan) RA: 23 hr, Dec: −60
Located between RA 22 hr 5 min–1 hr 25 min and Dec −57 — −75, Tucana contains most of the SMC, a galaxy close to our own of irregular class and partnered by the LMC in Dorado. The constellation contains several interesting nebulae and two globular clusters. Star α is a mag 2·8 K3 type at a distance of 62 light years.

Ursa Major (Great bear) RA: 11 hr, Dec: 50
This is possibly the most famous of all the constellations because its main geometry is associated with a form known as the 'plough' in modern times and the area of sky within its boundaries contains many interesting phenomena. Ursa Major occupies a position in the northern sky extending between RA 8 hr 5 min–14 hr 30 min and Dec 28—73. The constellation is bordered by Draco, Camelopardus, Lynx, Leo Minor, Leo, Coma Berenices, Canes Venatici and Boötes. It has a highly irregular outline and prominent stars are spread across its large area.

Taking the Plough stars in sequence, and moving in increasing hours of RA from the northernmost component of this well-known form: α, Dubhe, is a mag 1·81 A0 spectra double with component mags of 1·88 and 4·82 and 107 light years away; star β, Merak, is a mag 2·37 A1 type with absolute mag 0·5 at a distance of 78 light years; γ, Phad, is an A0 type of mag 2·44 (absolute mag 0·2) 90 light years away; δ, Megrez, is a comparatively faint member at mag 3·3; ε, Alioth, is the brightest member of the 'seven sisters' at mag 1·79 with an A0 spectra, absolute mag of 0·2 and it lies at a distance of 68 light years; star ζ, Mizar, is an A2 mag 2·06 double with component mags of 2·26 and 2·94 and the former is itself a double – the first spectroscopic binary to be discovered; and finally, η, Alkaid, at the comparatively remote distance of 210 light years, is a B3 type of mag 1·87 (absolute mag −2·1).

Four interesting doubles are located in the constellation: star θ is actually a double with mag 3·19 and 14; star ι has mag 3·12 and 10·8 components; star κ has component mags of 4·0 and 4·2; and star ο, Muscida, has components of mag 3·57 and 15. There are six Messier objects in the constellation (M81, M82, M97, M101, M108 and M109). M97 is the famous planetary Owl Nebula, looking like a staring owl, with the balance comprising galaxies in their own right.

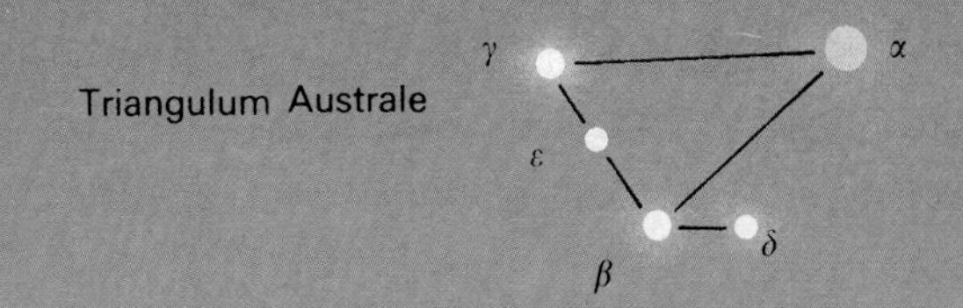
Triangulum Australe
γ
α
ε
β
δ

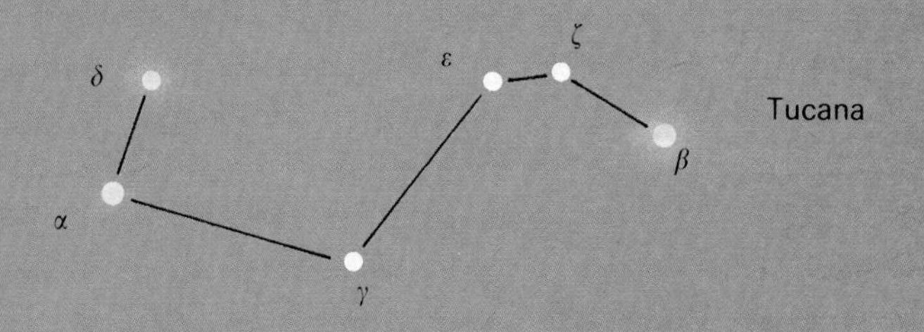
δ
ε
ζ
Tucana
β
α
γ

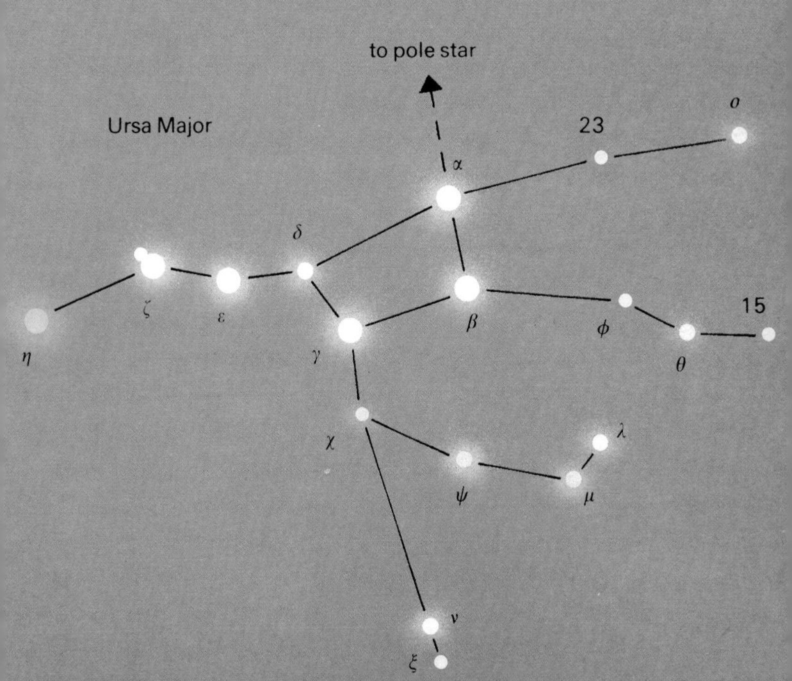
to pole star
Ursa Major
23
o
α
δ
ζ
ε
η
γ
β
ϕ
15
θ
χ
λ
ψ
μ
ν
ξ

Ursa Minor (Small bear) RA: 15 hr, Dec: 75
Ursa Minor was one of the originally named 48 constellations, and has the supreme distinction of capping the north celestial pole. Much of Ursa Minor seems nestled within the cradle of Draco but this constellation is unmistakably distinctive and having found the star nearest to the north celestial pole the rest is easy to recognize.

The pole star itself, or the star which today is most easily observed to be closest to the north celestial pole, is the celebrated α Polaris, a mag 1·99—2·1 variable of F8 type and an absolute mag of −4·6. Lying at a distance of 680 light years, the star is attended by a mag 8·9 companion. Star β, Kochab, is a mag 2·04 K4 type and star γ, Pherkad, is of mag 2·04 so that when Polaris, a notable Cepheid, is dim it is actually fainter than these stars.

Ursa Minor is surrounded by Draco, Cepheus and Camelopardalis and occupies a portion of the sky between RA extremes of 13 hr–18 hr and Dec 66—90. The 5 hr longitude in RA describes the main part of the constellation; the northernmost part is circumpolar.

Vela (Ship's sails) RA: 09 hr, Dec: −50
Vela is one of the three constellations formed from the original, and unwieldy, Argo Navis. It is bordered by Antlia, Pyxis, Puppis, Carina and Centaurus in an area between RA 8 hr–11 hr 5 min and Dec −37 — −57. The designation of component stars was set up when Vela was part of Argo Navis and the brightest star here is γ Velorum, a double with component mags of 1·88 and 4·31. Star δ, Koo She, is a mag 1·95 A0 type with a mag 5·1 companion. A third companion to δ is actually a spectroscopic binary viewed as a single mag 10 star.

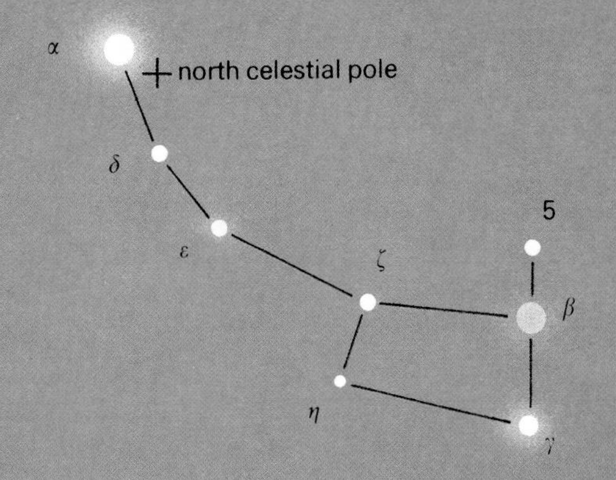
Ursa Minor
α
north celestial pole
δ
ε
ζ
5
β
η
γ

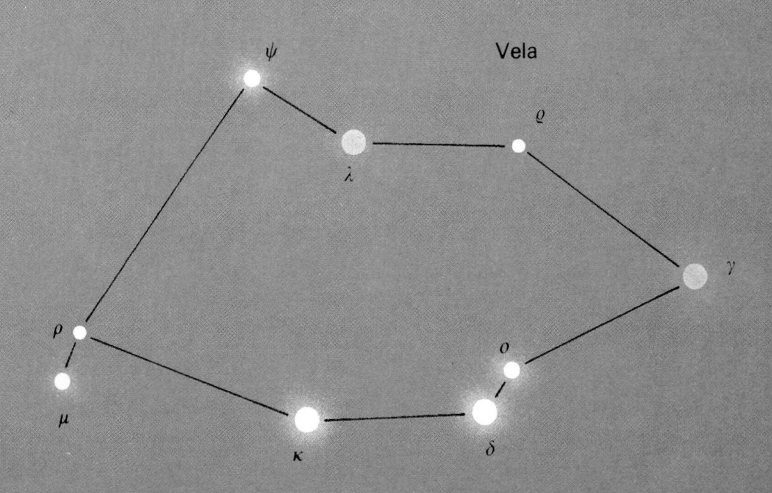
ψ
Vela
ϱ
λ
γ
ρ
o
μ
κ
δ

Virgo (Virgin) RA: 13 hr, Dec: 00
Consistently associated with female innocence and virginity, this constellation is one of the earliest to be named and was one of Ptolemy's original 48.

Virgo is a zodiacal constellation and it is bordered by Coma Berenices, Leo, Crater, Corvus, Hydra, Libra, Serpens and Boötes. Second only to Hydra in area, and much more conspicuous because of its squared area, Virgo lies between extremes of RA 11 hr 35 min–15 hr 10 min and Dec −22—15.

The most prominent star in the constellation is α, Spica, southernmost of the seven stars making up the geometrical form, with mag 0·91 and a location 220 light years from the Sun. The star is a variable with a mag 1·01 companion and is seen as a hot white object. Star γ, Postvorta, is a mag 2·76 F0 type with an absolute mag of −3·3 and is found to lie at a distance of only 32 light years. Far to the west of the constellation lies β, Zavijava, of mag 3·8, and to the north ε, Vindemiatrix, is a mag 2·86 star of spectral type G9 with an absolute mag of 0·6 at a distance of 90 light years.

Virgo lies adjacent to Coma Berenices and the northern border of Virgo contains many interesting nebulae – all external galaxies. Most notable are M58, M59, M60, M84, M87, M89 and M90. On the south-western border (with Corvus) the telescope will reveal the magnificent Sombrero Galaxy (M104) with M49 and M61 located between stars β and ε.

Volans (Flying fish) RA: 08 hr, Dec: −70
One of Bayer's constellations, Volans is more properly seen as an adjunct to Carina and this inconspicuous group is flanked by that constellation together with Pictor, Dorado, Mensa and Chamaeleon. It is found within an area between RA 6 hr 35 min–9 hr and Dec −64 — −75. None of the stars in this region of the celestial sphere is of significance or as bright as mag 3. The keen observer may be able to discern the mag 9 companion to ζ and, more easily, make out the double star γ.

Vulpecula (Fox) RA: 20 hr, Dec: 25
Vulpecula was one of the constellations named by Hevelius in 1690 and is located alongside Sagitta but is also bordered by Delphinus, Pegasus, Cygnus and Lyra. The form of the constellation is long and comparatively narrow, occupying a region of the celestial sphere located between RA 18 hr 55 min–21 hr 30 min and Dec 19—29.

The constellation is probably best served by the sinuous threads of the Milky Way that course through the centre of this area. The stars within Vulpecula are very faint and extremely inconspicuous and there is nothing of very great note in the constellation. The nebula M27 (NGC 6853) can be found to the south, close on the border with Sagitta. This is a planetary nebula and is widely known as the Dumbbell.

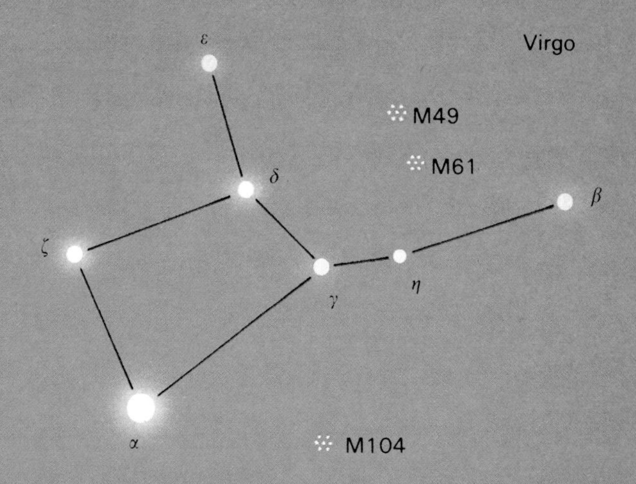
Virgo
ε
M49
M61
δ
β
ζ
γ
η
α
M104

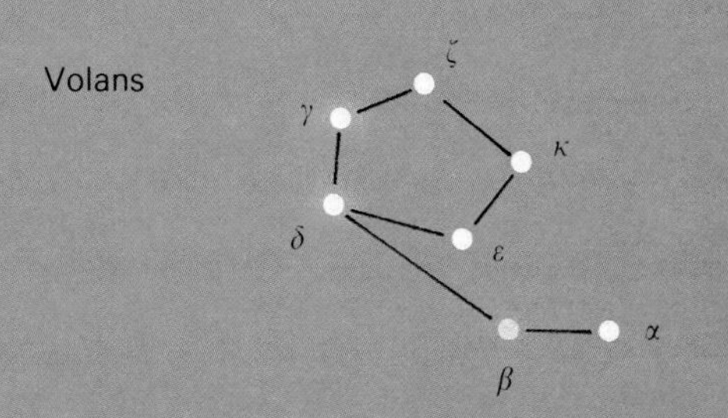
Volans
ζ
γ
κ
δ
ε
β
α

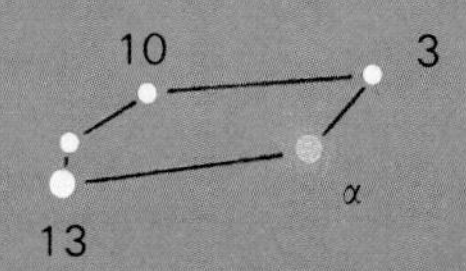
Vulpecula
10
3
α
13

Constellation notes

Cygnus: Further points of interest concerning this constellation are provided here so that the reader may appreciate the phenomena in this area and the observer can place Cygnus on the list of priority objects.

After α, Deneb and β, Albireo, the next brightest star is γ, Sadr, a mag 2·22 F8 type with an absolute mag of −4·6 and a measured distance of 750 light years. Star ε, Gienah, is a mag 2·46 K0 type with absolute mag 0·7 which lies at a distance of only 74 light years. A double star, seen as the mag 2·87 δ, Cygni, has components of mag 2·9 and 6·44 with the system measured to be 270 light years distant.

A particularly intriguing variable can be found about half way from γ, Sadr, to Albireo. It is a long-period variable (409 days) which changes from mag 4 to mag 14 and back again. The star, χ Cygni, is a classic example of the type of long-period variable which, although basically regular, show considerable fluctuations in the form of their light-curves.

An interesting double, called 61 Cygni, can be found between stars τ and υ. It has component mags of 5·3 and 5·9 and exhibits a very high proper motion of more than 80 km/s; this is the observed effect of its close proximity to the Sun, being only 11 light years distant. The star SS Cygni, a variable of mag 8—12, is also of interest and with this class of celestial phenomenon the amateur can provide invaluable information to professional astronomers. The temperamental fluctuations of variable stars are unique to specific examples, and careful measurement and analysis of particular cycles of luminosity flux can contribute valuable data about these remarkable characteristics.

Much of Cygnus lies, of course, in the Milky Way and even with limited optical aid the constellation gives the impression of endless stellar clouds receding into the distance in the central plane of the Milky Way. However, an interesting area, the Northern Coalsack, can be found here and it provides a forbidding contrast to the myriads of stars seen around it. The Veil Nebula and the North American Nebula are also in Cygnus and two open clusters, M29 and M39, complete the scenario of celestial splendour.

Orion: The magnificence of this constellation is hard to surpass and the proliferation of interesting celestial phenomena deserves more analysis than the cursory review contained in the main body of this section.

Star η Orionis can be found about one-third of the way along a line joining δ and β and is an eclipsing binary with stellar components of mag 3·59 and 4·98; the period of revolution about the common centre of mass is approximately eight days. A striking quadruple system is that of σ Orionis, just south-west of Alnitak. The four component stars are of magnitudes 4·0, 7·0, 7·5 and 10·0.

One of the most spectacular sights on the photographic plates of

Orion appears as a faint, hazy smear of light to all but the largest telescope. Nevertheless, it is precisely this type of phenomenon which beckons the amateur to bigger and better instruments. This is the Great Nebula in Orion, or simply the Orion Nebula, listed as M42 and M43 in the Messier catalogue.

Through the telescope, the Orion Nebula has a distinctly greenish tinge, probably caused by ionized oxygen atoms, but very large observatories and photographic plates record a magnificent array of white, red, purple and mauve hues. The brilliant white centre of the nebula seems to direct the eye down to a centre of raging energy. It is not hard to imagine that this broad expanse of gas, 16 light years across, is the spawning ground of new, hot stars and indeed the nebula is made visible by the light from these stellar breeding grounds.

The Trapezium, a close nest of four stars at the very centre, is recorded as θ Orionis with component mags of 6·0, 7·0, 7·5 and 8·0. This group of four stars is estimated to be about $3{\cdot}0 \times 10^5$ years old and it is quite easy to distinguish from the nebula which it illuminates. The extremely tenuous material of the gaseous nebula has a calculated density of only 10^{-17} g/cm^3 and consists mainly of hydrogen. The Orion Nebula is receding from the solar system at the rate of nearly 18 km/s although considerable variation exists in velocities measured in specific regions of the cloud.

Another rather interesting, certainly very different, observational feature in the Orion constellation is the famous Horsehead Nebula found just south of star ζ, Alnitak, itself a distant double with a system mag of 1·79. The Horsehead Nebula is listed in the Barnard catalogue as No. 33 and is found to be 5′ across in the sky, and literally resembles a horse's head. The nebula is a very dark patch of sky but a warm red glow from behind it and to the immediate north-west of Alnitak indicates the presence of a bright nebula beyond.

The Planets

Properties of the solar system

Despite the fact that neighbouring stars are barely a few light years distant, and in the grand cosmic scale of the galaxy hardly discernible as separate from the Sun, the planets, asteroids, comets and meteorites make the solar system a haven for carbon-based life forms.

The boundary of the solar system may be defined, in terms of physical effects, as the limit of the Sun's influence. One value is given by the extent of the heliosphere, the region swept by the solar wind. Its boundary, where the pressure of the outward flowing protons and electrons is balanced by incoming particles, probably lies at about $1{\cdot}5 \times 10^{10}$ km.

Alternatively, the gravitational radius of the Sun extends well beyond $1{\cdot}5 \times 10^{11}$ km and it is equally true that this should be regarded as a limit on the solar system. There is an interesting correlation between the outer limit of the heliosphere and the orbital configuration of the planets – they all lie within the heliosphere.

A common unit of distance measurement, most conveniently related to distances *within* the solar system, is the Astronomical Unit (AU). This is equivalent to the mean distance of the Earth's orbit about the Sun ($149{\cdot}6 \times 10^6$ km) and serves to clarify the relative distance of planets compared to the Earth–Sun distance.

The heliosphere can be said to extend out to some 50–100 AU while the gravitational envelope reaches more than 10^3 AU. This chapter considers the gravitational radius of the Sun to define the solar system and confines its comments to phenomena generated within that obtained value.

From **Fig. 59** it is immediately obvious that the nine known planets can be grouped into roughly two size categories: the four so-called terrestrial, or inner, planets and the four giant, or outer, planets. Pluto has been considered a former satellite possibly ejected from the orbit it originally held around Neptune (see p. 210).

Other properties categorize the two groups, notably the measured density values of the nine bodies. Here also the planets fall conveniently into two distinct groups to reflect intrinsic characteristics of elementary composition. The four terrestrial planets **(Fig. 60)** possess mean densities between $3{\cdot}9$–$5{\cdot}5$ g/cm^3, whereas the predominantly gaseous outer planets reveal mean density extremes of between $0{\cdot}7$–$1{\cdot}7$ g/cm^3. These values are obtained by dividing the mass of the planet by its measured volume: $\dfrac{m}{\frac{4}{3}\pi r^3}$

where m equals the planetary mass and r is the known radius. For example, the mean density of Venus is:

$$\frac{4{\cdot}87 \times 10^{27}}{9{\cdot}29 \times 10^{26}} = 5{\cdot}2 \text{ g/cm}^3.$$

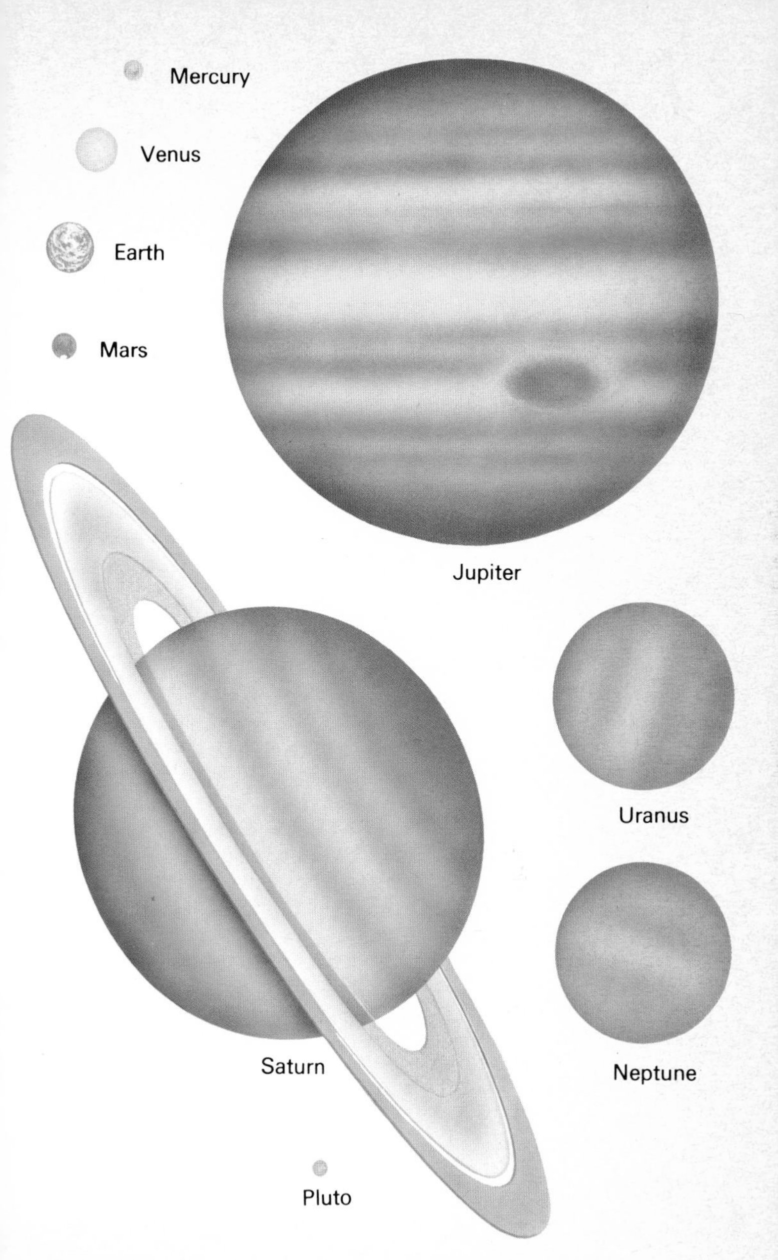

Fig. 59 The relative sizes of the planets in the solar system. Jupiter, Saturn, Uranus and Neptune are collectively known as gaseous, giant or Jovian bodies while the remainder are terrestrial bodies.

The implications of these figures are considered under relevant subsections but it is important to recognize the marked difference in the two groups. The four giant outer planets rotate on their polar axes with periods of between 10–16 h while the four terrestrial planets exhibit periods of between 1–243 days. Also, all the giant planets have many important satellites, whereas the terrestrial planets have few. Earth's Moon is a rather special case and will be considered later. Also, the ratio of satellite mass to primary mass is so great for the Earth–Moon system that it requires a special theory to explain its origin.

Finally, while the terrestrial planets contain about twice the Earth mass the combined outer planets equal 445 times the Earth mass. In other words, Mercury, Venus, Earth and Mars only possess 0·45 per cent of the planetary mass, while Jupiter alone accounts for 71 per cent, and together with Saturn claims 92 per cent of the total.

The importance of these two bodies is immediately obvious and evolutionary models of the solar system must concentrate on the predominance of Jupiter and Saturn, but this asymmetry between the planetary groups should be borne in mind throughout the chapter.

Planetary orbits were a profound source of confusion to the ancients, and Ptolemaic and Copernican theories were only a close approximation to orbital mechanics. Less than 400 years ago, Kepler introduced his three famous laws defining orbital motion and effectively set the scene for adequately describing the observed trajectories. As Kepler explained in his first law, the orbit of a secondary body about a primary describes an ellipse with the centre of mass of the primary occupying one focus of that ellipse. The second law states that the line joining the planet and the focus occupied by the primary will sweep out equal areas in equal periods of time, regardless of the planet's position in its orbit. Its velocity therefore varies around its elliptical orbit, being fastest when close to the primary. The third law states that the square of the orbital period is proportional to the cube of the semi-major axis.

Some consideration must now be given to the periods of revolution used throughout this book and expressed in terms of sidereal or synodic measurement. From the moving platform of Earth, the third planet from the Sun, the observed positions of the other planets must be compensated for in order to derive the actual period of the orbit. A full 360° passage round the Sun is known as the sidereal year which,

Fig. 60 Above: The relative scale orbits of the four inner planets. As viewed from Earth, when Mercury is on the far side of the Sun it is said to be at superior conjunction (B) and when an interior planet, in this example Venus, is between Earth and Sun it is said to be at inferior conjunction (A). An outer planet, in this case Mars, on the far side of the Sun would be said to be at conjunction. If Earth lies between the Sun and an outer planet the latter is said to be at opposition. Below: Kepler's second law states that the line joining the planet and the focus occupied by the primary will sweep out equal areas in equal periods of time regardless of the planet's position in its orbit.

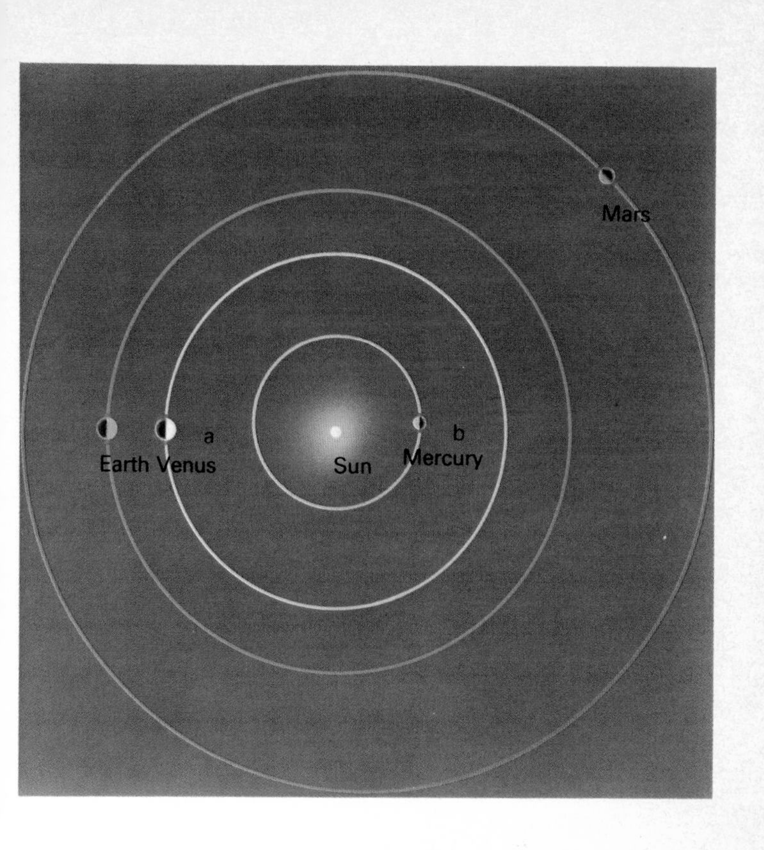
Mars
a
b
Earth Venus
Sun
Mercury

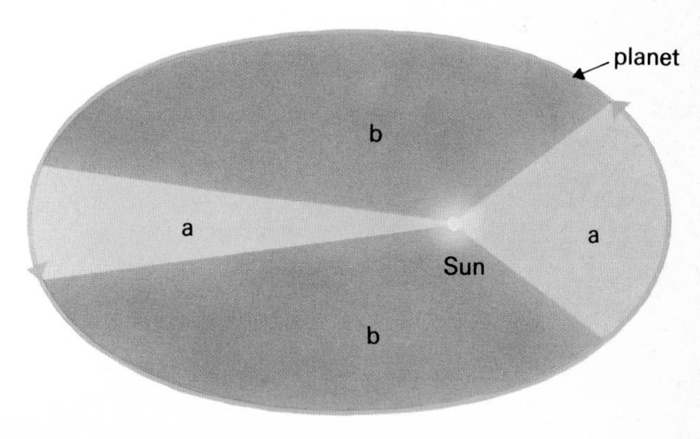
planet
b
a
a
Sun
b

for Earth, can be measured as the elapsed duration between successive passages of the Sun past a fixed point in space, usually the known position of a star. With the other planets, however, what is measured is the synodic period, or time after which the planet appears in the same position relative to the Earth and the Sun, perhaps between successive oppositions. In the case of an inner planet which overtakes the Earth, the synodic period is less than the Earth's sidereal year, while for outer planets it is greater. A planetary sidereal period is the time taken for a particular planet to make one full revolution around the Sun. **Figures 60** and **61** show the terminology applied to planetary positions.

Interior planets, those positioned between the Earth and the Sun, are said to be in inferior or superior conjunction when they lie on near or far sides of the Sun respectively. Exterior planets, with orbital radii greater than the Earth, are in conjunction when they are on the far side of the Sun and in opposition when their orbital motion places them on the opposite side of the Earth to the Sun.

These positions are extremely important for observation since an interior planet will be impossible to observe against the light of the Sun at superior and inferior conjunction. The most favourable position will therefore be obtained when the planet is midway between these locations. An exterior planet will be favourably situated for observation when it lies close to opposition.

It has been convenient to discuss planetary orbits as though they described circles about the primary body, although Kepler's first law adequately describes an orbit as an ellipse with one foci of the ellipse occupied by the prime mass. The measure of ellipticity is given as the distance between the two foci divided by the length of the major axis. This is described as the eccentricity of the orbit.

Finally, using the plane of the Earth's orbit as a reference, planetary inclinations are set against this arbitrary measure for comparison. Orbital planes set up by the other planets are measured in angles to the ecliptic, a term which, when hypothetically fixing the Sun, describes the Earth's movement against the stellar background. No planet exhibits an orbital inclination greater than 17·2° and in fact only one planet exceeds 7° (see below). The inclination of a planet's rotational axis is the angle between a line perpendicular to the specific orbital plane of that planet and the actual polar axis.

The Moon

With a magnitude of −12·7 at maximum illumination, the Moon has an observed average angular diameter of 31·09′, compared with the Sun's angular diameter of 31·98′, and sidereal and synodic periods of 27 days 7 h 43 min and 29 days 12 h 44 min respectively. Observation of particular features is usually undertaken during the phases when the terminator, the line separating lit and unlit portions, moves across

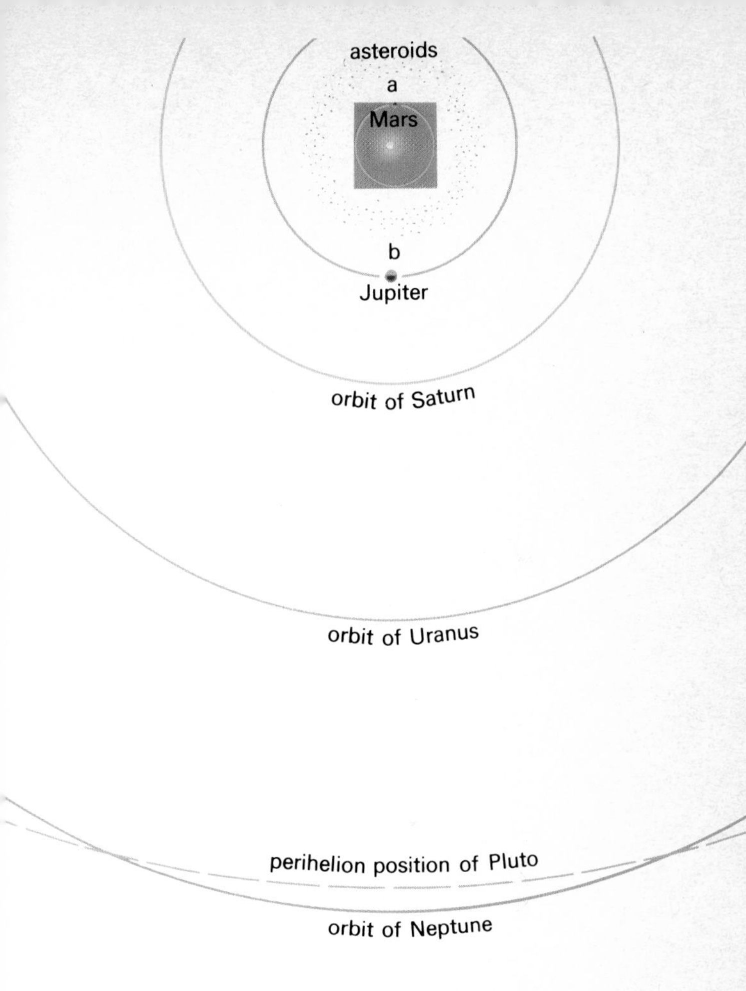

Fig. 61 The inner solar system, contained by the radius of Mars' orbit, is shown as area A in the small box. Between Mars and Jupiter lie the asteroids, a field of rocky debris, remnants from the beginning of the solar system. Note the two extremes for the orbital position of Pluto. The elliptical path of the planet brought it within the orbital radius of Neptune in 1978 and not for nearly 40 years will it resume its position as the outermost planet in the solar system.

orbit of Pluto (aphelion)

Fig. 62 Moon map 1 – the north-west quadrant

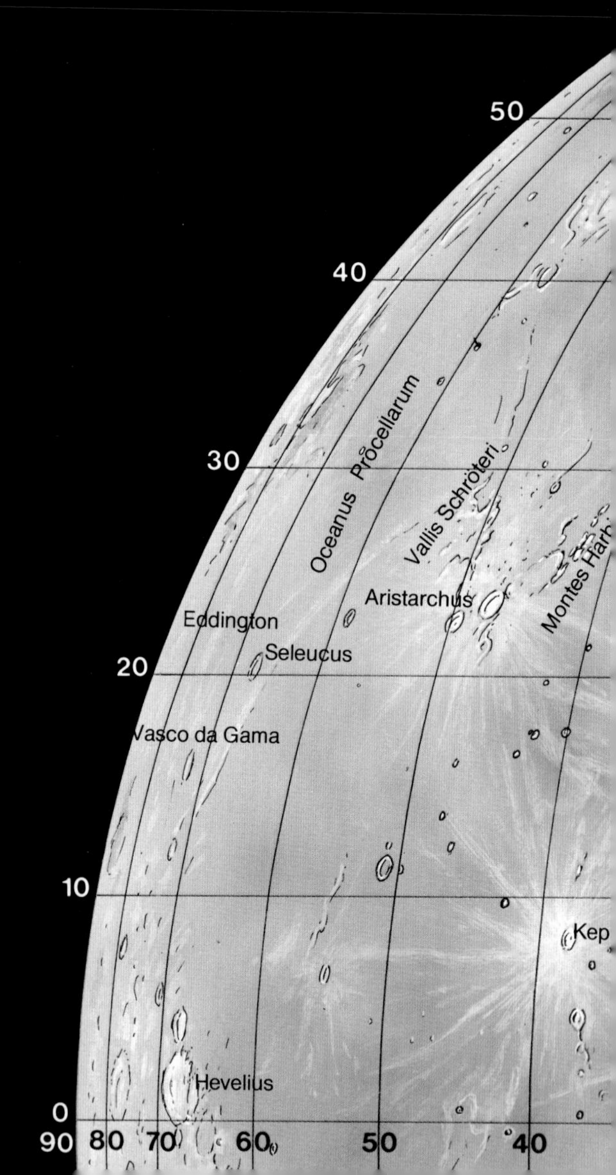

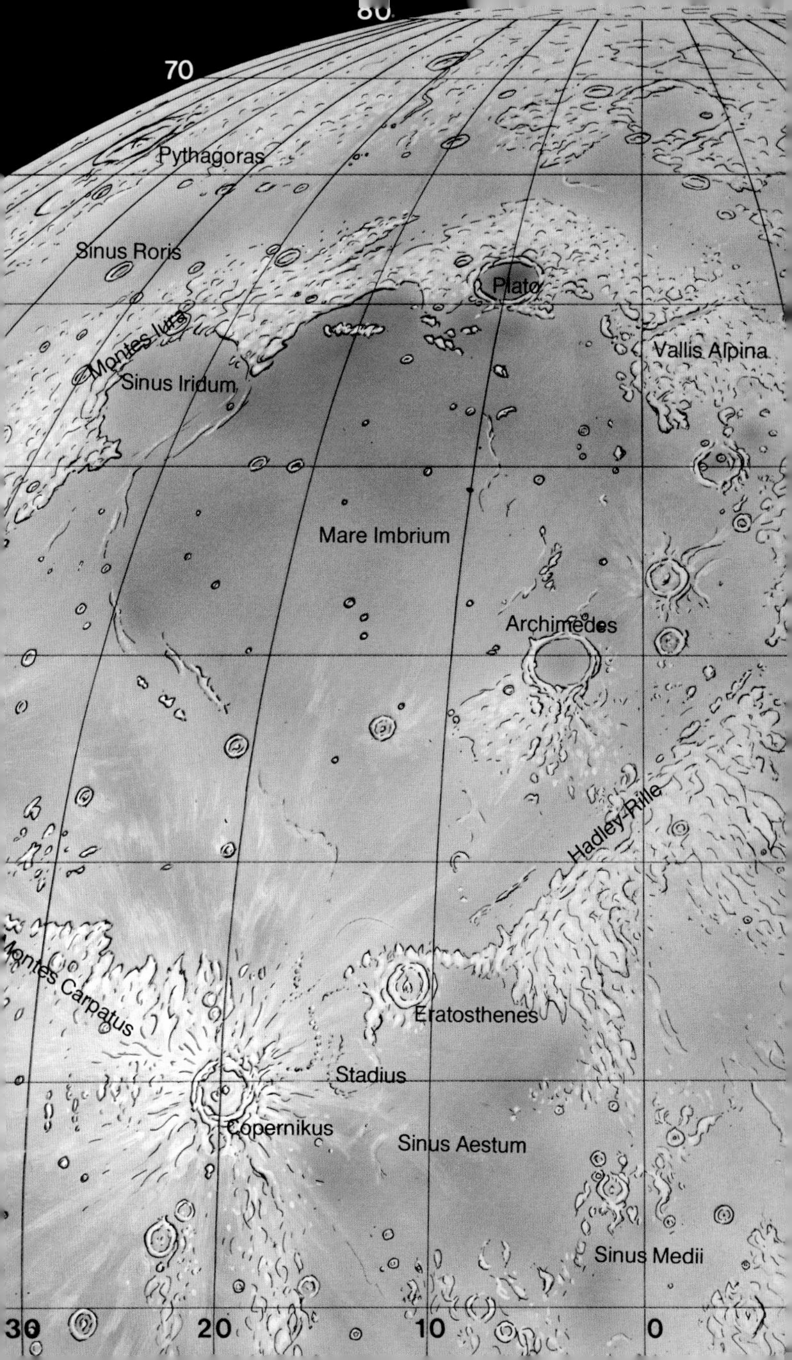

80
70
Pythagoras
Sinus Roris
Plato
Montes Iura
Sinus Iridum
Vallis Alpina
Mare Imbrium
Archimedes
Hadley-Rille
Montes Carpatus
Eratosthenes
Stadius
Copernikus
Sinus Aestum
Sinus Medii
30
20
10
0

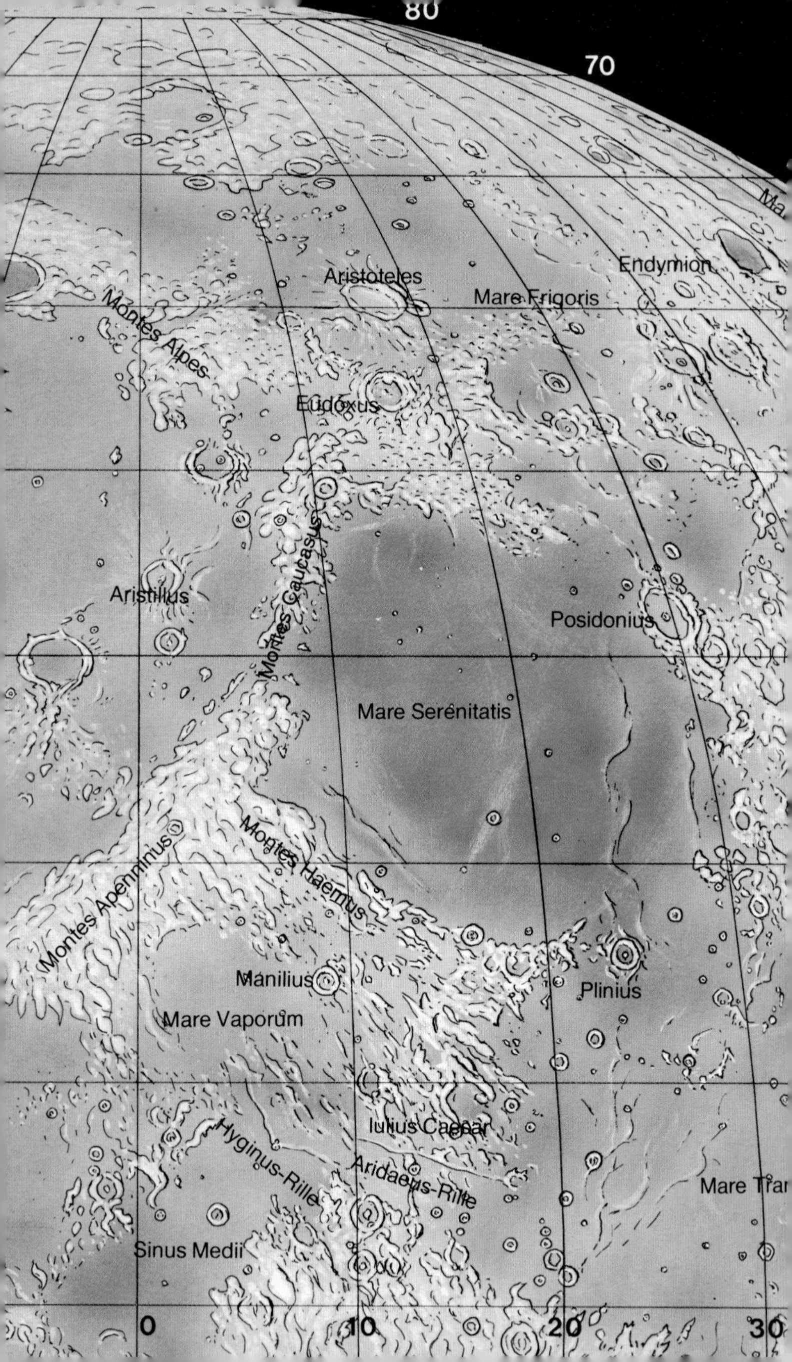
80
70
Ma
Endymion
Aristoteles
Mare Frigoris
Montes Alpes
Eudoxus
Montes Caucasus
Aristillus
Posidonius
Mare Serenitatis
Montes Haemus
Montes Apenninus
Manilius
Plinius
Mare Vaporum
Iulius Caesar
Hyginus-Rille
Aridaeus-Rille
Mare Tra
Sinus Medii
0
10
20
30

Fig.63 Moon map 2—the north-east quadrant

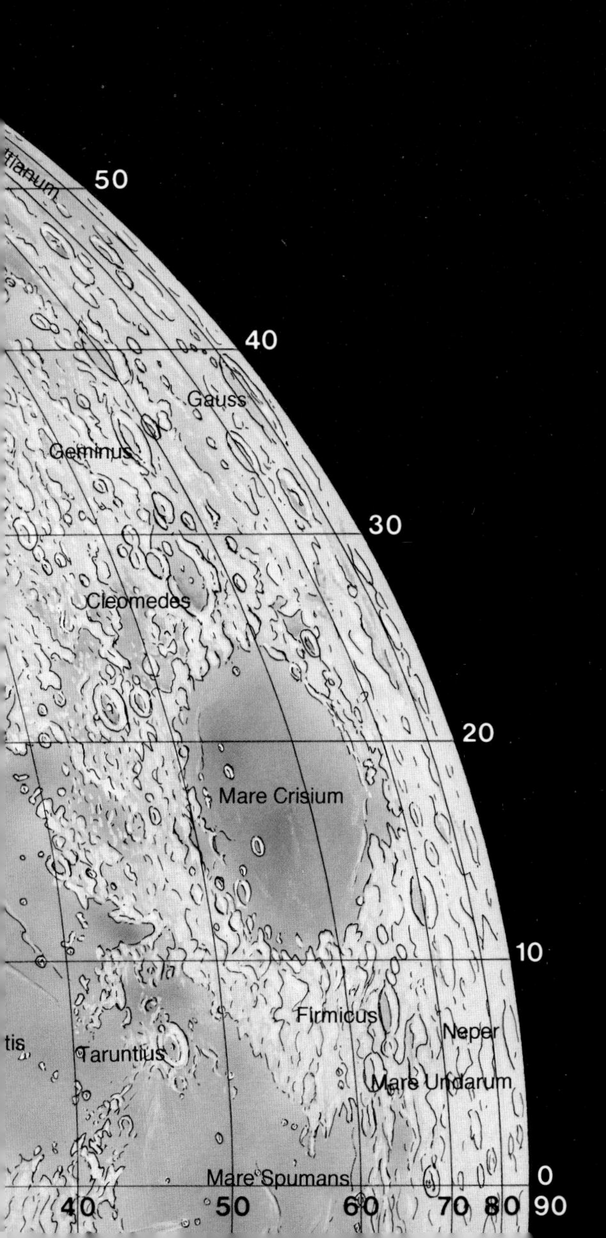

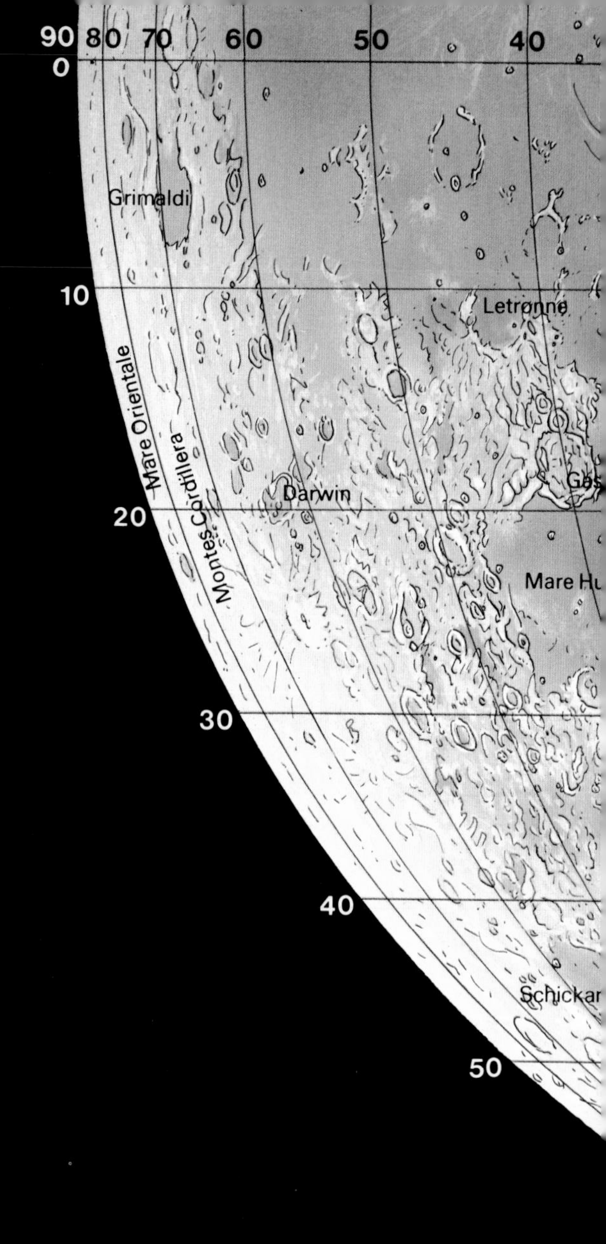

Fig.64 Moon map 3—the south-west quadrant

30
20
10
0
Sinus
Medii
Fra Mauro
Ptolemaeus
Alphonsus
Arzachel
Bullialdus
Mare Nubium
Purbach
Tycho
Maginus
Schiller
Clavius
70
80

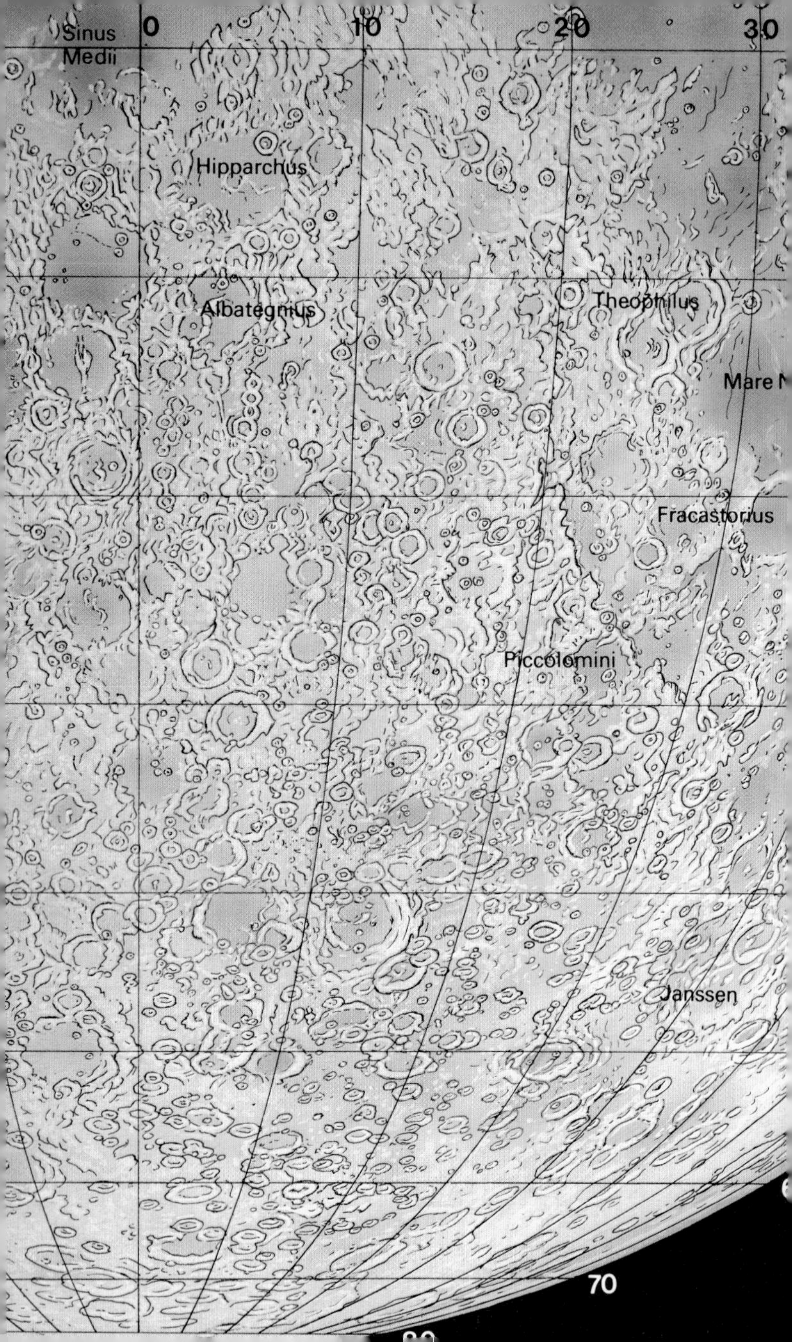

Sinus Medii
0
10
20
30
Hipparchus
Albategnius
Theophilus
Mare N
Fracastorius
Piccolomini
Janssen
70

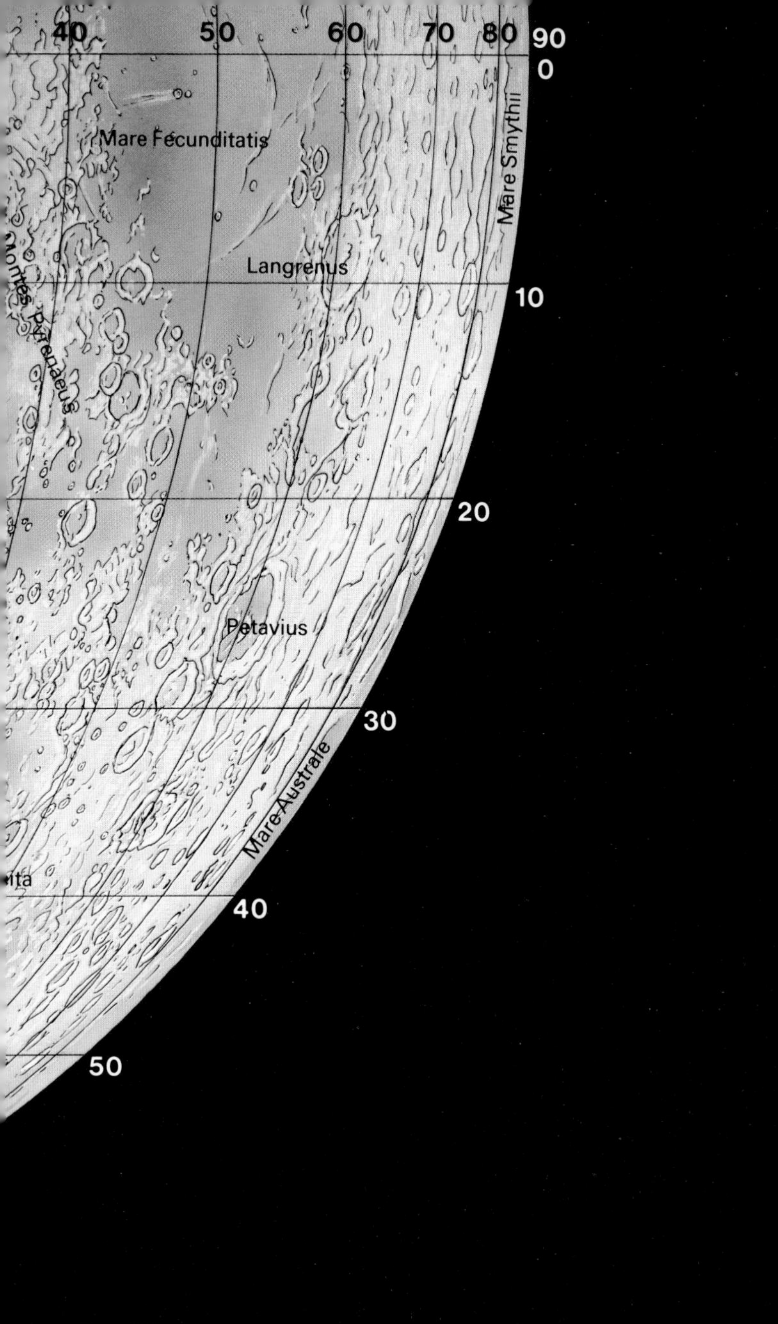

Fig.65 Moon map 4—the south-east quadrant

Index to moon maps

Feature	Map	Position	
Albategnius	4	12°S	4°E
Alphonsus	3	13°S	3°W
Archimedes	1	30°N	4°W
Ariadaeus Rille	2	7°N	11°E
Aristarchus	1	24°N	48°W
Aristillus	2	34°N	1°E
Aristoteles	2	50°N	18°E
Arzachel	3	18°S	2°W
Bullialdus	3	21°S	22°W
Clavius	3	58°S	14°W
Cleomedes	2	27°N	55°E
Copernicus	1	10°N	20°W
Darwin	3	20°S	69°W
Eddington	1	22°N	72°W
Endymion	2	55°N	55°E
Eratosthenes	1	15°N	11°W
Eudoxus	2	44°N	16°E
Firmicus	2	7°N	64°E
Fra Mauro	3	6°S	17°W
Fracastorius	4	21°S	33°E
Gassendi	3	18°S	40°W
Gauss	2	36°N	80°E
Geminus	2	35°N	57°E
Grimaldi	3	6°S	68°W
Hadley Rille	1	25°N	3°W
Hevelius	1	2°N	67°W
Hipparchus	4	6°S	5°E
Hyginus Rille	2	8°N	7°E
Janssen	4	46°S	40°E
Julius Caesar	2	9°N	15°E
Kepler	1	8°N	38°W
Langrenus	4	9°S	61°E
Letronne	3	10°S	43°W
Maginus	3	50°S	6°W
Mairan	1	42°N	43°W
Manilius	2	15°N	9°E
Mare Australe (Southern Sea)	4	37°S	89°E
Mare Crisium (Sea of Crises)	2	15°N	60°E
Mare Fecunditatis (Sea of Fertility)	4	5°S	50°E

Feature	Map	Position	
Mare Frigoris (Sea of Cold)	2	50°N	35°E
Mare Humboldtianum (Humboldt's Sea)	2	57°N	83°E
Mare Humorum (Sea of Moisture)	3	24°S	40°W
Mare Imbrium (Sea of Rains)	1	30°N	15°W
Mare Nectaris (Sea of Nectar)	4	15°S	35°E
Mare Nubium (Sea of Clouds)	3	20°S	14°W
Mare Orientale (Eastern Sea)	3	20°S	89°W
Mare Serenitatis (Sea of Serenity)	2	25°N	15°E
Mare Smythii (Smyth's Sea)	4	3°S	85°E
Mare Spumans (Foaming Sea)	2	1°N	65°E
Mare Tranquillitatis (Sea of Tranquillity)	2	10°N	35°E
Mare Undarum (Sea of Waves)	2	7°N	68°E
Mare Vaporum (Sea of Vapours)	2	14°N	3°E
Montes Alpes (Alps)	2	45°N	2°E
Montes Apenninus (Apennines)	2	22°N	3°E
Montes Carpatus (Carpathian Mountains)	1	15°N	24°W
Montes Caucasus (Caucasus Mountains)	2	34°N	9°E
Montes Cordillera (Cordillera Mountains)	3	24°S	79°W

Feature	Map	Position
Montes Haemus (Haemus Mountains)	2	20°N 8°E
Montes Harbinger (Harbinger Mountains)	1	28°N 41°W
Montes Jura (Jura Mountains)	1	45°N 37°W
Montes Pyrenaeus (Pyrenees Mountains)	4	15°S 43°E
Neper	2	7°N 83°E
Oceanus Procellarum (Ocean of Storms)	1	30°N 60°W
Petavius	4	25°S 61°E
Piccolomini	4	30°S 32°E
Plato	1	51°N 9°W
Plinius	2	15°N 24°E
Posidonius	2	32°N 30°E
Ptolemaeus	3	10°S 3°W
Purbach	3	25°S 2°W
Pythagoras	1	65°N 65°W

Feature	Map	Position
Schickard	3	44°S 54°W
Schiller	3	52°S 39°W
Seleucus	1	21°N 66°W
Sinus Aestum (Seething Bay)	1	12°N 8°W
Sinus Iridum (Bay of Rainbows)	1	44°N 32°W
Sinus Medii (Central Bay)	1, 2, 3, 4	0° 0°
Sinus Roris (Bay of Dews)	1	52°N 48°W
Stadius	1	11°N 14°W
Taruntius	2	6°N 46°E
Theophilus	4	12°S 26°E
Tycho	3	43°S 11°W
Vallis Alpina (Alpine Valley)	1	50°N 3°W
Vallis Rheita (Rheita Valley)	4	40°S 48°E
Vallis Schröteri (Schröter's Valley)	1	26°N 51°W
Vasco da Gama	1	15°N 85°W

the feature of interest. The low angle of illumination casts prominent shadows that accentuate the undulating surface. Prominent features suitable for observation can be found in the maps **(Figs 62–65)**.

Under ideal conditions Transient Lunar Phenomena (TLP) have been observed **(Fig. 66)**. These imply that a subsurface mechanism is at work releasing gaseous products which, in the vacuum of space, temporarily obscure surface features until dissipated. In the light of geophysical analysis by Apollo space vehicles, it seems likely that fission decay of unstable isotopes may be producing these gases. This type of moon observation can be rewarding for amateur groups.

The Moon lies at a mean distance of $3{\cdot}84 \times 10^5$ km from the Earth, and exhibits synchronous rotation, always presenting the same hemisphere to the observer. However, with an orbital eccentricity of 0·043–0·0668, an inclination of 5·1°, and various other factors, the Moon appears to rock, causing additional areas of both latitude and longitude to become visible from Earth at different times. In total up to 59 per cent of the surface can be seen due to this effect, which is known as libration.

Since the Moon's orbit lies at an angle of 5·15° to the ecliptic, occultations of the Sun and Moon (eclipses) are only possible if conjunction and opposition respectively occur at the intersection of the two planes. The Sun's corona can only be observed during a solar eclipse due to the near equality of the two apparent diameters **(Figs 67** and **68)**. At other times, when the Moon is waxing or waning, reflected Earth light dimly illuminates the dark side of the hemisphere facing the Earth.

With a radius of $1{\cdot}738 \times 10^8$ cm and a mass of $7{\cdot}35 \times 10^{22}$ kg, $m/(\frac{4}{3}\pi r^3)$ produces a mean density for the Moon of 3·34 g/cm^3, or less than 0·6 that of the Earth. The small mass causes the surface gravity to be slightly less than one-sixth Earth value. The low escape velocity of 2·38 km/s means loss of atmospheric gases so that the environment of the Moon is, essentially, a vacuum.

Viewed through the telescope, the Moon can be divided into two type areas: maria and highland terrain. The darker maria surface has a

Fig. 66 (below) Fissures and vents in the outer layers of the lunar crust release trapped gases (X) which temporarily obscure surface features over large areas. These TLP are sometimes observed by students of selenography. **Fig. 67 (above right)** An eclipse of the Moon occurs when the Earth passes between the Sun and the lunar sphere, but only when the planes intersect at the time of opposition. **Fig. 68 (right)** A solar eclipse requires the Moon to pass between the Earth and the Sun and for the orbital planes to intersect at the time of conjunction.

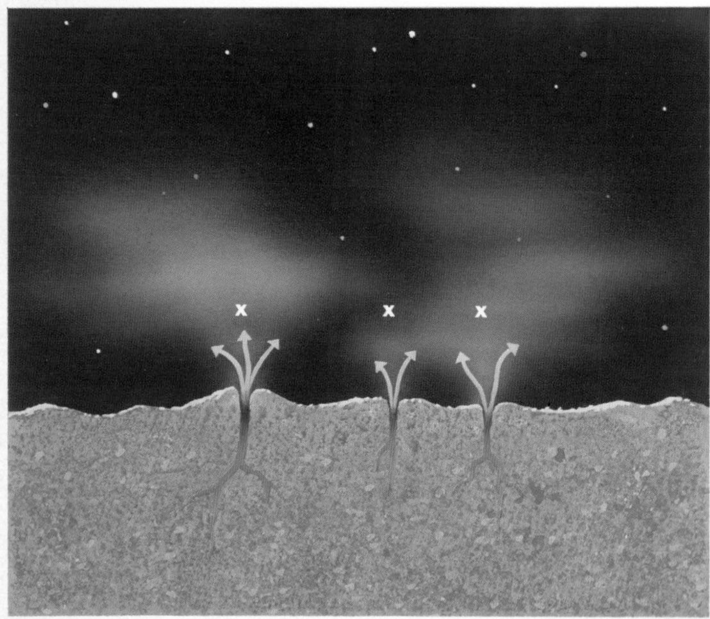

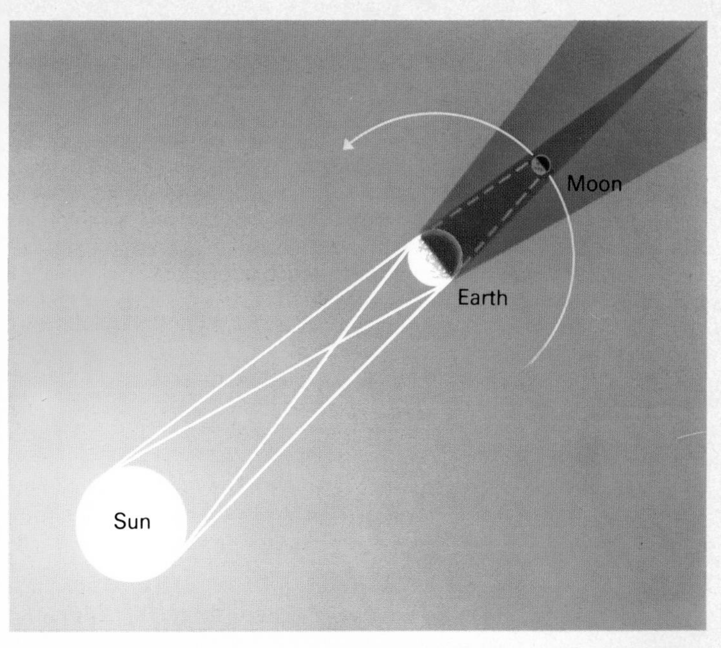
Moon
Earth
Sun

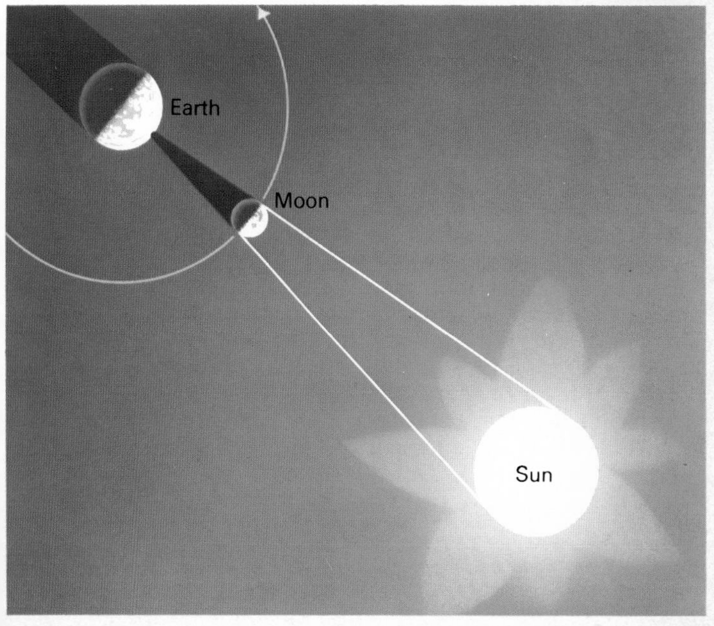
Earth
Moon
Sun

younger age than the lighter-coloured 'highland' material but the latter does not necessarily constitute regions of higher elevation **(Fig. 69)**. Superimposed on the two surfaces the observer will recognize craters, mountain chains, ridges and channels.

From the Earth, about 500 000 craters can be seen through the most powerful telescopes but there is no lower limit on the physical size of these features; several are more than 200 km in diameter, with the largest (Clavius) 230 km across. However, the vast majority are less than 100 km in diameter with the frequency increasing with decreased size. The absence of light-scattering effects due to the lack of air molecules on the Moon creates the impression of great depth when comparatively low-angle sunlight cuts across a crater rim and fails to strike the floor.

Generally, craters less than 40 km across have a ratio of wall height to diameter of 1 : 10, but very large craters may have ratios as low as 1 : 50 and wall slopes are rarely greater than 30°. This ratio change is due to the mechanism that formed the crater and also the continual erosion from mass wasting, electrostatic transportation and internal seismic activity. Several large craters exhibit a cluster of mountain peaks in the centre which is of considerable importance to the theories of crater origin.

It would seem that most craters were formed by impact and such a central thrust cone is also most characteristic of the rebound processes observed in massive explosions on Earth. Considerable efforts have been expended to locate volcanic craters with no positive indication of their existence. Several craters exhibit radial arrays of secondary, or ejecta, craters produced by upper surface material hurled out from the point of impact at great speed on ballistic trajectories. Rim or wall material comes from progressively deeper regions affected by the impact, and the terraces within the walls of Copernicus and Tycho are typical of mass wasting and slump phenomena adjusting the surface material to a particular angle **(Fig. 70)**.

Most isolated mountain chains are the surviving remnants of walls enclosing craters more massive than any seen with the telescope. Such colossal excavations are called basins and most have been extensively flooded with a basaltic lava which has subsequently dried and produced the maria regions **(Fig. 71)**. In places these vast lava sheets are 25 km thick and show pronounced wrinkle ridges – evidence of structural contraction on earlier, solidified extrusions. Prominent among these basins are Mare Serenitatis, Mare Nectaris, Mare Nubium and, where sheets of lava merge to cover vast areas, Mare Tranquillitatis and Oceanus Procellarum. Mare Imbrium is flanked to the east and south-east by the Montes Apenninus, the

Fig. 69 The composition of the Moon can be considered using either of two criteria: the composition and distribution of rocks (into core, mantle and crust as shown in the upper view), or the boundaries determined by seismic profiling (with an asthenosphere and a lithosphere as shown in the lower view).

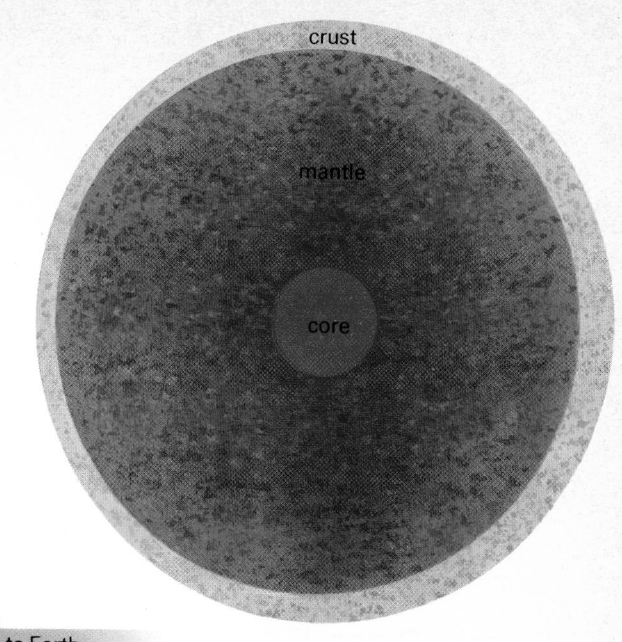
crust
mantle
core

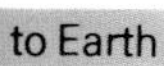
to Earth

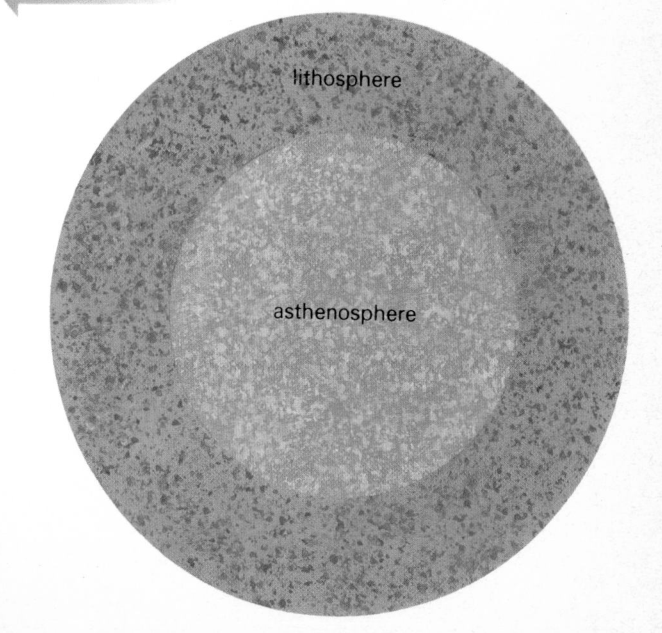
lithosphere
asthenosphere

remaining wall of its once magnificent rim.

Several craters of moderate size and comparatively recent origin display prominent rays extending several hundred kilometres across the lunar surface overlaying other craters and surface features **(Figs 72 and 73)**. These are best seen at full Moon when the Sun angle makes crater observation virtually useless.

Many of the ridges and other features of the lunar surface appear to be due to volcanic processes. Apart from the valleys caused by faulting of the rocks, the Moon also has winding channels known as sinuous rilles. Such rilles seem to be collapsed lava tubes formed when hot magma drains away from beneath the surface; later the roof collapses, debris accumulates on the floor and rille evolution comes to an end **(Fig. 75)**.

Studies by manned and unmanned lunar exploration have revealed the stages of lunar evolution. The Moon was formed along with the rest of the planets, probably about $4{\cdot}7 \times 10^9$ years ago, consisting of a homogeneous mixture of nickel iron, iron sulphide, magnesium silicates and deposits of uranium, potassium and thorium. Shortly thereafter, chemical separation caused the lighter elements to rise to the surface (a process accelerated by the residual heat from accretion) forming a crust of aluminium silicates, 60 km and 100 km thick on the near and far side respectively.

This asymmetric crust profile possibly indicates an early Earth–Moon association and it is interesting to note that the Moon's centre of mass is offset in the direction of Earth by about 3 km. Considerable melting occurred as radioactive elements decayed and magmatic basalts sent aluminium, calcium and potassium to the crustal region with magnesium and iron settling towards the centre; this light anorthositic highland material forming the outer crust. However, throughout this period, which probably lasted $0{\cdot}5 \times 10^9$ years, the chemical composition of the Moon's interior was altering, evolving towards its current state.

For about $0{\cdot}8 \times 10^9$ years after its formation, the Moon was continually bombarded by massive meteorite-like chunks of debris which pulverized the surface and that considerably remodelled the outer crust by deep penetration of the impact shock waves. These planetesimals excavated all the major basins although it should be noted that the majority occurred on the near side of the Moon.

About $3{\cdot}9 \times 10^9$ years ago, the rain of massive boulders had excavated the last of the big basins and only the smaller craters (less

Fig. 70 (above right) Random impacts in the early phase of the Moon's evolution produced large basins several hundred kilometres across. The material that formed the arcuate rims came from the deepest part of the basin while other debris was thrown out to form secondary craters surrounding the main impact zone. **Fig. 71 (right)** About 2×10^8 years after the last major basin had been formed, basaltic lava rose to the surface and flooded most of the floor area.

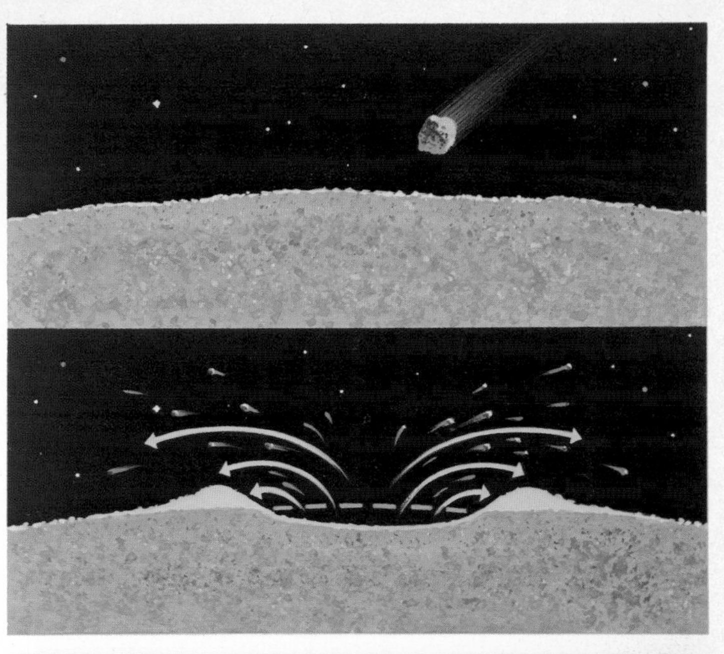

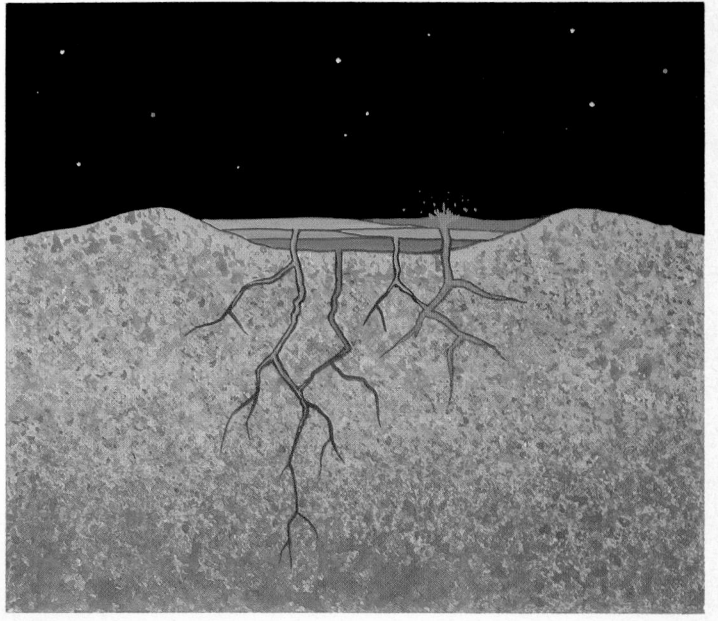

Fig. 72 (above left) The Moon today provides visual evidence of the major basaltic intrusions which are seen as dark patches on the lunar sphere. **Fig. 73 (left)** Highland areas preserve much of the original crust but successive bombardment by random debris in the solar system has covered the primordial surface with numerous craters. **Fig. 74 (above)** Fine particles overlay impact events and ejecta blankets.

than 500 km in diameter) were still being formed from random impacts. A short while thereafter, the thermal energy from uranium, potassium and thorium decay brought basaltic lavas to the surface, filling low basin floors in successive waves lasting in some cases up to $0{\cdot}4 \times 10^9$ years. The maria areas were completely formed by $2{\cdot}9 \times 10^9$ years ago but local extrusions complicated the stratigraphy with ash, or fumarole, deposits.

Impact craters continued to be formed and even the crater Tycho, 87 km in diameter, is only 10^7 years old, confirming the present likelihood of more large scale impacts. The Moon has essentially remained unchanged for 3×10^9 years and the interior is very inactive, there is the possibility of a molten core and moderate gaseous emission from time to time. Tidal interaction with the Earth generates significant seismic activity when the Moon passes through perigee in its orbit, and the resulting moonquakes occur at depths of several hundred kilometres.

The Moon is gradually receding from the Earth and will reach a point where it takes 60 days to complete one revolution. By that time the Earth will have slowed to synchronous rotation and will itself take 60 days to rotate on its polar axis; the Sun will shine for 30 days in any one location followed by 30 'days' of night while the Moon appears stationary over a fixed point on the Earth. However, this is not likely to occur much before the Sun becomes a giant star.

Observation of the Moon is potentially rewarding during the low point in its orbit because many events are triggered by the Earth which could lead to particularly exciting results. For instance, seismic activity is more likely to release gases at this time than any other and a comparatively small telescope should reveal even moderate emissions.

Mercury

Mercury, the innermost planet of the solar system, lies in an orbit of between $45{\cdot}9 \times 10^6 - 57{\cdot}9 \times 10^6$ km (0·31–0·47 AU) which is particularly unfavourable for observation of surface features due to the Sun's proximity. At maximum elongation, when the planet lies at apparent extreme east or west positions in its orbit, it is never more than 28° from the Sun and is therefore visible only in the opening or closing minutes of daylight. It is possible to see the planet in daylight but the glare from the Sun severely hampers useful observation. As a result it was impossible to secure any significant information about the planet, apart from its orbital geometry, until a spacecraft was sent to Mercury in 1974.

It had been known for some time that Mercury orbits the Sun with sidereal and synodic periods of 88 days and 116 days respectively. Up to a decade ago astronomers believed the planet was in syn-

Fig. 75 Collapsed lava tubes like this in the Hadley region have left channels in the lava that flooded basin area.

chronous rotation with the Sun, taking 88 days to make one full rotation about its polar axis; in other words a 'day' on Mercury was equal to its 'year'. By 1965, radar reflections from Earth endorsed increasing suspicion that the planet was not in synchronous rotation and did in fact take about 59 days to make one 360° rotation. Moreover, the observed radio flux from both lit and unlit sides showed that temperatures were too evenly distributed to imply a perpetual darkness for the anti-solar hemisphere.

Refined measurements revealed a planetary 'day' of 58·7 days displaying a unique 3:2 ratio for orbital and rotation periods. This means that Mercury has a spin-orbit coupling which provides for three 360° rotations in exactly two planetary revolutions about the Sun. Temperatures at the surface are the product of solar radiation and like the Moon the planet has virtually no atmosphere (see below).

The 3:2 spin-orbit coupling has a profound effect on temperatures and this is directly related to the 0·206 orbital eccentricity which is the highest eccentricity of any planet except Pluto. Since the orbital angular velocity exceeds the spin velocity at perihelion (closest approach to the Sun) the Sun would appear to reverse its motion across the sky before the planet moved away from perihelion. This would slow the orbital velocity below the spin velocity and return the apparent movement of the Sun to its normal direction.

At perihelion, longitudes 0° and 180° alternately face the Sun receiving 2·5 times the solar irradiation received at longitudes 90° and 270° which only face the Sun at aphelion (farthest point from the Sun). The mean temperature extremes at the surface are 350° C and −170° C which reflect the absence of an atmosphere for effectively transferring heat to the dark side. Nevertheless, the temperatures would be more extreme in respective values if Mercury were indeed in synchronous rotation with the Sun.

In 1974, when the first accurate observation of the planet was made from the US Mariner 10 spacecraft, instruments detected a thin trace of helium sustained at a surface pressure not exceeding 10^{-9} mb. This gas may in fact be a peripheral effect of the close proximity of the planet to the Sun and it is uncertain whether this is trapped helium from the parent star or a surface emission from the interior of Mercury. Also, the planet was observed to possess a very small dipole magnetic field (varying between 350–700 gamma) aligned along the polar axis but only 1/100th as intense as that in the Earth's interior.

From this it can be inferred that Mercury has experienced the same cycle of chemical differentiation as that undergone by the Earth and the mean planetary density of 5·4 g/cm³ implies a large iron core, probably contributing as much as 80 per cent of the planetary mass.

Fig. 76 (above right) With a surface reminiscent of the lunar topography, Mercury is thought to contain a large iron-rich core giving it a high mean density. **Fig. 77 (right)** Mercury bears evidence of major impacts in its early phase but there is no indication of the extensive lava sheets that characterize Moon basins.

iron-rich core

With only 0·055 times the Earth's mass in a volume 0·06 times that of the Earth, Mercury has a greater abundance of heavy elements than any other planet in the solar system, giving it a surface gravity 0·37 times that of the Earth **(Fig. 76)**.

Within a sphere of 2440 km radius, Mercury should have a core of about 1900 km radius. The inclination of Mercury's orbit to the ecliptic is 7°, greater than any other planet with the exception of Pluto and this afforded added complexity for the trajectory required of a visiting space vehicle. Nevertheless, the first, and only, planetary explorer to visit the planet used the gravitational influence of Venus to accelerate it on to Mercury in 1974 and, because of the unique nature of the spacecraft's orbit about the Sun, it performed three close passes across the surface at 176-day intervals. From those reconnaissance fly-by's came the first pictures of the planet's topography showing it to be Moon-like with basins, craters, extruded lava sheets and rilles **(Fig. 77)**.

Many of the larger craters exhibit the same structure and profile as lunar craters of similar dimensions. Craters with diameters greater than 10 km are seen to possess a central peak rising from a flat floor, although the rims are somewhat more subdued with respect to the diameter than similar structures on the Moon. Lava flows are less prominent than those on the Moon and several large basins up to 680 km in diameter are devoid of basaltic fill material in the quantities observed on the lunar surface.

It appears likely that Mercury was first differentiated into an iron core with a thin silicate mantle and rigid crust. Continual bombardment from meteorite-like bodies and crustal volcanism remodelled the surface area over wide regions and major impacts scoured massive basins similar to those found on the Moon **(Fig. 78)**. Also, the sudden termination of massive impact events was followed by lava flows but, unlike the Moon, these appear more spasmodic and less confined to basin impact areas.

The final phase involved occasional cratering by impacts, volcanic flow activity and, probably, minor tectonic displacement caused by subcrustal readjustment from tidal interaction with the Sun. This may have continued to the present day, but it is of course impossible to observe any of these features or changes from the Earth.

Venus

Venus is generally referred to as Earth's 'sister planet' but it is only the similar volume which qualifies it for this distinction. With the exceptionally low orbital eccentricity of 0·007, the planet deviates only $1{\cdot}6 \times 10^6$ km from its aphelion distance of 109×10^6 km (0·72 AU) and presents sideral and synodic periods of 224·7 and 583·9 days

Fig. 78 Bright, rayed craters testify to the sustained bombardment from fragmented bodies.

respectively. Like Mercury, Venus is best observed at maximum elongation when it lies about 51° away from the Earth–Sun line but, unlike the innermost planet, it reflects a considerably higher percentage of light and this gives it a magnitude value of −4·2 compared with −0·2 for Mercury. Principally, this is due to the dense layers of cloud surrounding Venus and the greater surface area of the visible phases from Earth; the planet is the brightest body in the solar system apart from the Sun and the Moon.

The Venusian atmosphere is an opaque concentration of carbon dioxide in rapid motion about the planet, probably driven by thermal convection in the lower regions. About 65 km above the surface, the atmosphere becomes stable and for an additional 20 km of altitude the circulatory patterns predominate with the cloud tops moving round the planet in a period of 100 h. Radar reflectivity measurements of the surface undulations, however, indicate a true planetary rotation period of 243·2 days and a mechanism has yet to be found which will satisfactorily explain the high velocities measured in the upper atmosphere.

Surface pressure is about 90 times that of the Earth with a temperature of about 485° C – making Venus the hottest planet in the solar system. Even though Venus is twice as far from the Sun as Mercury, it is considerably warmer at the surface due to the re-radiation of absorbed heat in the infrared regions of the spectrum. The longer wavelengths are absorbed by the clouds and gases present in the lower atmosphere and are prevented from direct escape, hence the accelerated green-house effect **(Fig. 79)**.

Carbon dioxide constitutes about 97 per cent of the atmosphere due to the high surface temperature. Crustal rocks liberate carbon dioxide at the surface some of which is broken down into carbon monoxide and oxygen; the latter probably moves through a cycle of reactions culminating in sulphuric acid, hydrochloric acid and hydrofluoric acid. The lower regions of the atmosphere contain very little water, probably about 0·05 per cent, and in the upper regions molecular abundances are altered by ultraviolet radiation from the Sun.

Until late in 1975 when two Russian spacecraft landed on Venus (surviving a temperature that melts lead) it had been impossible to obtain any photographs of the surface due to the opaque nature of the atmosphere **(Fig. 80)**. The two panoramic sequences **(Figs 81 and 82)** show that fractured boulders litter the area in great profusion with much evidence of thermal stress. On a larger scale, radar mapping from Earth and orbit reveals a varied terrain, with widespread plains, craters, and high, mountainous areas.

Venus is not unlike the Earth, as it may seem at first, and its physical characteristics provide a mean radius of 6052 km (95 per cent that of the Earth) with a mean density of 5.2 g/cm³, although its orbital inclination of 3·4° is third after Pluto and Mercury. The abundance of carbon dioxide is about the same as that in carbonate rocks on Earth and were it not for the increased energy from the Sun, Venus may well have emerged to be like our own planet. It is necessary, however, to explain how the temperatures were first generated to remove the carbon dioxide gas from surface rocks. One theory suggests that Venus was originally more like the Earth with a cooler, thinner atmosphere and substantial quantities of water. Gradually, solar heat raised temperatures to the boiling point of water, releasing more water vapour and trapping thermal energy in the infrared bands. Once set on this course, temperatures would soar to the point where carbon dioxide was liberated from rock carbonates to the observed level. If this was the case, there may have been a period when very simple micro-organisms developed and, if permitted to evolve to the hard-shell stage, fossils could conceivably await detection by some sophisticated space vehicle. However, there are serious problems imposed by this model of Venus' evolution not least of which must take account of the depleted water-level today, and caution must be applied to theoretical conjecture until more precise information is obtained.

Fig. 79 The atmosphere of Venus is predominantly carbon dioxide with a surface pressure nearly a hundred times that of the Earth; at an altitude of about 50 km, pressures are about the same as those of Earth at sea-level

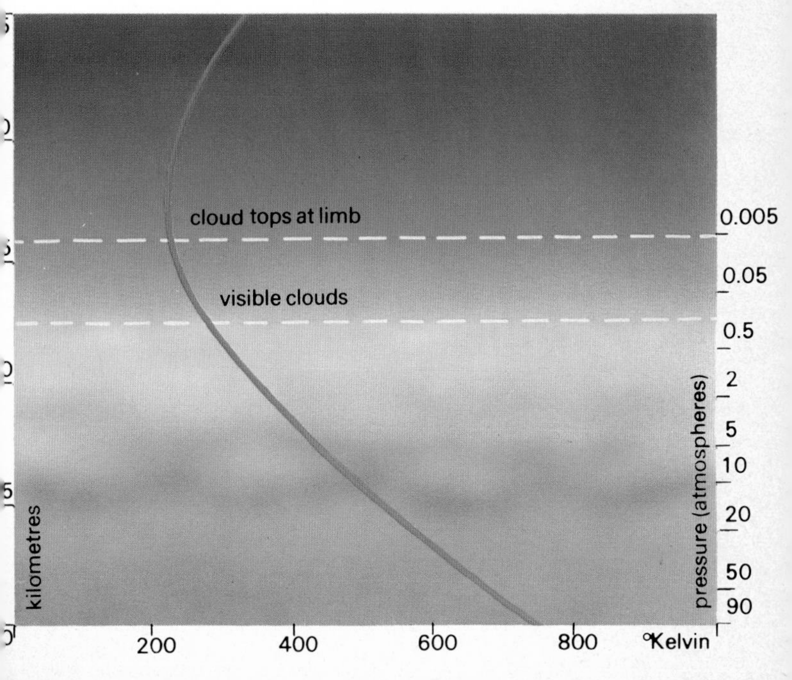

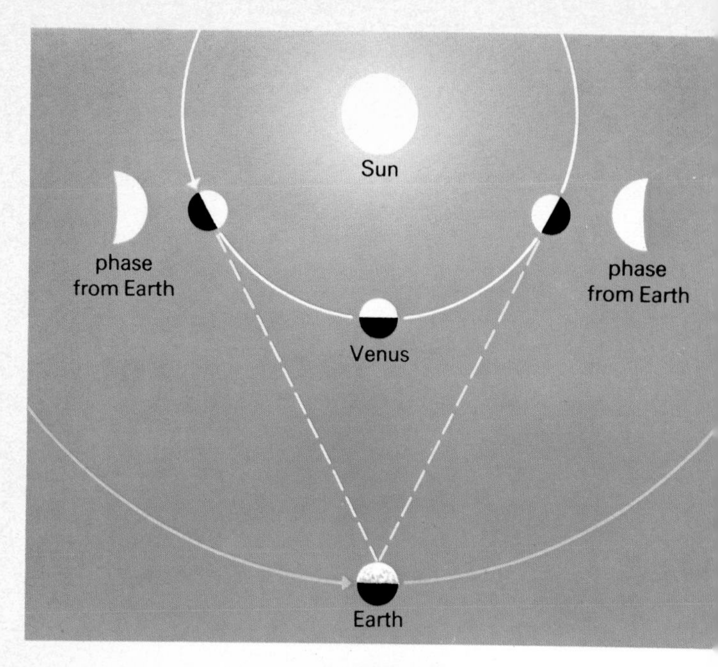

Sun
phase
from Earth
phase
from Earth
Venus
Earth

Fig. 80 (left) The phases of Venus are shown here with a simulated view of the planet as it would appear in a large telescope. **Fig. 81 (top)** Seen in ultraviolet light, the atmosphere contains banded clouds that move around the planet every 100 h. **Fig. 82 (above)** Views of the surface of Venus taken by a Soviet unmanned spacecraft reveal volcanic rocks; light curved area is part of the spacecraft.

Mars

No other body has undergone more revision of its possible models than the nearest exterior planet to the Earth – Mars. For several decades it was thought that the so-called Red Planet supported an Earth-like environment, and up to the mid-1930s popular opinion accepted the idea that intelligent life had evolved to a significant level.

Mars orbits the Sun with its aphelion at 249·1 × 10^6 km and perihelion at 206·7 × 10^6 km, presenting an orbital eccentricity of 0·093 and sidereal and synodic periods of 687 and 780 days respectively. Inclined at 1·9° to the ecliptic, the orbit places Mars on the outer edge of the Sun's ecosphere (see p. 62) making it a potential life-bearing planet and outermost representative of the four terrestrial planets occupying the inner solar system. Mars reflects less light per unit area than any other planet except Mercury and this gives it a magnitude of −1·98.

The distance between Earth and Mars at opposition varies according to their respective positions on the two orbital ellipses and close approaches occur at approximate 15-year intervals. An especially close approach occurred on 10 August 1971 when Earth passed within 56·2 × 10^6 km of Mars; not since 5 September 1877 had the planets been as close as this. However, oppositions in January 1978 and February 1980 provide separation distances of 97·8 × 10^6 and 101·7 × 10^6 km respectively. The next exceptionally close approach occurs on 28 September 1988 when Mars will lie just 58·4 × 10^6 km beyond Earth. It is at times of closest approach that Mars is best situated for detailed observation, and the planet itself is the subject of continual speculation regarding atmospheric and surface phenomena.

With a polar axial tilt of 25·20°, which is very similar to Earth, Mars rotates in just 24 h 37 min 23 s and moves through seasonal cycles in a 23-month period. As an example of the protracted seasons, spring and summer in the northern hemisphere last 194 and 178 Mars days respectively versus 93 and 94 days for respective seasons on Earth. The variation in Mars' seasonal duration is a product of the elliptical orbit; orbital velocity differs by 4·5 km/s between aphelion and perihelion, modifying the transit time through the four seasonal periods.

The planet has an equatorial radius of 3397 km with an oblateness (the difference between equatorial and polar diameters divided by the diameter at the equator) of 0·005. Mass and volume are 0·107 and 0·15 times the Earth respectively, displaying a calculated mean density of 3·9 g/cm^3. This is considerably lower than the other three terrestrial planets but similar to the Moon's. Gravity at the surface is 0·38 times that of the Earth, almost exactly the value calculated for

Fig. 83 Seen from an approaching spacecraft, Mars displays intriguing surface features that upon closer examination reveal significant events on a colossal scale.

Fig. 84 Map of Mars to show the principal surface features.

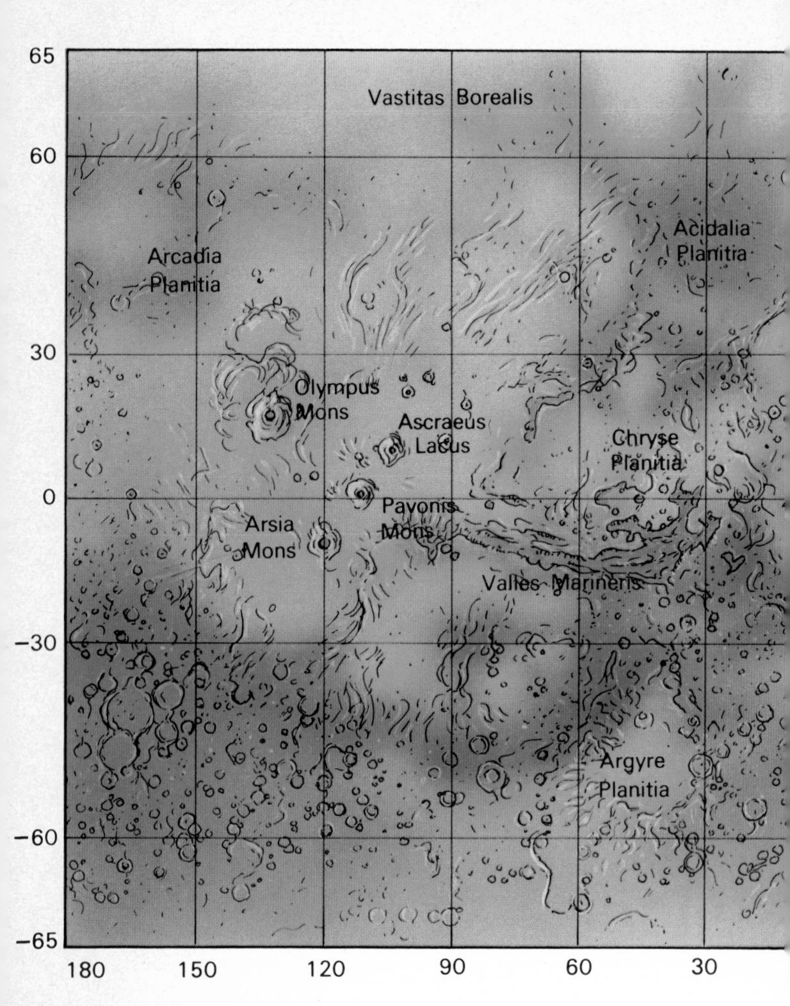

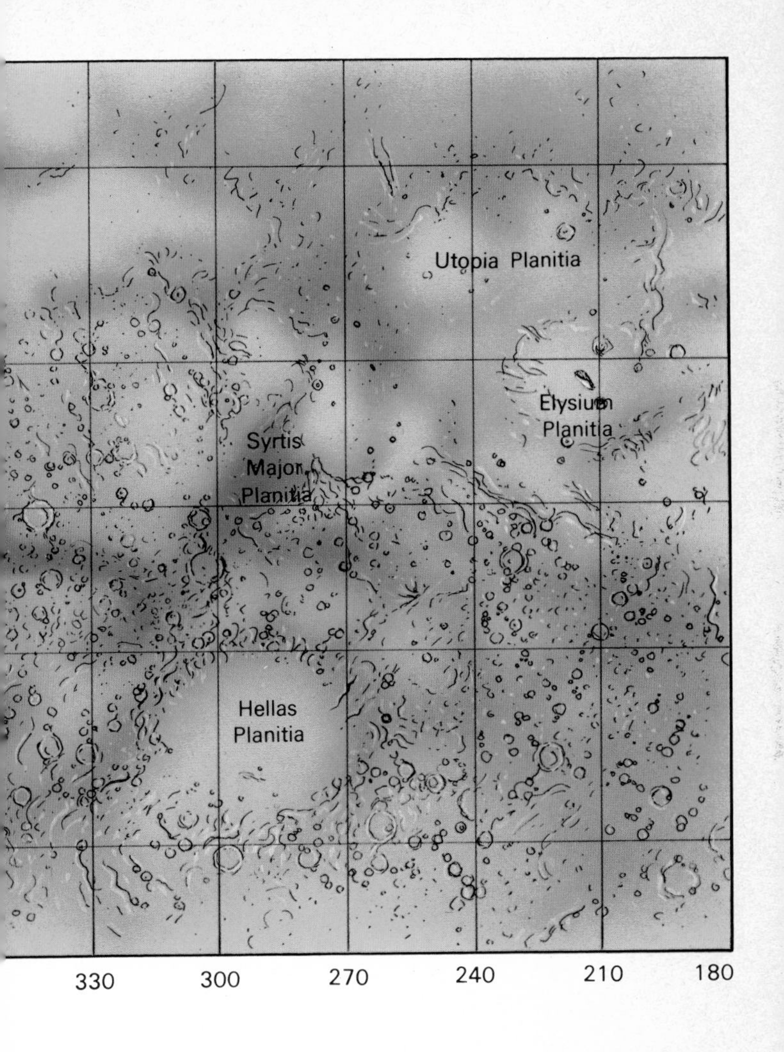
Utopia Planitia
Elysium Planitia
Syrtis Major Planitia
Hellas Planitia
330
300
270
240
210
180

Mercury which is considerably more dense per cubic centimetre but much less massive.

There is an indication in the asymmetry of the planet that Mars has not experienced the amount of differentiation that is thought to have occurred in Earth and Mercury. This could be related to the absence of a magnetic field; Earth and Mercury both possess dipole fields.

Surface features can be divided into five primary categories: cratered terrain, volcanic plains and basins, volcanic mountains and calderas (collapsed vents), canyons and eroded faults and sedimentary modification **(Fig. 83)**. The cratered terrain represents the oldest, preserved topographical structure and confirms a uniform distribution of early planetary bombardment from debris left over at the accretion of the planets. It is now known that some time during their history all the bodies in the inner solar system were exposed to crustal pulverization on a massive scale **(Fig. 84)**.

Volcanic plains and basins record the second phase of evolution. The basins were dug by impact from large bodies; the Hellas region displays a diameter of 2000 km, testimony to a meteoroid between 100–200 km in diameter puncturing the crust more than $3{\cdot}5 \times 10^9$ years ago. (It is interesting that all planets seem to display evidence that massive bombardment lasted for only $0{\cdot}6 \times 10^9$ years after the formation of the solar system about $4{\cdot}7 \times 10^9$ years ago.) Sheets of lava were periodically laid down over vast areas to line the basins and establish a geophysical cycle which today provides enormous crater-free, desert-like areas of totally obliterated topography.

In one area of Mars, the Tharsis region, huge volcanic cones rise from an uplifted plateau. Olympus Mons **(Fig. 85)** is more than 600 km across and 25 km high with a caldera about 80 km across. Vertical cliffs 2 km high ring the periphery indicating atmospheric and dust erosion on a wide scale. A major tectonic event has rent the surface just south of the equator and produced a 5000 km long canyon up to 140 km wide and nearly 5 km deep **(Fig. 86)**.

Most important, however, are the regions of polar and sedimentary modification. Several areas at mid-northern latitude contain sinuous channels identical to dried river beds which together with glacial scrape marks indicate a very different environment earlier in the planet's history. The polar caps today consist of water ice covered with carbon dioxide frost during the winter **(Fig. 87)**.

The atmosphere of Mars is essentially carbon dioxide with indications of carbon monoxide, argon and ozone. Surface pressure is only 5–7 mb and temperatures rarely rise above freezing point although noon temperatures at the equator reach about 18°C in summer. Mars has two moons, Phobos and Deimos, orbiting the

Fig. 85 (above right) Olympus Mons, the largest volcanic mountain on any planet in the solar system, is more than 600 km across and 25 km high with a caldera 80 km across. **Fig. 86 (right)** East of the Tharsis ridge that supports massive volcanic mountains, Valles Marineris cuts a deep incision in the crust.

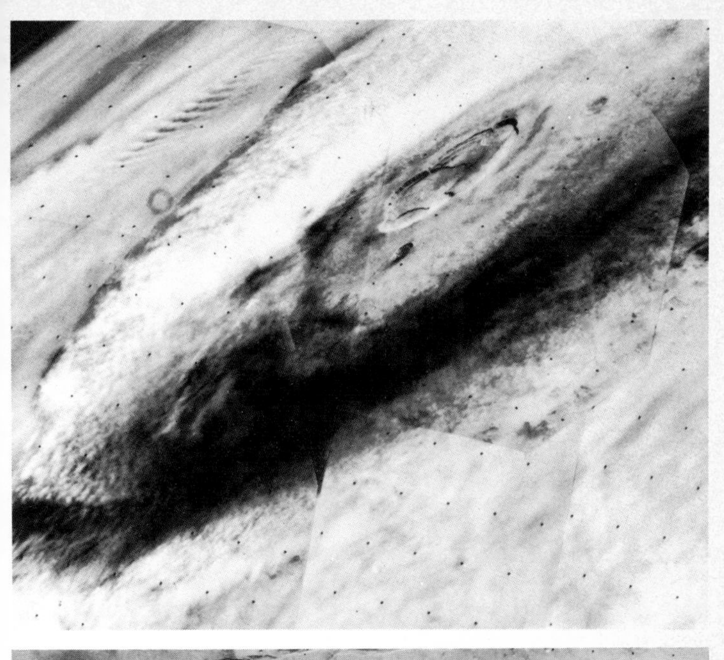

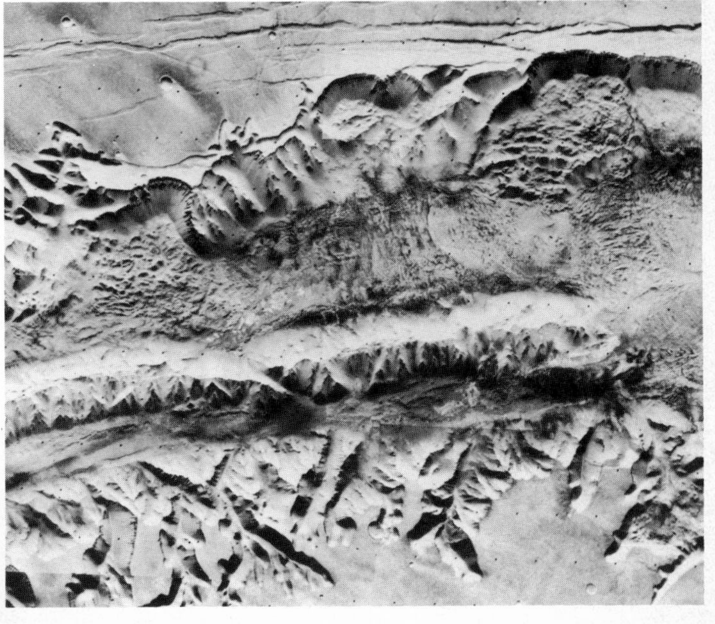

Fig. 87 Views of Mars taken from the surface by the Lander spacecraft of the Viking programme revealing many types of rock, most of which can be attributed to a volcanic origin. **(This page)** The bottom sequence of photographs shows a

surface sampler trenching an area 20 by 23 cm. The rock in the foreground is 20 cm high.

Fig. 87a The crater Arandas with outpourings of material reminiscent of experimental craters in waterlogged ground.

Fig. 87b The surface of Mars as seen at the site of Viking Lander 1 in Chryse Planitia. The red coloration of the rocks is largely due to various iron oxides.

planet at distances of 6034 and 19 950 km respectively **(Fig. 88)**. Phobos has a synodic period 7·7 h and conducts three revolutions of the planet in a single Mars day while Deimos, conducting a full revolution in 30·3 h, has a synodic period of 5·5 Earth days. Both moons are small, irregular-shaped rocky bodies pulverized by primordial bombardment bearing many craters; Phobos is about 16 × 23 km and Deimos is only 13 × 19 km in size.

Observationally, Mars is a jewel in the solar system with many seasonal and cyclic phenomena visible from Earth. The dust storms which can be observed from Earth usually originate over the southern highlands and may spread to cover practically the whole Martian surface. Waves of darkening and colour changes were once associated with the possible growth of vegetation, but now that more is known about the planet it seems unlikely that such advanced life forms have ever emerged. Nevertheless, it is just possible that simple life mechanisms could still exist on Mars, and the planet is the major contender among the planets for ever having had any form of life.

It has been suggested that conditions amenable to life are cyclic throughout Mars' history, and that biological evolution could lie dormant between environmentaly favourable periods. If that is the case it may be possible to observe primitive links in the chain of life. Fossilized evidence might be found in areas where sedimentary deposits appear to have played a major role in the evolution of the planet. However, much about Mars remains enigmatic.

Jupiter

Jupiter orbits the Sun far beyond the planet Mars and lays claim to several historic landmarks in science. Four large satellites of Jupiter became the first objects to be discovered through the telescope and, later, measurement of the orbital motion of one of Jupiter's moons produced the first qualitative calculation to determine the speed of light.

The orbit of the planet, inclined a mere 1·3° to the ecliptic, has aphelion and perihelion values of $815{\cdot}7 \times 10^6$ and $740{\cdot}9 \times 10^6$ km respectively (4·95 and 5·45 AU), displaying an eccentricity of 0·048. Jupiter is the innermost, and most magnificent example, of the four gaseous outer planets which are so very different from the four terrestrial-type ones found in the inner solar system. Its volume is 1316 times that of the Earth yet its mass is only 317·9 times greater, rendering a mean density of 1·3 g/cm^3.

Jupiter is found to have a high oblateness value of 0·06 and the

Fig. 88 (above right) The Martian moons, Phobos and Deimos, are probably captured asteroids. Seen from the surface, Deimos would appear to move round Mars in the opposite direction to Phobos, due to the 30·3-h period. **Fig. 89 (right)** Jupiter, largest planet in the solar system, is primarily composed of hydrogen.

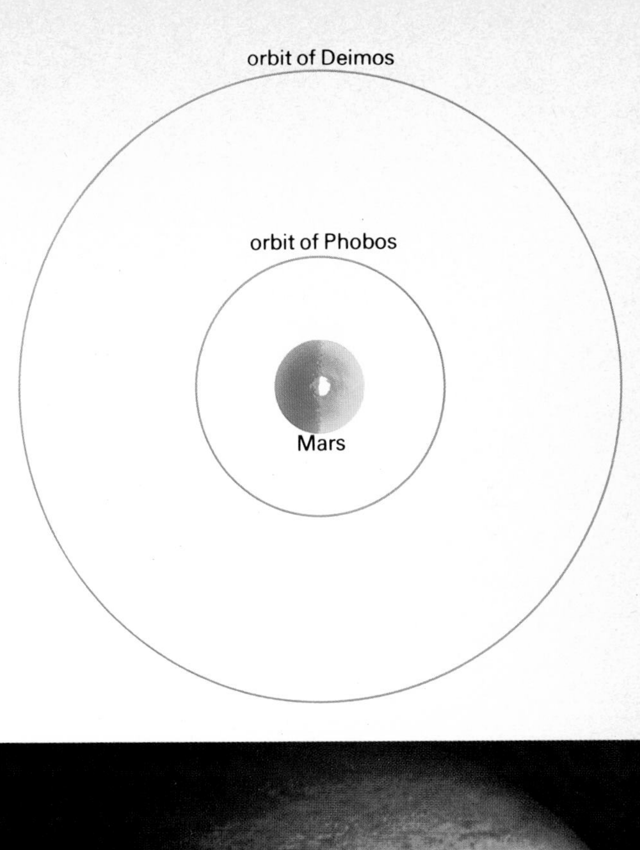
orbit of Deimos
orbit of Phobos
Mars

entire planet is visibly compressed at the poles. With an equatorial radius of 71 400 km, Jupiter rotates on a polar axis tilted 3·08° to the orbital plane in various periods according to latitude. Therefore, the planet must be thought of more as a liquid body than a comparatively rigid sphere.

The 20° of latitude centred on the equator rotate in a period of just 9 h 50 min 30 s. However, north and south of this band a rotational period of 9 h 55 min 41 s is accepted as a standard mean for the various periods, some longer and some shorter, which are observed from atmospheric phenomena. The polar regions, with a radius of 66 700 km, just 93 per cent of the equatorial value, are a confusion of different rotational velocities and this is seen in the configuration of the upper atmosphere.

Jupiter takes 11·86 years to make one sidereal revolution and because it is so distant from the Earth its synodic period, 399 days, is more closely matched to the period of the Earth's orbit than any of the interior planets. The most prominent feature on Jupiter is the system of belts and zones girdling the planet along lines of latitude with spots and markings superimposed on the light and dark bands. The lighter coloured zones appear to be the product of rising atmospheric columns with clouds of ammonia and ammonia hydrosulphide crystals. The currents flow out of these convection cells and down into the cooler, dark belts 20 km lower **(Fig. 90)**.

Most of Jupiter's atmosphere lies in the outer 1000 km of the planet and here the ratio of hydrogen to helium is 4:1 with less than 1 per cent made up from heavy elements. The banded structure of the Jovian atmosphere is a result of the enormous rotational speed – about 35×10^3 km/h at the equator. Wind speeds of 600 km/h are reached at the edges of the belts and zones. Opposing currents cause great wind shear and result in gigantic eddies being formed.

Most of Jupiter's weather is formed in a 70-km-thick layer with upper and lower pressures of 500 and 4500 mb respectively. The Great Red Spot has formed in one of the rising cells of warm atmosphere. This is thought to be the product of super-convection with the top of the 40 000 × 13 000-km convection cell sitting 8 km above the surrounding cloud deck. Below this atmospheric region, compressed hydrogen and helium are found to be in the ratio 88:12 and well mixed down to a depth of about 1000 km; the gaseous hydrogen is compressed to a liquid at a temperature of 2000° C. At a depth of 3000 km the pressure is $9{\cdot}3 \times 10^7$ g/cm^3.

About 22 000 km below this region, temperatures reach 11 000° C and pressures are up to $3{\cdot}1 \times 10^9$ g/cm^3. At this point the liquid molecular hydrogen is turned into liquid metallic hydrogen (temperatures are too high for it to become a solid) and from here for a further 45 000 km temperatures and pressures build up to 30 000° C and $1{\cdot}03 \times 10^{11}$ g/cm^3 respectively. It is thought likely that a central rocky core exists containing 15 Earth masses and this would provide

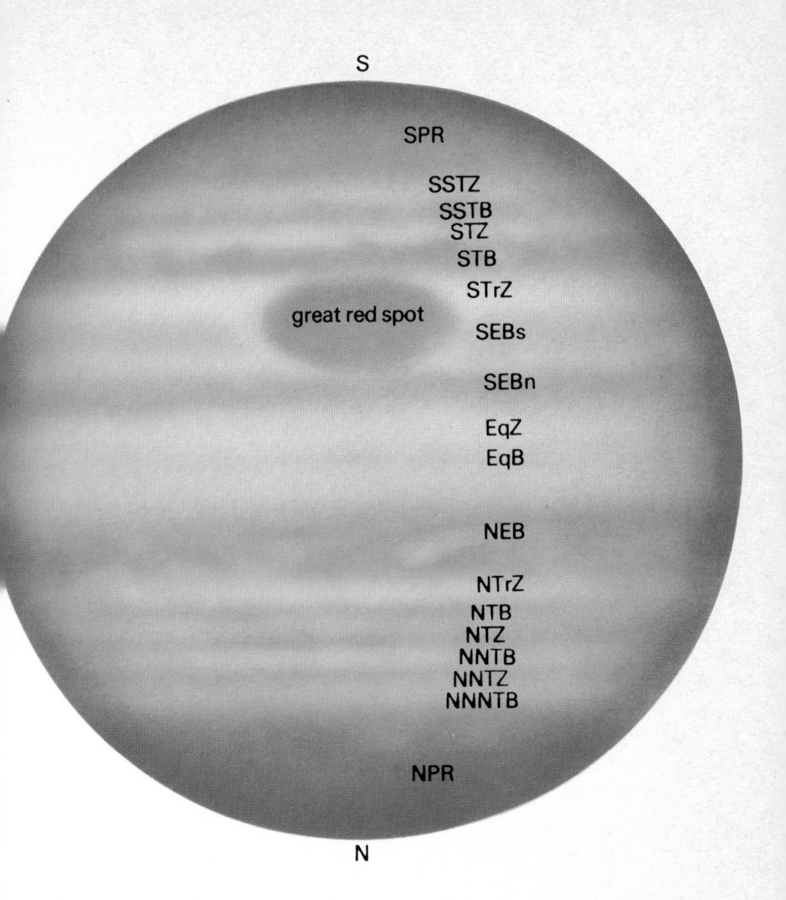

Fig. 90 The prominent belts and zones are important indicators of the turbulent conditions within the Jovian weather system.

a mean hydrogen-helium mass distribution ratio of 7:3 for the entire planet **(Fig. 92)**. In summary, Jupiter must be considered as a fluid planet with a predominance of hydrogen being transformed to a liquid at a depth of 2 per cent and into metallic hydrogen at a depth of 37 per cent.

As if Jupiter were not magnificent enough in the telescope there are invisible qualities to its environment that dominate the solar system. These are the magnetosphere and the radiation belts associated with it. Like a giant comet with its tail pushed back by the solar wind (high-speed protons and electrons). Jupiter's magnetosphere **(Fig. 93)** is about $14{\cdot}5 \times 10^6$ km across and probably stretches $1{\cdot}5 \times 10^9$ km back into the solar system. The interaction

Fig. 91 (below) Voyager 2 photograph of Jupiter. One of the 'barges' can be seen in the North Equatorial Belt. The Great Red Spot is out of the picture to the right. **Fig. 92 (right)** Above: Beneath the atmosphere, gaseous hydrogen is compressed to a liquid and then to a solid giving the element a metallic quality. It is possible that a small rocky core lies at the very centre of Jupiter. Below: A cross-section through the outer layers of the planet showing the continuous gas convection cycle.

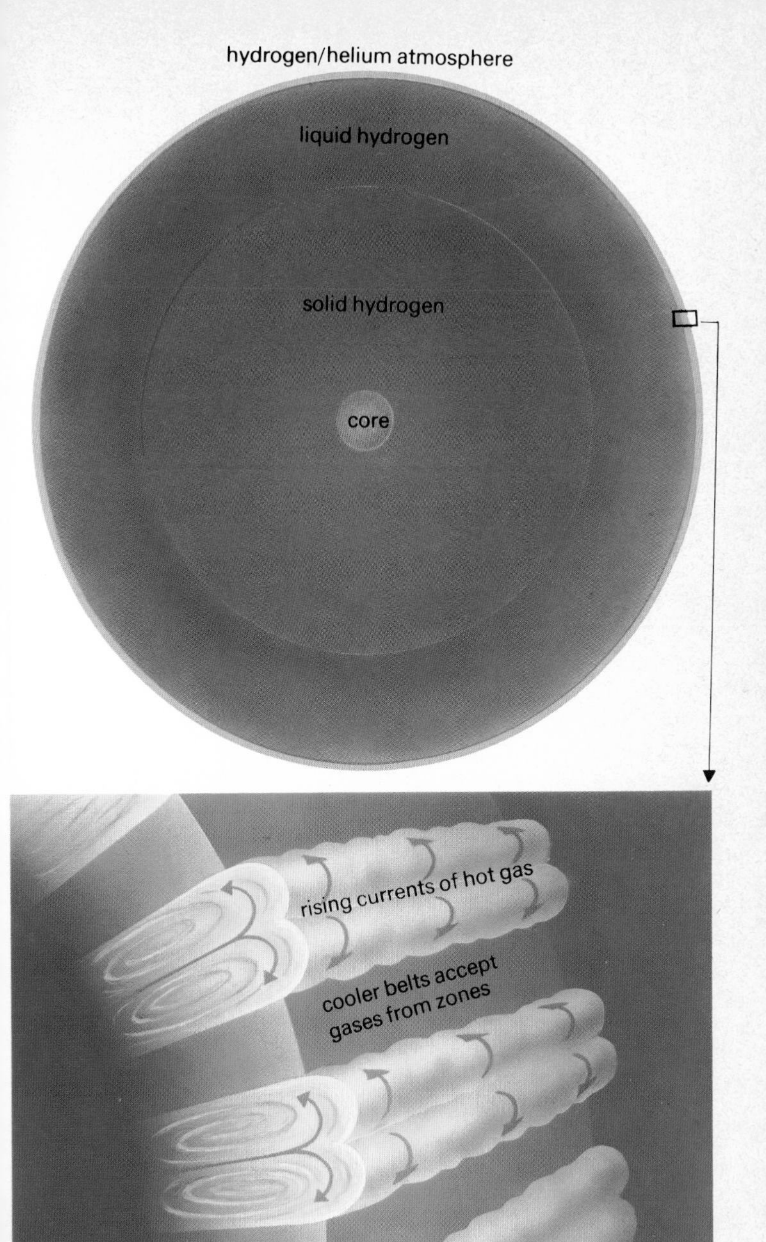
hydrogen/helium atmosphere
liquid hydrogen
solid hydrogen
core
rising currents of hot gas
cooler belts accept
gases from zones

between the solar wind and the magnetosphere sets up a bow shock wave which stands apart from the magnetosphere and this magnetopause is a transition region for the particles flowing round the belts of trapped radiation. The bow shock wave pulsates in and out according to fluctuations in the pressure exerted by the solar wind.

Jupiter's magnetosphere consists of three separate zones: the inner zone extends from the planet itself out to a distance of some $1{\cdot}44 \times 10^6$ km; the second zone, or current sheet, carries electrified particles out to a distance of up to $4{\cdot}3 \times 10^6$ km; and the outer zone lies between $4{\cdot}3 \times 10^6$–$6{\cdot}4 \times 10^6$ km and sends high speed electrons spiralling up the solar wind back towards the Sun. Space vehicles in the vicinity of Mercury have detected these particles which, to achieve such penetration up-wind of the solar radiation, require a power input exceeding 10^{11} W.

The magnetic field is slightly offset from the centre of the planet and is found to be tilted 11° to the Jovian equator, wobbling up and down through 22° as the planet rotates on its axis. The strength of the field is 17 000 times that contained in the Earth's magnetosphere with total energy 20 000 times the Earth's radiation belts. Field intensity varies between 4·2–14·8 Gauss (the Earth's magnetic field has an intensity of 0·35 Gauss at the surface). It it were visible from the Earth, Jupiter's magnetosphere would appear as a disc four times the diameter of the Sun or Moon. In all respects it is the most important natural laboratory in the solar system.

One of Jupiter's moons, Io, operates like a power switch by funnelling magnetic field lines between itself and the planet developing an electric potential of 4×10^5 V across the surface to alternately switch an electric current on and off. Measurements of the radio signals from the flux tube linking Jupiter and Io detect an energy level of 10^{13} W.

Apart from being the most massive planet in the solar system with an extreme example of magnetosphere/solar wind interaction, Jupiter supports a satellite system comprising four large moons the size of terrestrial planets and twelve confirmed other smaller bodies **(Fig. 94)**.

Five Jovian moons lie at distances between $1{\cdot}82 \times 10^4 - 1{\cdot}88 \times 10^6$ km. The innermost moon, Amalthea, is only 75 km in radius but the other four are much larger; Io, 1818 km; Europa, 1533 km **(Fig. 95)**; Ganymede, 2608 km; and Callisto, 2445 km. These are known as the Galilean satellites, after Galileo discovered them in

Fig. 93 (above right) The inner and outer magnetic fields are surrounded by the magnetopause (a) which in turn sets up a bow shock wave (b) against the impinging radiation from the stream of protons and electrons in the solar wind (c).
Fig. 94 (right) This image of Europa was taken by Voyager 2 from a range of 241,000 km. The complex pattern on its surface is thought to have been caused by fracturing of an icy mantle.

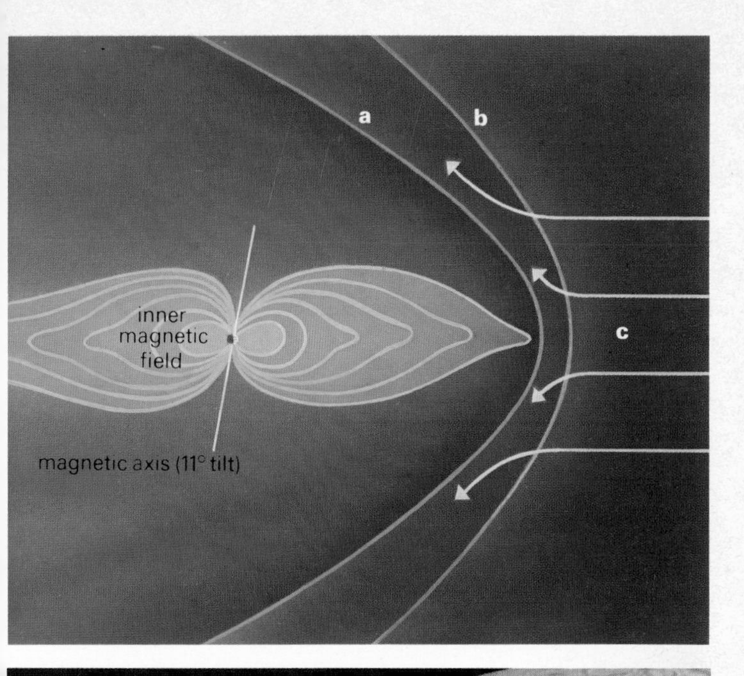
a
b
c
inner magnetic field
magnetic axis (11° tilt)

1609; their density values are 3·6, 3·0, 1·9 and 1·8 g/cm^3 respectively. From these data it can be inferred that Jupiter's Galilean satellites reflect the primordial geochemistry at condensation and much can be learned about the early history of the solar system from studies of these important satellites. Also, it is interesting to compare the density and diameter ratios with the general characteristics of the solar system.

Voyager images showed Io to be the most volcanically active object in the Solar System. There is some rocky lava, but the volcanoes also eject molten sulphur and sulphur dioxide, which give the satellite its reddish colour. The heating of the interior arises from tidal 'squeezing' by Jupiter as Io's orbit varies slightly because of gravitational forces exerted by the other satellites. Europa proved to be very smooth, and almost without impact craters. It probably consists of a rocky core surrounded by layers of water and ice, the latter about 100 km thick.

Ganymede and Callisto have very much lower densities, so they probably consist of mostly ices, surrounding rocky central cores. Ganymede's surface varies very greatly. There are dark plains and lighter, younger material, as well as strange 'grooved terrain' with long parallel furrows about 1 km deep and 15 km in width. All this shows that the satellite was geologically active at some time in its lifetime. Callisto has even more craters than Ganymede, but these are shallow, probably because the surface, being ice, gradually flowed and flattened out any features. One large impact created the set of concentric rings known as Valhalla, which are about 3000 km across, and larger than any similar features on the planets or the Moon.

The other satellites are very small and insignificant. The largest of them orbit in two groups of four **(Fig. 94)** well outside the Galilean satellites. The outer group orbits in a retrograde direction unlike all the other satellites, and may consist of captured bodies similar to minor planets.

The two Voyager probes obtained many exciting results when examining Jupiter, including the discovery of a narrow ring system about 3·56 × 10^8 km in diameter, and around 30 km thick, as well as of a small satellite orbiting just outside the ring. Aurorae and powerful lightning flashes were also observed on the planet itself.

Saturn

Probably the most intriguing view an observer will remember is a first look at the magnificent rings of the planet Saturn; an unparalleled view in this solar system at least **(Fig. 96)**.

Saturn resides in a 29·46-year orbit of the Sun with aphelion and perihelion values of 1·507 × 10^9 and 1·347 × 10^9 km (10·07 and 9·0 AU) respectively presenting an eccentricity of 0·056. With an orbital radius nearly twice that of Jupiter, an observer on Saturn would see the Sun as a disc only one-tenth the diameter viewed from Earth and little more than one per cent of the luminosity.

Fig. 95 The four Galilean moons of Jupiter. Only Europa (top right) is smaller than Earth's moon; Io (top left) is larger while Callisto (bottom right) and Ganymede (bottom left) are larger than Mercury.

The synodic period is 378 days and the orbit is inclined 2·5° to the ecliptic. Like Jupiter, Saturn presents a yellow appearance from the Earth and shows light and dark zones and belts like its companion, albeit not as well defined. Inclined 26·7° to the plane of its orbit the planet is seen to have a measured oblateness of 0·1 with an equatorial radius of 60 450 km. A remarkably low mean density of 0·7 g/cm³ means that Saturn would literally float in a sea of water, and within a volume 755 times the Earth, the planet contains 95·2 Earth masses. It is very difficult to obtain a precise measurement of the rate of rotation but careful observation of discrete surface features indicates a value of approximately 10 h 14 min in the equatorial region.

Saturn must be very similar to Jupiter, its low density implying a predominantly hydrogen chemistry, and calculations indicate that the planet contains 80 per cent hydrogen, 18 per cent helium and 2 per cent heavier elements. Like Jupiter, the main body of the planet must be assembled from molecular hydrogen transitioning to metallic hydrogen at great depth and observations of the atmosphere indicate trace quantities of methane and ammonia. It seems quite likely that Saturn has a dense, heavy element core probably containing 15–20 Earth masses within a radius of 12 000 km from the centre.

Thanks to spacecraft images, we now know that Saturn's impressive system of rings consists of many thousands of narrow ringlets encircling the planet. The particles in the rings are very small, the largest being no more than perhaps 10 m across, while the smallest are little more than dust. They probably consist of water ice with a mixture of other materials. The dynamics controlling these particles are very difficult to understand, but it seems that certain small satellite bodies, only a few kilometres across, help to contain them in their orbits and 'shepherd' them into seven major rings, only three of which are easily visible from Earth. The A ring lies between about 136 500 and 121 000 km, separated by Cassini's Division from ring B, the brightest of the three, lying between about 117 500 and 92 200 km. Inside the B ring is the very faint C ring, sometimes known as the Crepe Ring, reaching down to about 73 200 km. A narrow ring D exists even closer to the planet, and outside ring A we find faint, narrow rings F, G and E.

Saturn has a large number of satellites: seventeen have been confirmed and another six are thought to exist. Three small bodies, Atlas, Prometheus and Pandora orbit just outside ring A – Atlas acts to give a sharp outer edge to that ring. Then come two satellites, Janus and Epimetheus, roughly 200 and 120 km across, respectively, in almost the same orbit at 151 400 and 151 500 km. Every four years the innermost one catches up with the slower outer one, they swing round one another and change orbits.

Mimas, 392 km across, orbits at 186 000 km from Saturn. It is heavily cratered unlike the next moon, Enceladus, 510 km in diameter, which orbits at 238 000 km and has large smooth areas on its surface, rather

Fig. 96 A composite Voyager picture (in true colour) of Saturn, also showing three satellites, Tethys, Dione and Rhea. The black spot on the southern hemisphere of the planet is the shadow of Tethys. Some of the 'spokes' can be seen (particularly in the top left) on the bright B ring.

Fig. 97 A close-up view in false colour of two of Saturn's rings – the C ring (blue in this picture) and the B ring (orange). Seen from this close, the two rings are clearly made up of many smaller ringlets, of which more than 50 are evident in this frame.

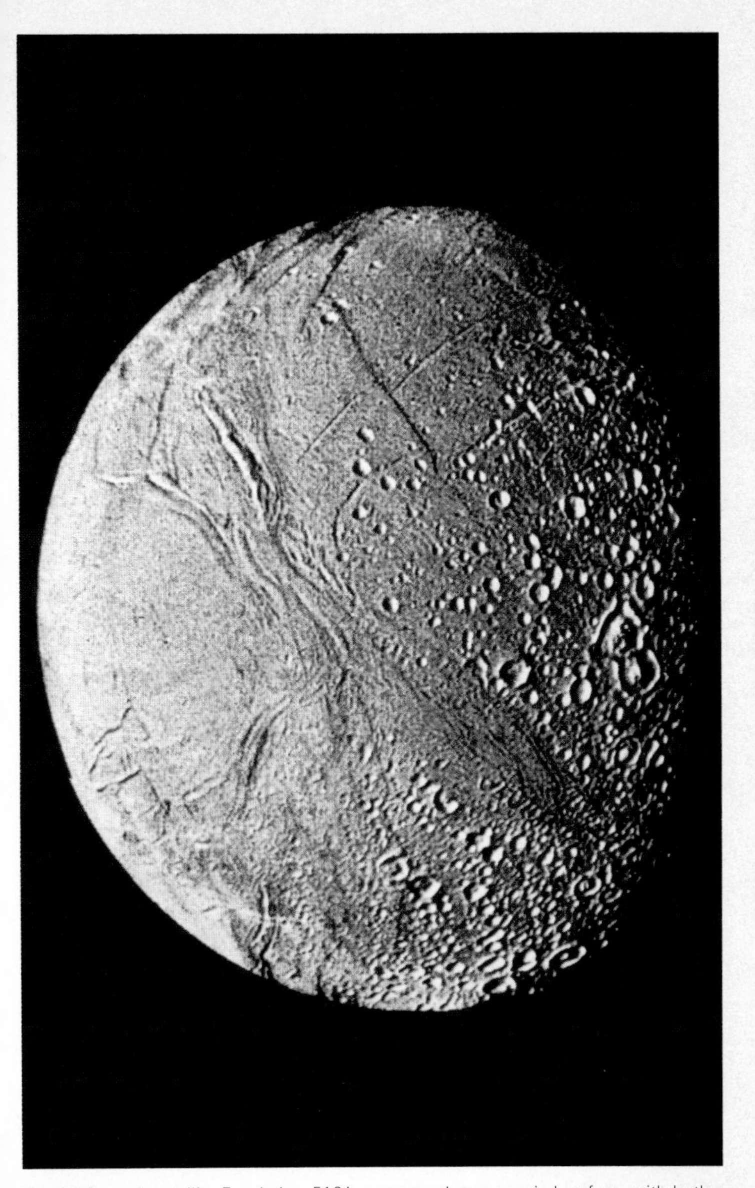

Fig. 98 Saturn's satellite Enceladus, 510 km across, shows a varied surface, with both craters, wrinkle ridges and smooth areas. The surface has obviously been modified, but the satellite is too small to have any interior source of heating, so the processes involved remain a mystery.

like Jupiter's Ganymede. Beyond Enceladus at 295 000 km is Tethys, 1060 km across, heavily cratered but with low relief, probably indicating that it has an icy interior. Two small satellites, Telesto and Calypso, share Tethys' orbit, gravitationally locked approximately 60° ahead of and behind the large body. The next satellite, Dione, at 377 000 km, has a similar small body, Helene, in its orbit. Dione and Rhea (orbiting at 527 000 km) show wispy markings, probably produced by ice deposits on the surface.

The largest satellite, Titan, at 1 222 000 km from Saturn, is the size of a small planet, being 5150 km in diameter. It probably consists of half ice and half rocky material. It is the only satellite in the Solar System with a dense atmosphere, and the surface pressure is even higher than on Earth. The atmosphere consists mainly of nitrogen but some methane is present and there may be liquid methane on the surface.

Hyperion, orbiting 261 000 km outside Titan, is an irregular body roughly 410 × 260 × 210 km, reddish in colour, probably consisting of dirty ice. Iapetus at 3 561 000 km from Saturn, is odd: half is white, and the other half is reddish black. Finally, far outside the other moons, at 12 954 000 km is little Phoebe, about 220 km across and probably a captured asteroid.

Uranus

It is impossible to say when Man first observed the motion of planets across the stellar background. However, one thing is certain, until 1781 only the six planets out to and including Saturn had been recorded. In that year the renowned astronomer, Sir William Herschel, discovered Uranus.

Orbiting the Sun in 84·01 years, Uranus has a synodic period of 370 days, aphelion and perihelion values of $3{\cdot}004 \times 10^9$ and $2{\cdot}735 \times 10^9$ km (20·08 and 18·28 AU) respectively and an orbital eccentricity of 0·047. With a mass of 14·6 and a volume of 67 times the Earth, the planet displays a mean density of 1·3 g/cm^3, rather less than the value obtained for Jupiter. Because of this it is considered to have a core of 1–5 Earth masses surrounded by hydrogen, helium, methane and ammonia.

Uranus has a diameter of 50 800 km and a very strong magnetic field (50 times that of the Earth). Its thick atmosphere is almost featureless and consists of methane, hydrogen, acetylene and ammonia. Uranus has at least fifteen icy satellites, nine of which are less than 100 km in size. Another, Puck, is only 170 km across. Of the larger moons, Oberon is heavily cratered, but Titania and Ariel show signs of considerable geological activity, which is even more pronounced on Miranda. This small moon, 485 km across, has strange areas of parallel grooves and ridges, unlike anything seen anywhere else. Umbriel, the remaining moon, is very dark and almost featureless.

A peculiarity of Uranus lies in its polar tilt. While the planetary orbit

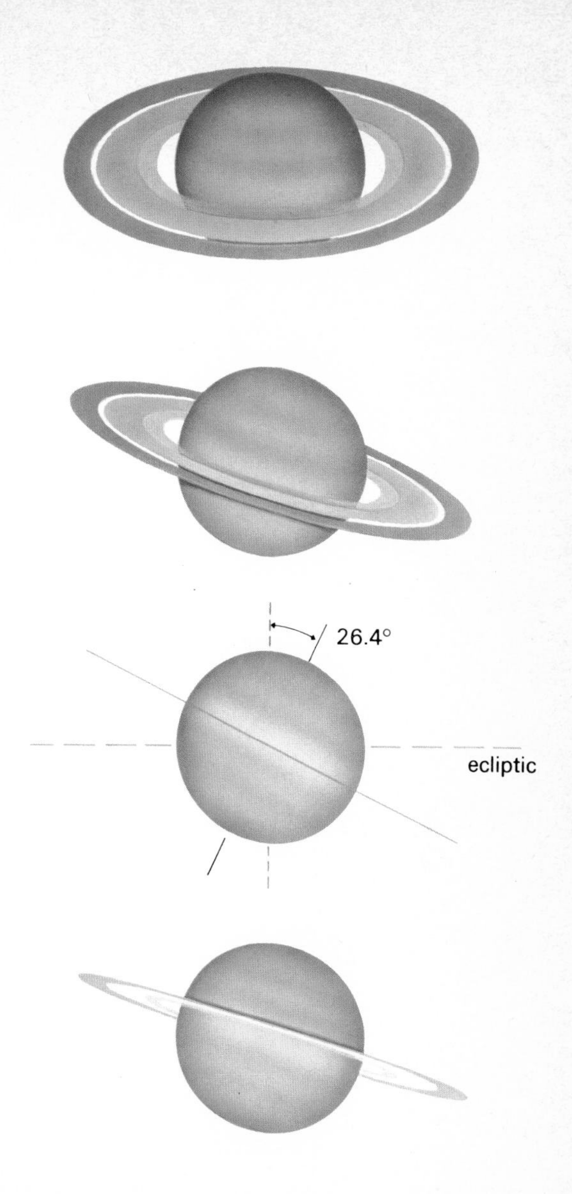

Fig. 99 The orbital path of Saturn presents a changing orientation of the plane of the planet's rings; edge on, the rings are invisible and the ring plane is 26·4° to the ecliptic because of the planet's polar tilt.

about the Sun has an inclination of 0·3° to the ecliptic, less than any other planet in the solar system, the polar axis is inclined fully 82·1° and the planet spins on this axis in a period of approximately 10·8 h. Moreover, Uranus spins in a retrograde direction to the other planets, that is, if the pole above the orbital plane is considered to be north, it spins in the opposite sense to all other planets in the solar system except Venus.

Sometimes, using the direction of rotation to establish the criteria for measuring angular tilt, Uranus is said to be in a posigrade state of rotation with an inclination of 97·9°. Either way it is a unique phenomenon. Since the planes of the satellites' orbits are in the plane of the planetary equator they are similarly inclined with respect to Uranus' orbit. **Figure 101** shows the repercussion of this phenomenon: in 1966, the north pole of Uranus pointed in the direction of orbital motion; in 1987 (shortly after the Voyager 2 fly-by) the south pole pointed towards Earth; in 2008 the south pole will point in the direction of travel; and in 2029 the north pole towards Earth.

There are ten very dark rings around the planet and unlike those of Saturn these are very narrow, ranging from 2 km to about 100 km in width. There may be a number of small, unseen satellites that keep these rings in place.

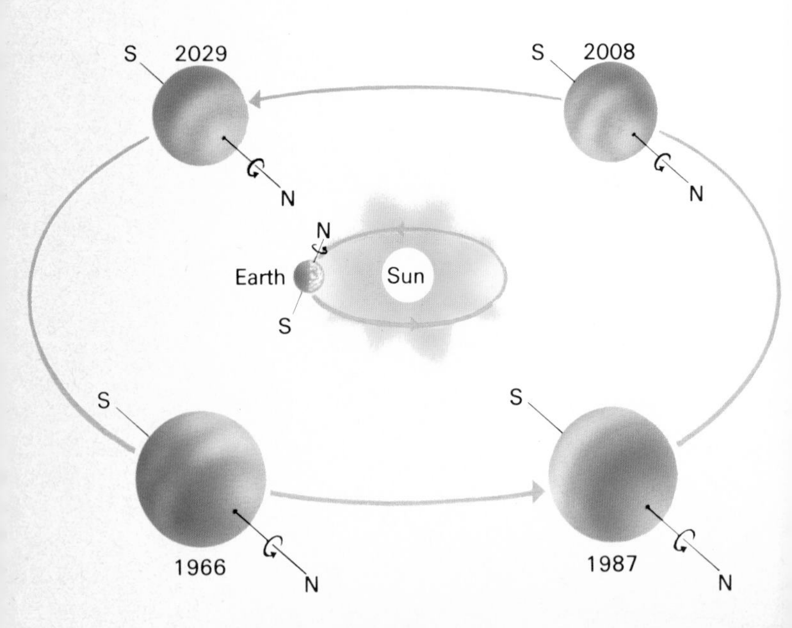

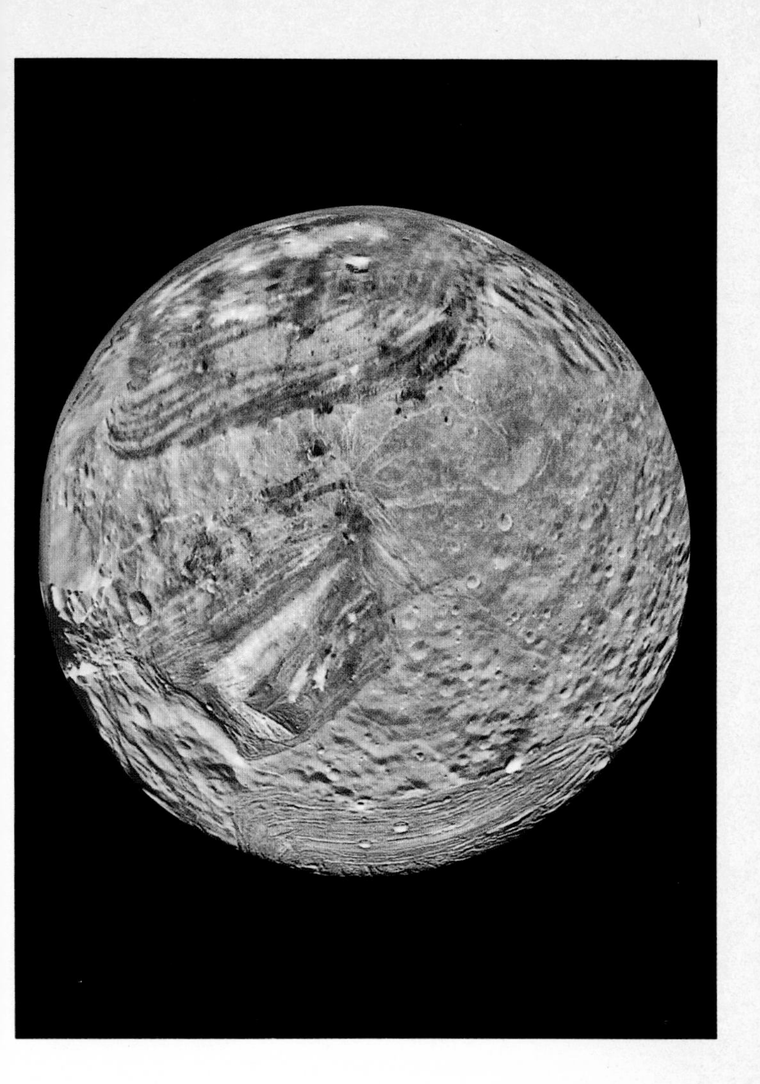

Fig. 100 (above) Uranus' moon Miranda is probably the strangest satellite in the Solar System. Just visible on the limb are sheer cliffs about 15 km high. The extraordinary 'circus' and 'chevron' patterns are thought to have arisen during separation of ice and rock mixtures within the satellite. **Fig. 101 (left)** The peculiar polar tilt of 82·1° provides an opportunity to view both equatorial and polar regions of the planet from Earth. In 1987 the south pole faced Earth, in 2008 the equator and in 2029 the north pole will be visible.

Neptune

Neptune was discovered in 1846 by Galle acting on the advice of Le Verrier who had calculated the position of an eighth planet from observed perturbations in the orbit of Uranus. Neptune lies in a 1·8° orbit varying between $4{\cdot}537 \times 10^9$–$4{\cdot}456 \times 10^9$ km (30·3–29·7 AU) respectively with an eccentricity of 0·009. The sidereal and synodic periods are 164·8 years and 367 days respectively. In these distant reaches of the solar system the temperature is −220° C and the Sun is only 1/900th as bright as seen from Earth with a diameter of 1·0′, or 1/30th the apparent size from Earth.

Because of the enormous distance separating Earth from Neptune it is very difficult to ascertain the planet's physical characteristics **(Fig. 102)**. However, mass and volume are apparently 17·2 and 54 times the Earth values respectively and the oblateness factor of 0·02 is based upon a radius of 24 300 km. The density is, therefore, 1.7 g/cm³ and because of this the planet is thought to be of similar composition to Uranus with an atmosphere of hydrogen, helium and ammonia and the largest proportion of methane yet detected on any planet. At

Fig. 102 (right) Neptune, as seen from the surface of its largest moon, Triton, would carry similar banding to that observed on Uranus. **Fig. 103 (below)** The two moons of Neptune (to scale). (a) Triton is thought to have a diameter of 6000 km, making it the largest satellite in the solar system; (b) Nereid has a diameter of about 280 km.

−220°C methane and ammonia will exist in a solid state.

Neptune has two moons, Triton and Nereid with orbits of 355×10^3 and $1{\cdot}3 \times 10^3$–$5{\cdot}57 \times 10^6$ km respectively **(Fig. 103)**. Triton is inclined 20° to the planet's equator, itself tilted at 28·8° to the planet's orbit. Its orbital motion is retrograde and it has been suggested that within 10^7–10^8 years the satellite will decay into the atmosphere of Neptune. With an approximate radius of 3000 km and a mean density of about 4 g/cm³ Triton may possess an atmosphere. Nereid exhibits an inclination of 28° and appears to have a radius of only 140 km. Its eccentricity, at 0·75, is unique in planetary satellite systems.

Pluto

Pluto was discovered in 1930 due to perturbations in the orbits of Uranus and Neptune. Orbital values exhibit aphelion and perihelion values of $7{\cdot}375 \times 10^9$ and $4{\cdot}425 \times 10^9$ km (49·2 and 29·5 AU) respectively, for an eccentricity of 0·25 and a 17·2° inclination to the ecliptic, both latter values being the highest for any planet. Pluto takes 247·7 years to perform one revolution and exhibits a synodic period of 366·7 days. Therefore, Pluto actually moves within the orbit of Neptune at perihelion and since 1978 has lost the honour of being the outermost planet; however, the orbital nodes of Neptune and Pluto are displaced in the plane of the ecliptic and they cannot collide.

It has been impossible to obtain a definitive value for Pluto's size but the very accurate occultation measurements of a star have narrowed the conjecture by putting a positive upper limit of 2900 km on its radius. Assuming it to be fully this size, the calculated perturbations in the orbits of Uranus and Neptune require a mass which would dictate a mean density of 60 g/cm³ within a volume established by the cube of the 2900 km radius. Clearly this is most unlikely and the conclusion is inevitable: Pluto does not contain sufficient mass to account for the observed perturbations **(Fig. 104)**.

This view was confirmed by discovery in 1978 of a satellite, Charon, about 1300 km in diameter and orbiting around 20 000 km from the planet. Both Pluto and Charon have very low densities, not much more than that of water, so probably consist mainly of ices. Pluto has a very low mass, perhaps only two thousandths of that of the Earth. Suggestions have frequently been made that Pluto is an escaped satellite of Neptune and that this could account for the planet's small size and mass, and also for the exceptional orbit of Triton, this being the result of interaction between the former satellites, which led to Pluto's escape. However, there have always been many objections to this theory, and the discovery of a satellite to Pluto seems to make it quite unacceptable, as it is difficult to devise any mechanism which would not disrupt the Pluto–Charon system. Roughly every 124 years Pluto and Charon undergo mutual eclipses. These have recently shown that Pluto's surface has mottled ice deposits.

Fig. 104 Pluto resides in the cold, outer regions of the solar system and its lone position among the gaseous outer planets is unique. It may be an escaped moon of Neptune and if so it is the only known example of a natural satellite freed from planetary bondage. Here, it is silhouetted against the stars of the Milky Way. The Sun appears as a star-like point; it gives only 1/1500th the light Earth receives although it is still brighter than a full Moon.

Minor planets

Classification of bodies within the solar system must, by necessity, adopt an arbitrary set of criteria. Such is the case with the minor planets, or asteroids, orbiting the Sun between Mars and Jupiter **(Fig. 105)**. Classification of a group of fragmented planetesimals began in 1801 with the discovery of the asteroid Ceres, found to be the largest member of several hundred sizeable 'boulders'. Ceres has the largest diameter of any asteroid (955 km) and next in size are Pallas, 520 km, and Juno, 240 km.

These three bodies would seem to be primordial debris from a very early phase of planetary formation but there are at least 2000 asteroids regularly charted and observed. As might be expected, frequency increases with decreased mass and it is impossible to place a lower limit on the size. Nevertheless, it is unlikely that fragments with less mass than 10^{-13} g would be found. Much of this highly fragmented material is the product of collisional bombardment (as against gravitational accretion) and displays substantial modification from the very early chemistries.

Most of the asteroids lie in a band between 1·8–4·0 AU from the Sun in comparatively eccentric orbits at inclinations of 0°–10° to the ecliptic. However, several 'maverick' bodies are in elliptical orbits with perihelion values as close as 0·19 AU to the Sun (perihelion for the innermost planet, Mercury, is 0·3 AU), or 28×10^6 km.

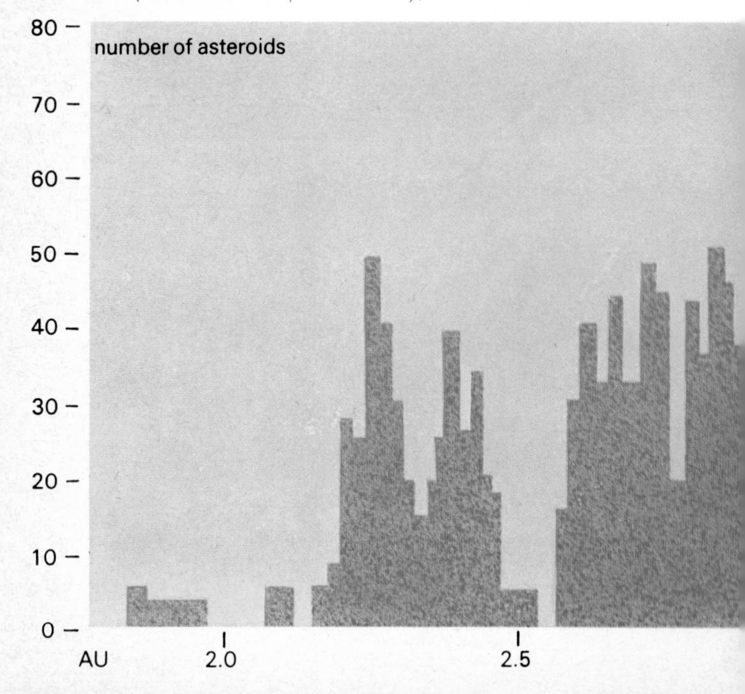

Evolution of the solar system

Although there are many conflicting theories concerning the early phase of planetary evolution, it is possible to state a few broad truths that have been confirmed by astronomical and space-borne observations. It is most unlikely that the solar system is much older than the currently accepted value of between $4{\cdot}7 \times 10^9$–$4{\cdot}8 \times 10^9$ years and the condition of the terrestrial planets confirms a period of intense planetary bombardment at least up to $3{\cdot}9 \times 10^9$ years ago; all four inner planets exhibit crater scars.

It appears that the Earth has the most tectonic activity, although it is very possible that Venus is also active, forming crustal plates, and so on. There may be some volcanism on Mars, but major evolution has ended, and Mercury is essentially inert. The atmosphere on Venus resembles that of Earth some $3{\cdot}0 \times 10^9$ years ago, before carbon dioxide became locked into carbonate rocks. Both Venus and Mars are unlikely to be the sites of any biological development.

The outer planets, especially Jupiter, may yet pass through further periods of contraction but their evolution is essentially spent and it is with Mars and Venus, closest to Earth, that further changes are likely.

Fig. 105 The asteroids are grouped in orbits between Mars and Jupiter, and this population density chart shows the majority to lie between 1·8–4·0 AU from the Sun.

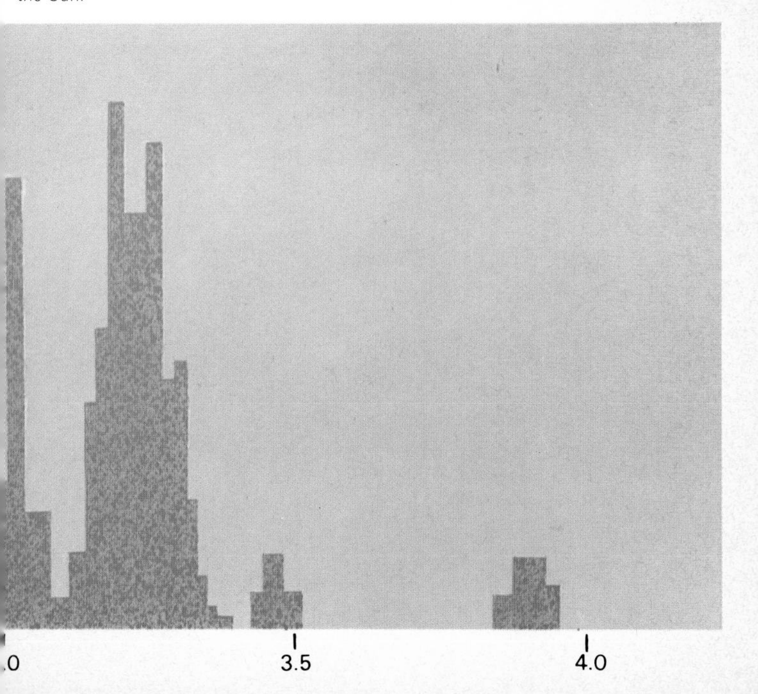

Comets and Meteorites

History and designation

The word 'comet' is a close approximation to the Greek word for hair and in times of astronomical ignorance, a comet with its sweeping tail emanating from a spherical head, was thought to be similar to human hair blowing in the wind. Chinese astronomers, though, likened it to a broom sweeping the constellations on successive nights.

For several thousand years comets were synonymous with catastrophe, their presence implying a portent of famine, earthquake or fire. Throughout the Roman era and the Middle Ages comets were thought to be either true celestial objects or vaporized emissions from Earth's atmosphere. This is all the more confusing since early celestial observation had charted the peculiar motion of the planets and even cursory observation of a comet would have shown it to be a truly celestial object, revolving with the heavens and rising and setting as the Earth revolved on its axis.

It was not until the late 16th century that Brahe measured the position of a comet from widely separated observatories and came to the conclusion that it was following a path through the solar system at a distance more remote than that of the Moon. Working on this hypothesis, Halley attempted to develop the idea that comets were orbiting the Sun in like manner to the planets. In 1682, he drew upon 70 years of telescopic work from various contributors and developed a theory based on his own observations of a comet in that year.

The comet he observed had been seen in 1607 and before that, apparently, in 1531. In fact the same comet can now be traced back to 86 BC and perhaps even further. Basing his calculation on Newton's new theory of gravitation, he predicted that the comet would return in 1758. It did appear then, 16 years after Halley's death, and has since been named in his honour. Halley's Comet appeared again in 1985–86, when it was the subject of intense investigation both from Earth and by several spacecraft.

Gradually, through such work, fears were allayed and comets came to be respected as celestial phenomena of astronomical importance. Designation systems were established, employing any one of three categories, to record their discovery and sequence their appearance.

Comets which are discovered or rediscovered are given the year of discovery, followed by the order in which they are recorded for that year. For instance, the first new comet to appear in 1980 will be designated 1980a followed by 1980b, 1980c and so on through to 1981 when the small letter suffix will start again. When sufficient information about the orbit of the 'new' comet becomes available, through sustained observation and tracking, a second designation is applied; a Roman numeral indicates the order, for that year, in which the comet passes perihelion, its closest approach to the Sun. For instance 1980n (the 14th comet discovered in 1980) may become 1981 II (the second comet to reach perihelion in 1981). The third designation category simply employs the name of the person, or persons, who actually discovered the comet, for example, Seki or Arend-Roland. In the latter case the pre-hyphen name is the person who discovered the comet (Arend) and the post-hyphen name honours the person who first observed it on its next passage (Roland).

Structure

The structure of a comet can be divided into four sections: nucleus, coma, tail and halo **(Fig. 106)**. The nucleus and the coma together form the head of the comet which is most usually observed, in small telescopes at least **(Fig. 107)**. From this a tail of varying length develops as the object approaches and recedes from the Sun. The halo is not visible from the Earth and it is only with the advent of space-borne observatories capable of recording the ultraviolet bands of the spectrum that such a phenomenon has been detected.

Fig. 106 The nucleus of a comet (a) produces the coma (b) when the tiny body comes within effective range of the solar wind. This, in turn, produces the characteristic tail (c). Observations in the ultraviolet portion of the spectrum have identified a large halo (d) of hydrogen gas surrounding the head. (Arrow indicates direction towards the Sun.)

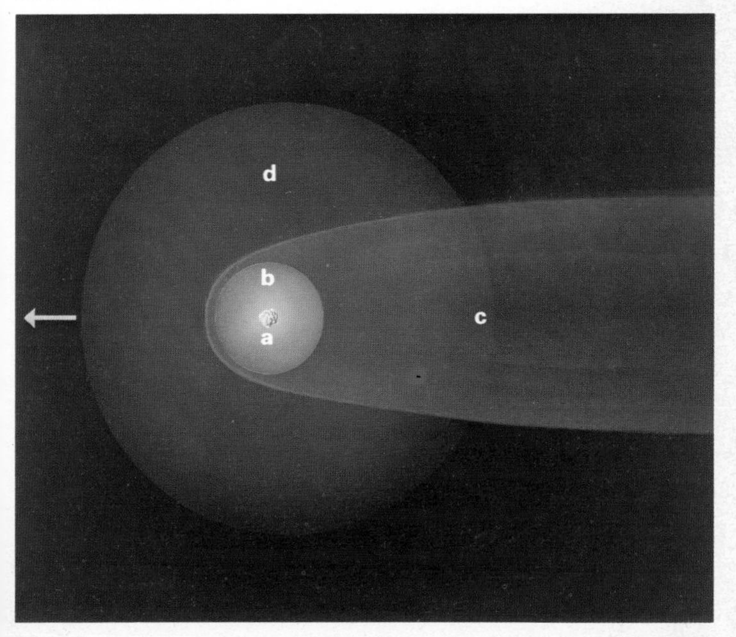

Owing to its small size, even large telescopes can resolve no detail within the nucleus. It may sometimes split into two or more parts, perhaps during a close solar passage, and the separate bright components within the coma may be visible even with small telescopes. The nucleus is, of course, the only permanent part of the comet and can be thought of as the reactive core from which all other phenomena are directly produced. Although the visible comet is indeed a splendid and observationally rewarding object, it is perhaps important to emphasize its transient nature. It is the nucleus itself, unobserved for much of its travels, that contains the information which is most important to the scientific community. Only by measuring the nature of the coma and the tail can the astronomer understand the composition of the nucleus.

For much of the time, the nucleus is a dormant structure and it is only when its orbit carries it to within effective range of the Sun that the coma and the tail emerge. The mass of the nucleus is so low that it has not been observed to influence measurably any other body in the solar system, although there have been recorded observations of cometary passage through the satellite system of Jupiter. Considerable effects were noted in the comet's orbit; none were observed in the motion of the moons.

It has been known for many years that comets have very low masses when compared with other bodies (even small planetary satellites) in the Solar System. With the return of Comet Halley in 1985–86, and the various spacecraft missions that investigated it, it became possible, at long last, to confirm that comets have extremely low densities, perhaps as low as one-quarter of that of ice. They seem to consist of loose collections of ice and dust particles with a very dark, outer crust that contains a large amount of organic (i.e. carbon-based) compounds. The Giotto probe's images of Comet Halley showed it to be an irregularly shaped body that was approximately 15 × 8 km in size. The ices were evaporated by solar heating as the comet approached perihelion, releasing jets of gas and dust through holes in the crust. Water, carbon-dioxide, and carbon-monoxide ices predominate, but there is surprisingly little methane. Some dust is similar to ordinary silicate rocks, but other particles are unknown minerals.

Comets lose part of their mass (perhaps 1%) at each return, so they gradually fade and become inactive. This means that Comet Halley cannot have been orbiting the Sun ever since the formation of the Solar System but must be a recent phenomenon.

On successive passes of the perihelion more and more material is left in protracted accumulations strung out along the orbital path. The predicted lifetime of a comet (or number of perihelion passes) must take account of the strength of the solar wind, the actual perihelion distance, the mass of the nucleus and the material remaining available for future vaporization. When the comet approaches to within about 2 AU, solar radiation impinging upon the nucleus vaporizes the ices and develops a diffuse cloud around the nucleus known as the coma.

Many elements and molecules have been identified and inferred, including hydrogen, carbon, oxygen, many metals, water, methane, ammonia, and hydrogen and methyl cyanides, as well as silicate dust. When the comet reaches the inner solar system the coma develops rapidly and may have a radius of $10^3 - 2 \times 10^5$ km **(Fig. 108)**. Although not all comets have tails, these usually develop near the Sun, and consist of gases and dust driven from the head by pressure of solar radiation and by bombardment from the solar wind. They may reach lengths of as much as 2×10^8 km.

The gases and dust usually form two distinct types of tail. The tiny dust particles spread out behind the comet in its orbit and often give the appearance of wide, fan-shaped or curved tails. As the dust only reflects sunlight, dust tails appear yellowish in colour. The dust particles may later give rise to distinct meteor showers if the Earth passes close to the comet's orbit. The gas, or plasma, tail is controlled by the energetic particles in the solar wind, and appears as a narrow, straight, bluish tail that points away from the Sun **(Fig. 109)**.

The first radio waves from a comet were detected in 1973 coming

Fig. 107 Comets have distinct features that make them attractive subjects for telescopic observation.

from the region of the dust cloud, indicating the presence of methyl cyanide and hydrogen cyanide found in the spiral arms of galaxies. The halo surrounding the coma is a hydrogen envelope visible only in ultraviolet light and this can be as large as 2×10^7 km in radius, a phenomenon caused by the irradiation of the head.

Orbits

The origin of comets is subject to much speculation, but it seems likely that they originate in a shell lying well outside the known boundary of the solar system. Stellar perturbations affect their early orbits and send the bodies in towards the Sun. The eccentricity of the orbit is about 1·0, that of a parabola, which is theoretically 'open' so that the comet may never return to the inner solar system. When the eccentricity is less than 1·0, the comet is definitely periodic, and will return, albeit perhaps after a very long interval **(Fig. 110)**.

Periodic comets are fairly arbitrarily divided (at 200 years) into those with long and short periods. About 60 per cent of periodic comets are of long period, and follow elongated orbits. Planetary perturbations (particularly by Jupiter) can alter orbits and periods, and even cause the comet to be ejected from the solar system on a hyperbolic orbit.

Fig. 108 (below) The familiar dust tail is often accompanied by a plasma tail curved by the magnetic lines of the solar wind. **Fig. 109 (right)** This negative image of Comet Halley is a composite of two photographs taken on 10 March, 1986. The relatively featureless dust tails curve away to the left (North), while the much more complex gas tail extends towards top right.

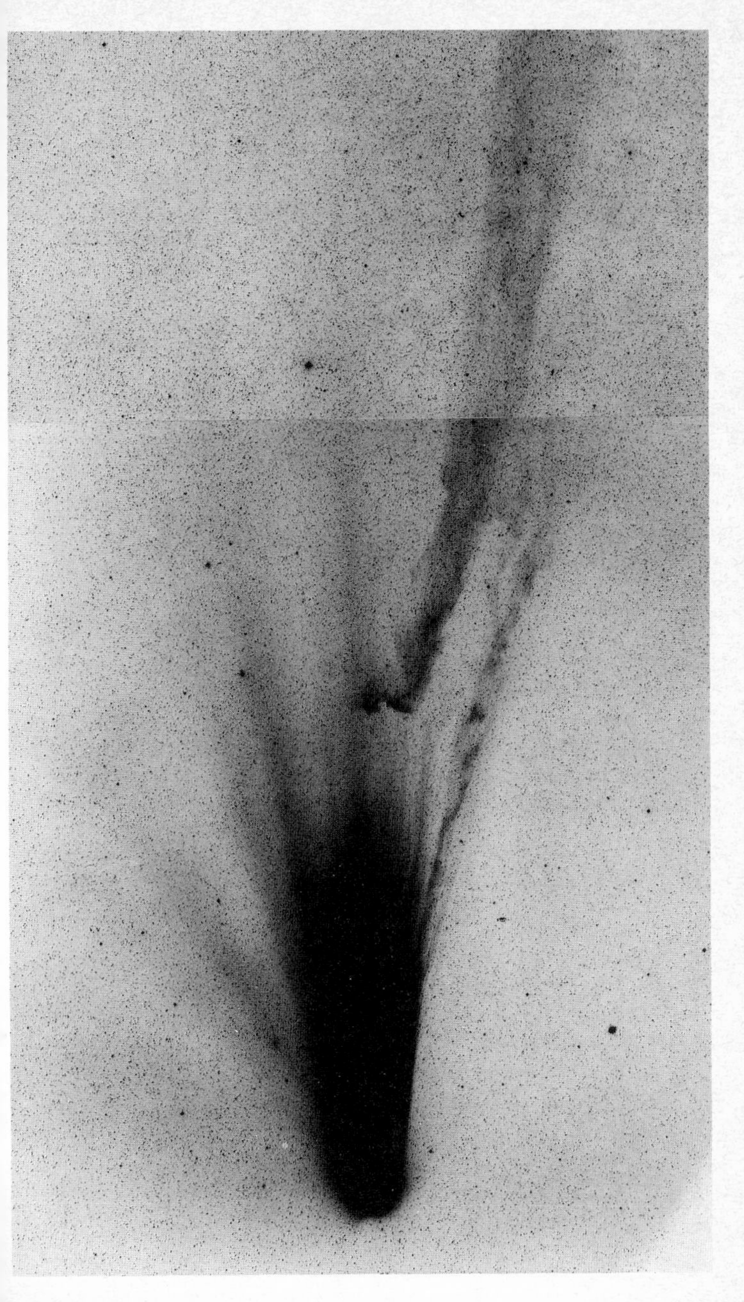

Generally, the orbits display eccentricities close to unity indicating an origin on the periphery of the solar system with an orbit constrained by the gravitational envelope of the Sun. It would be impossible for any new cometary body to pass unperturbed through to perihelion, and the modification of orbits by the planetary masses is primarily responsible for determing the period of the orbit. Further indication of the remote origin of proto-comets comes from a study of the inclinations of long-period types where the random angles to the ecliptic imply a halo unrelated to the plane of planetary formation.

An example of the interrelationship of remote, dormant, existence and outer planet perturbation can be found in the calculated trajectories for comet Kohoutek. Originally in a 4×10^6-year-period orbit at a distance of 5×10^4 AU ($7{\cdot}48 \times 10^{12}$ km), Kohoutek was displaced by a local star into a deflected orbit from which it became attracted to the Sun. This occurred 2×10^6 years ago and the comet reached perihelion in December 1973 before moving back out into the solar system.

It has been perturbed by the planets into an orbit with perihelion and aphelion values of $2{\cdot}09 \times 10^7$ and $5{\cdot}15 \times 10^{11}$ km respectively with a period of $7{\cdot}5 \times 10^4$ years. In the case of Kohoutek, the comet has been removed from its initial orbit far beyond the outermost planet, brought to a very close (0·14 AU) perihelion and deflected to a long-period orbit. The importance of this comet, as with all first approach comets, lay in the pristine and unvaporized condition of the nuclear material which, when released as a coma and a tail for

Fig. 110 (below) No two comets exhibit identical characteristics and each one produces phenomena directly related to the comparative abundance of materials in their composition. **Fig. 111 (right)** Short-period comets are dramatically influenced by the presence of other bodies in the solar system.

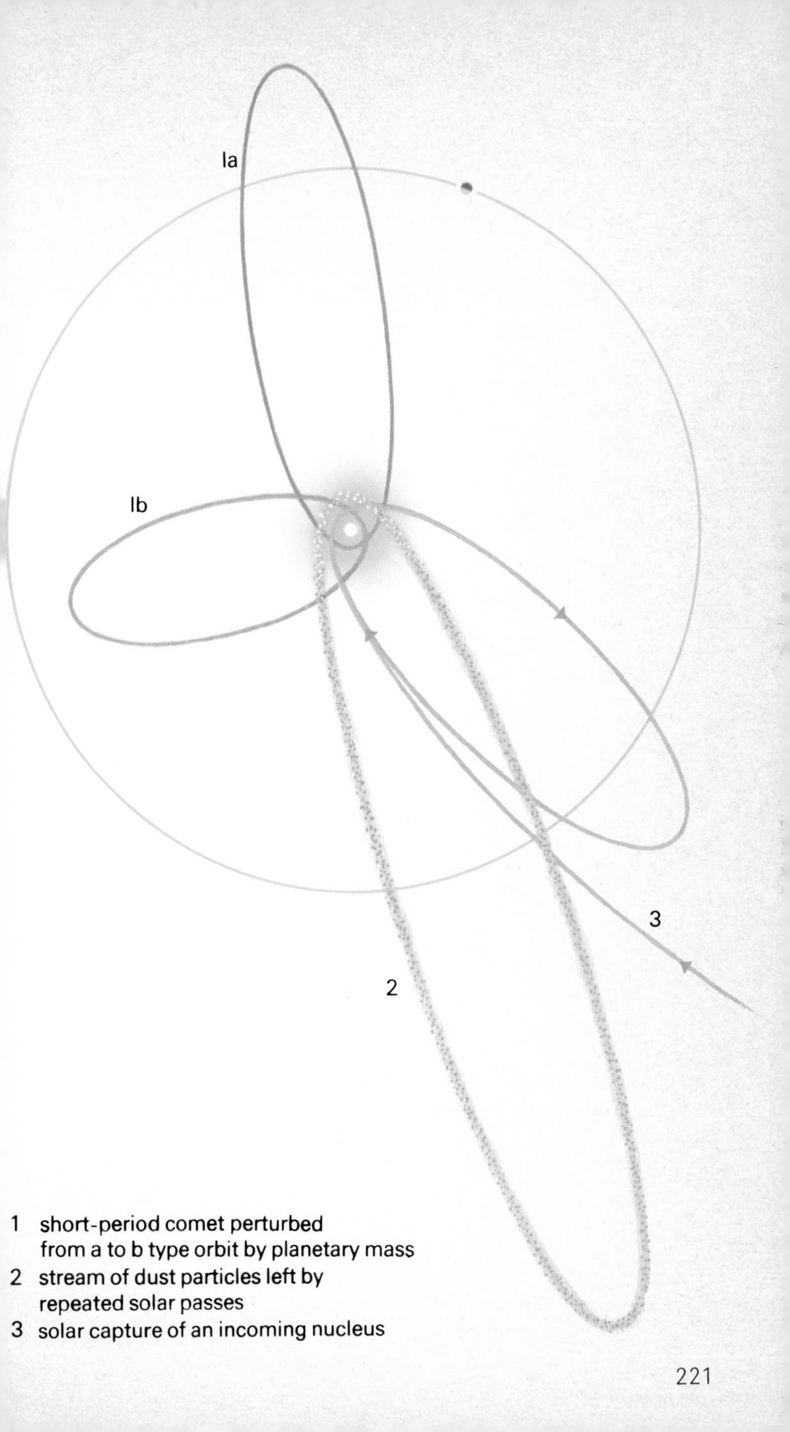

1 short-period comet perturbed from a to b type orbit by planetary mass
2 stream of dust particles left by repeated solar passes
3 solar capture of an incoming nucleus

the first time in 1973, enabled astronomers to measure residual material from the very earliest years of the solar system.

The second group of comets belong in short-period orbits of between 3–3 × 10^2 years taking them beyond Jupiter at aphelion and in to a very close solar pass at perihelion **(Fig. 111)**. They are nearly all inclined less than 20° to the ecliptic and several examples lie within Mars and Jupiter or within Jupiter and Saturn radii from the Sun. Halley's comet is an exception in one respect since unlike most short-period comets it is in a retrograde orbit inclined 17·8° to the ecliptic. Most short-period comets have aphelia close to the planet Jupiter, whose dominant mass has succeeded in playing a major role in setting up the geometry of short-period orbits.

Perihelia come very close to the Sun and comet Encke is an example, passing just 0·3 AU (4·49 × 10^6 km) from the Sun in its 3·3-year orbit. One reason for Halley's prominent appearance is that in its 76·2-year orbit it comes from 35·3 AU to within 0·6 AU of the Sun at high velocity and is influenced by massive radiative forces emanating from the solar surface.

Observation of comets is a particularly rewarding occupation for amateur astronomers. It is essential to record such pertinent characteristics as brightness, rate of motion and celestial position and report these to the recognized astronomical bodies of the country concerned from where the information can be distributed to professional astronomers around the world.

Comet observation can have important repercussions for man and beast alike. Although the probability of contact between Earth and a sizeable comet is very low, if the two bodies were to experience a head-on collision, the relative speed of 80 km/s would give rise to an enormous explosion and disastrous effects for the region concerned. It is thought likely that the nucleus of a small comet struck the Earth on 30 June 1908 in the Tunguska river valley area of Siberia. Trees were felled within a radius of 48 km and if the comet had been of a more massive type the consequences would have been disastrous.

Meteors

Several times throughout the year, Earth apparently sweeps through a cloud of celestial debris with the result that friction with the atmosphere causes the incoming particles to glow and streak across the sky leaving a bright tail of ablated fragments. In accepted nomenclature the particle is termed a meteoroid before entering Earth's atmosphere and a meteor when it is seen to interact with atmospheric molecules.

In order to establish the point of meteoroid origin, whether inside or outside the solar system, it is necessary to measure the geocentric velocity (a combination of the speed within the solar system and the Earth's orbital motion). To extract the velocity of gravitational attraction (11·2 km/s) it is necessary to use the following equation:

$$v^2 = Vm^2 - (11{\cdot}2)^2$$

where Vm is velocity measured in km/s and v the geocentric velocity.

Most meteoroids exhibit geocentric velocities of 15–50 km/s but some are observed as high as 72 km/s and since the Earth is moving at nearly 30 km/s the maximum heliocentric velocity can be no more than 72 − 30 = 42 km/s in a head-on collision. Since the heliocentric escape velocity is $\sqrt{2} \times 29{\cdot}8$ km/s (the mean orbital velocity of the Earth) the derived value of 42·1 km/s agrees well with the concept of captured origin and the meteoroids are, therefore, members of the solar system. Careful review of meteor orbits reveals a direct association with comets; the Orionids are in the orbit of Halley's comet, the Beta Taurids associated with comet Encke and the Leonids **(Fig. 112)** with comet 1866 I.

Because of the compact nature of a comet's nucleus only solar pressure could cause disruption, and the layout of meteoroid clouds undoubtedly owes much to the low mass qualities of the nucleus. Further evidence suggesting a cometary origin lies in the calculated densities which have a mean value of about 0·4 g/cm^3, considerably less than the 3 g/cm^3 of most minerals. The icy nature of a comet's

Fig. 112 The Leonid shower arises from the fragmented debris of a comet, broken apart by close passage around the Sun. Each year, Earth passes through this stream of particles and the spectacular showers caused by friction with atmospheric molecules are a frequent subject of amateur observation.

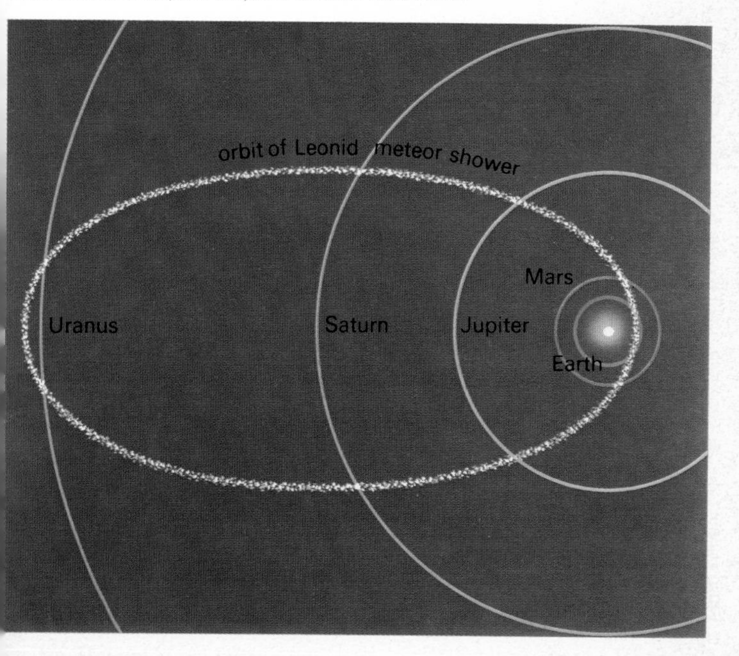

head fits well with this hypothesis. Meteoroids, generally, are between 0·8–5 mm in size.

Much needs to be understood about these shooting stars and two methods of analysis, optical and radar, are particularly useful. In the visual examination of trajectories, two wide angle cameras are trained on the same area of the sky but from vastly different locations up to 100 km apart. When a meteoroid penetrates the atmosphere its track is recorded against the same stellar background, and from the different observed trajectories at the two locations it is possible to compute height, velocity and path of the incoming particle.

During the day, it is possible to record pertinent aspects of the meteoroid's trajectory either: by using radar to detect ionization of the atmosphere during the period of intense heating upon descent, and thus record the change in frequency to calculate the Doppler shift; or by careful analysis of reflections from the ionization trail. Meteors are usually visible between 40–130 km. Careful observation of a meteor shower will reveal a radial 'starburst' effect as though the ionized tracks are pouring through one constellation and setting up a spoked-wheel of light in the sky. This is, of course, an optical illusion caused by perspective and lack of compensation for the immense distances separating the constellation of stars from the meteoroids.

Meteorites

A particularly large meteoroid is called a fireball which is characterized by exceptional brightness, while a larger fireball is called a bolide and is frequently associated with sparks, explosive disintegration and large bangs. When an extraterrestrial object enters the Earth's atmosphere and survives disintegration it is called a meteorite, totally unrelated to meteoroid showers which, as indicated, are generally associated with the orbits of comets and probably consist of nuclei broken into fragments under solar pressure.

Meteorites, unlike meteors, occur randomly and with no detectable preference for geographical location. On the contrary, their encounter with the planet occurs with such frequency that is is believed a meteorite falls to Earth every 2 days. The largest of these are associated with Meteor Crater in Arizona, USA **(Fig. 113)**. Here, a 1·2-km crater 175 metres deep and probably 3×10^4 years old was excavated by vaporization of a $9 \times 10^6 - 9 \times 10^7$ kg mass. (Because of the explosive disintegration often associated with meteorites it is common for a single specimen to break up into several fragments.)

In 1868 more than 10^5 fragments fell at Pułtusk in Poland, and in Holbrook, Arizona more than $1{\cdot}4 \times 10^4$ came down in 1912. On occasions like this the meteorite breaks up at comparatively low altitude, usually to the accompaniment of a cacophony of bangs and flashes, and the individual components follow similar trajectories, falling to Earth within a small region.

Meteorites can be divided into three types: iron, stony-iron and stony **(Fig. 114)**. More than 60 per cent of all meteorites are of the stony type, 30 per cent are of the iron type and the balance consists of stony-irons. The largest meteorite of the iron type is the Hoba meteorite which fell in South West Africa weighing $5{\cdot}4 \times 10^4$ kg, while the largest stony type is the Chinese Kirin Province meteorite weighing 1770 kg; specimens of the stony type are more brittle than iron types and tend to break up in the atmosphere.

Stony meteorites are further divided into chondrites, with small spherical inclusions, and achrondites, or those without inclusions. Chondrites account for about 85 per cent of all stony meteorites, and the carbonaceous chondrites contain evidence of organic matter (hydrocarbons, fatty acids and amino acids) which does not, however, indicate the presence of life on the parent body.

Three types of iron meteorite have been classified with varying quantities of nickel, and the type with between 6–14 per cent nickel exhibits the familiar Widmanstätten pattern when it is polished and prepared. This effect is particularly associated with slow cooling from the molten state and would seem to indicate a period of slow change in the chemistry of the metal.

The stony-iron category includes examples that contain iron within stone minerals and also metals embedded in silicate stone. The chemistries of the two components within the iron and the stone fragments seem to indicate a common ancestry to the separate iron

Fig. 113 Meteor crater in Arizona bears testimony to the unexpected arrival of random debris.

and stony categories with common origin and local morphology.

The questions concerning the origin of the meteorites are unanswered, as they were a decade ago, but there is unanimous agreement that they are genuinely extraterrestrial and originate from within the solar system. Moreover, there is evidence to support the view that meteorites belong to the same class of object as represented by the asteroids and that both result from the breakup of a planetary body or the group of bodies that would have accreted to planetary volume but for some unforseen cosmic quirk of fate.

The sequence following through stony, stony-iron and iron meteorites seems analogous to the chemical differentiation from radioactive heating deep within an accreted body of planetary size. Silicates would rise, metals would sink and a combination of the two would be found at some intermediate depth. If the meteorites are laid out in class sequence, it is possible to construct a picture of planetary formation, thermally induced differentiation and subsequent breakup from an unknown mechanism.

Tektites

Unlike meteorites, tektites (from the Greek word for molten) are a group of objects whose chemical composition is 70–80 per cent silica with a mean density range of between 2·3–2·5 g/cm^3. They are of precise shape, between 2–4 cm in size and apparently restricted to geographically preferred sites. Found mostly in the South China Sea, Australia and Tasmania, tektites are thought to have been formed either from the impact of a large meteorite, or to have arrived on Earth from a fragmented comet or the ejected material of a lunar impact; but they may not be of extra-terrestrial origin **(Fig. 115)**.

Based on measurements of radioactive decay most tektites are between $0{\cdot}3 \times 10^6$–$3{\cdot}0 \times 10^7$ years old. Meteorites, of course, are considerably older with an age range of from $4{\cdot}0 \times 10^9$–$4{\cdot}5 \times 10^9$ years, very close to the origin of the solar system. However, in argument against the terrestrial hypothesis it should be known that tektites contain less than 150 ppm of water (versus 6000 ppm or so for igneous rocks on Earth) and the mechanism by which the water could have been extracted is not easy to obtain; it is a very difficult process to extract water from molten glass.

At the other end of the scale, large impacts such as those which sculptured the major craters and small basins on the Moon are believed to have been preserved in certain areas. These are known as astroblemes, and although the Earth has undergone much geophysical modification which has virtually obliterated evidence of planetismal impact, there are certain shield areas that appear to contain the eroded depressions of major impact early in the Earth's history.

In summary, more than $1{\cdot}6 \times 10^7$ kg of material is collected by the Earth each year and many falls provide interesting assignments for amateur observation, particularly the meteor showers.

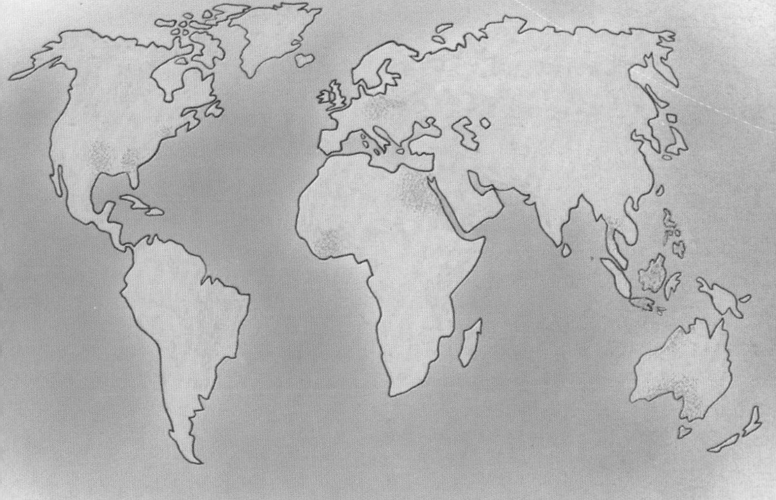

Fig. 114 (top) Meteorites are among the oldest known bodies in the solar system and can be grouped into one of three categories: stony, iron or stony-iron.
Fig. 115 (above) The distribution of tektites on the surface of the Earth is seen to favour selected regions.

The Galaxies

The Milky Way

On a clear night it is possible for the unaided eye to see about 2750 stars in each hemisphere. These are close to the solar system, and form just a small part of our own galaxy. In the northern hemisphere only one hazy 'star' visible to the naked eye does not belong to our galaxy, and is a galaxy in its own right – the Andromeda nebula or M31 (from Messier's catalogue of such objects). In the southern hemisphere the two Magellanic Clouds are small independent galaxies, close companions to our own. The Sun is just an insignificant member of the vast number of stars which form the population of the Milky Way galaxy.

The Milky Way, so called because it resembles a smudged pattern of flowing milk, is visible proof of the dense central plane of a disc-like accumulation of stars. The galaxy itself is more than a collection of stars, however, and careful study of its content and characteristics is the key to theories concerning the very origin of the universe itself.

The galaxy, or indeed any galaxy, contains stars, diffuse patches of neutral or ionized gas, dust particles and clouds of interstellar debris and high-energy particles known as cosmic rays. As mentioned earlier, most stars can be divided into at least two broad population classes, quite apart from the many divisions and subdivisions which describe their chemical composition, temperatures, sizes and densities.

In general terms, the Population I category can be considered to occupy positions making up the central disc of the galaxy and the arms which appear to radiate like curved spokes from the nucleus. Population II stars are less flattened in their distribution and are represented by the halo stars and globular clusters, with orbits at high inclinations to the central plane of the galaxy. Population I stars account for 83 per cent of the galactic mass and Population II types provide the balance of 17 per cent **(Fig. 116)**.

It is interesting to note that regional accumulations of specific population types upsets this homogeneous balance, and in the vicinity of the Sun less than 5 per cent of the stellar attendants belong to the Population II type. Since galactic matter in stars and dust clouds orbits the nucleus, unlike the situation presented by a rigid wheel where the outer section rotates faster than inner components, it is possible to measure the relative velocity (proper motion) of nearby stars and determine the approximate mass of the sum total.

The Sun is about 10 kpc (1 kpc = $3{\cdot}26 \times 10^3$ light years) or $3{\cdot}09 \times 10^{17}$ km from the centre of the galaxy, a distance equal to $3{\cdot}26 \times 10^4$ light years. Assuming that the Sun orbits the galaxy at a comparatively uniform distance from the centre it is possible to derive a mean rotational velocity of $2{\cdot}5 \times 10^2$ km/s.

From $2\pi(3{\cdot}08 \times 10^{17}\,\text{km})$ the period of the Sun's orbit can be calculated thus:

$$P = \frac{2\pi(3{\cdot}09 \times 10^{17}\,\text{km})}{250}\text{s} = 2{\cdot}5 \times 10^{8}\ \text{years}$$

where P equals the period. From this it is only necessary to modify the equation used to determine masses of binary star systems (page 49):

$$M_1 + M_2 = \frac{a^3}{P^2}$$

where M indicates masses of the components, a is the semi-major axis and P is the period, to obtain a rough estimate of the galactic mass. For this, the equation may be rewritten as:

$$M = \frac{a^3}{P^2}$$

where M is the mass of the galactic nucleus (in solar masses), a is the orbital radius in AU and P the period in years. Then:

$$M = \frac{(2{\cdot}06 \times 10^{9})^{3}}{(2{\cdot}5 \times 10^{8})^{2}}$$

Fig. 116 The central disc of the galaxy contains many young, Population I type stars while the older, Population II material is represented by halo stars in highly inclined orbits, whose trajectories are shown here by the arrowed paths.

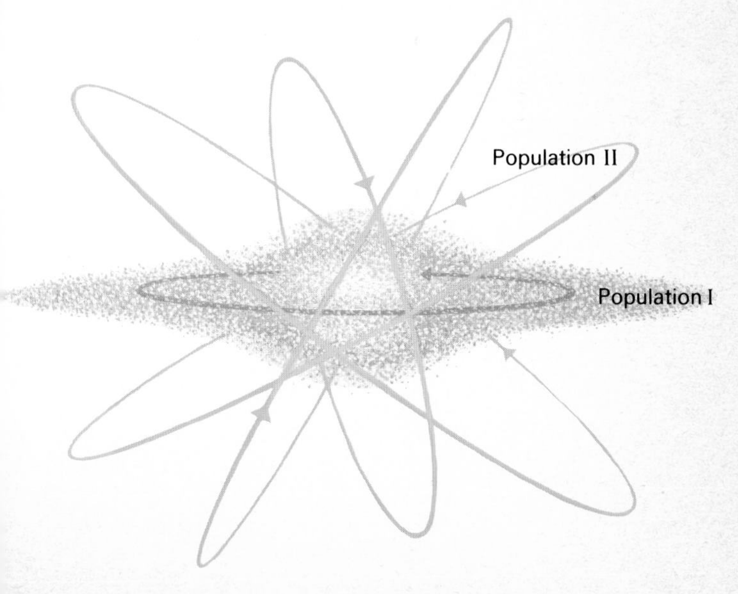

The result, $1{\cdot}4 \times 10^{11}$ solar masses, is in agreement with more sophisticated calculations which show that the total galactic mass lies between this figure and $2{\cdot}0 \times 10^{11}$ solar masses.

It has often been stated that the galaxy contains 10^{11} stars, which together with the above figure might be thought to easily yield the average mass of individual stars. However, such a conclusion should be treated with caution as the total number of stars in the galaxy is rather uncertain. It is impossible to say precisely how many black dwarfs, neutron stars or black holes may exist throughout the galaxy. However, the result of our equation does give an indication of the amount of mass which is to be found inside the solar orbit.

Structure of the galaxy

Since the stars are the only means by which the galaxy can be illuminated for visual observation it is from careful measurement of bright sources, including novae, that the dimensions of the galaxy have been obtained. The galactic disc is believed to have a nucleus 3 kpc thick and to be about 1·0–1·5 kpc thick in the region of the Sun. As stated above, the galactic centre is 10 kpc from the Sun and, while it is only with extreme difficulty that the outer 'edge' of the central plane can be observed, it appears that the distance from the Sun to the outer periphery is about 4·5 kpc. The galaxy is, therefore, about 14·5 kpc in radius, a value which shows it to have a thickness-diameter rato of approximately 1:10.

These dimensions are related to the nucleus, disc and arms which comprise the flattened structure of the main galaxy and the less discernible globular clusters (see p. 46) are observed to be accumulations of stars moving in high inclination orbits above and below the central plane of the galaxy. Gobular clusters represent stellar condensations from a phase in galactic evolution when the contraction of gas and dust permitted star formation beyond the limits of the galaxy as it exists now.

These clusters of stars are found to have radii of between 20–200 pc (65–650 light years) and can number up to a million individual stellar components which, although apparently in contact, probably have mean separation distances of no less than 2·5 light years. The globular clusters enclose the galaxy in a diffuse halo and have been important for determining the distance to the centre of the galactic nucleus by careful analysis of the orbital trajectories.

The spiral arms

Galactic arms pose major problems of interpretation. If arms were to be formed by some process, it would be expected that the differential rotation rates of material at various distances from the galactic centre would always tend to disperse the stars and destroy the structure of the arms. Such destruction of spiral arms by dispersion will occur in just 10^8 years, which is two orders of magnitude below the calculated age of the galaxy.

It could be assumed that the material in the galaxy has recently become aligned to give the unique structure which can be detected today, and that although the material is tending to disperse, the arms are just a chance occurrence. However, it would be unreasonable to expect other galaxies to be similarly affected. Yet countless galaxies can be seen with spiral structures, and it would seem that all those which can possess such features, do in fact have them. It is quite impossible to believe that they could all be affected by chance at the same time. The only other reasonable conclusion is that there is some mechanism which can maintain spiral structures for very long periods of time.

Careful analysis of many close galaxies shows the arms to contain interstellar material and young, massive stars. Stellar associations in the arms of our own galaxy are less than 2×10^7 years old, yet clusters of older stars are distributed at random throughout the galactic disc. This is a further indication that stars in the galaxy orbit the centre with lower velocities at greater distances from their centre of gravity. Indeed, star clusters will move out of the spiral arms at the rate of 0·5 kpc in 5×10^7 years; hence the youthful age of stars currently observed to lie within the arms, since insufficient time has passed for them to migrate from the spiral structure.

It is obvious that forces other than gravitation are at work to sustain the geometry of the spiral arms. However, acceptance of the rigid nature of the arms implies that the outer components are moving at a relative speed exceeding the escape velocity at that region of the galaxy. In other words, the kinetic energy of the velocity on the outer periphery of the galaxy would exceed the potential energy of the whole galactic mass.

One possible solution to this anomaly lies in the theory that the spiral arms are magnetic tubes holding ionized (positive or negative charged atomic) particles which in turn create a filamentary 'web' collecting dust moving out radially from the galactic nucleus. The accumulated dust clouds provide sufficient gravitational attraction to prevent the grains from spiralling up the arms and reaching escape velocity to be lost from the galaxy for ever.

The arms appear to funnel gas clouds but the observed flow rate, at a velocity of 50 km/s at the galactic nucleus and 7 km/s at the position of the solar system, is insuffícent to prevent the arms from winding back in on the central region of the galaxy. Since the magnetic spirals would tend to create loops, a mechanism must be operating to sustain the arcuate spirals observed in our own and so many other galaxies. A loss of one solar mass of disc material (dust grains) per year from the central region of the galaxy would be sufficient to unwind the arms in direct relation to the gas pressure within them and this would create a more dynamically symmetrical flow tube for the magnetic forces. In summary, the arms tend to loop back on the galactic nucleus, but outward surges of disc material unwind the loop as a result of the gas pressure **(Fig. 117)**.

Galactic nebulae

Observationally, some of the most intriguing categories of galactic phenomena are the various types of nebulae, originally catalogued as diffuse patches of luminous material distributed both inside and outside our own galaxy. Nebulae are now divided into extragalactic and galactic sources, the former being galaxies in their own right and the latter comprising clouds of non-luminous gas and dust lit to varying degrees of intensity by stars inside their voluminous envelopes.

There are two basic types of galactic nebulae, those which are accumulations of gas and dust brought together by magnetic and gravitational forces and where stellar formation may be taking place (see p. 53), and those which are the result of phases of stellar evolution. The latter include planetary nebulae – shells of material shed at a red-giant stage – and remnants of supernovae such as the magnificent example of the Crab Nebula in Taurus.

The Great Nebula in the constellation Orion is one of the more magnificent accumulations of gas and dust which is brightly lit by atomic excitation from stars embedded within and around the envelope. The presence of dark nebulae testifies to the lack of intrinsic luminosity in all types of dust and gas accumulation; only when lit by stars or related mechanisms can it be seen as a bright, albeit generally diffuse, cloud of matter. Absorption values in dark nebulae generally range between 25–90 per cent and this is a result of their composition and density. Apart from absorption, the elongated dust grains also give rise to polarization of starlight. The dark nebulae are concentrated along the spiral arms, appearing in the central plane of the Milky Way. The most notable examples are the Coalsack in Crux, and those in Taurus, Orion and Monoceros.

Apart from gas and dust, most nebulae contain various molecular compounds which may range from simple substances with only two atoms, such as carbon monoxide, to much larger molecules like ethyl alcohol, which has nine atoms, or to even more complex compounds. Most molecules have been detected by observations in the microwave region of the spectrum, and every year seems to bring identification of new and more complicated types, so that the total number now exceeds fifty. The densest concentration of molecules is observed in the Saggitarius B2 cloud close to the galactic nucleus, which contains about 10^6 solar masses of material.

Due to the flattened nature of the galaxy it is impossible to obtain a view of the dense central region or areas on the opposite side to the

Fig. 117 The spiral arms of the galaxy are theorized to funnel gas clouds from the nucleus along spiralling magnetic lines holding ionized particles. In this way, material is transported from the nucleus to the arms.

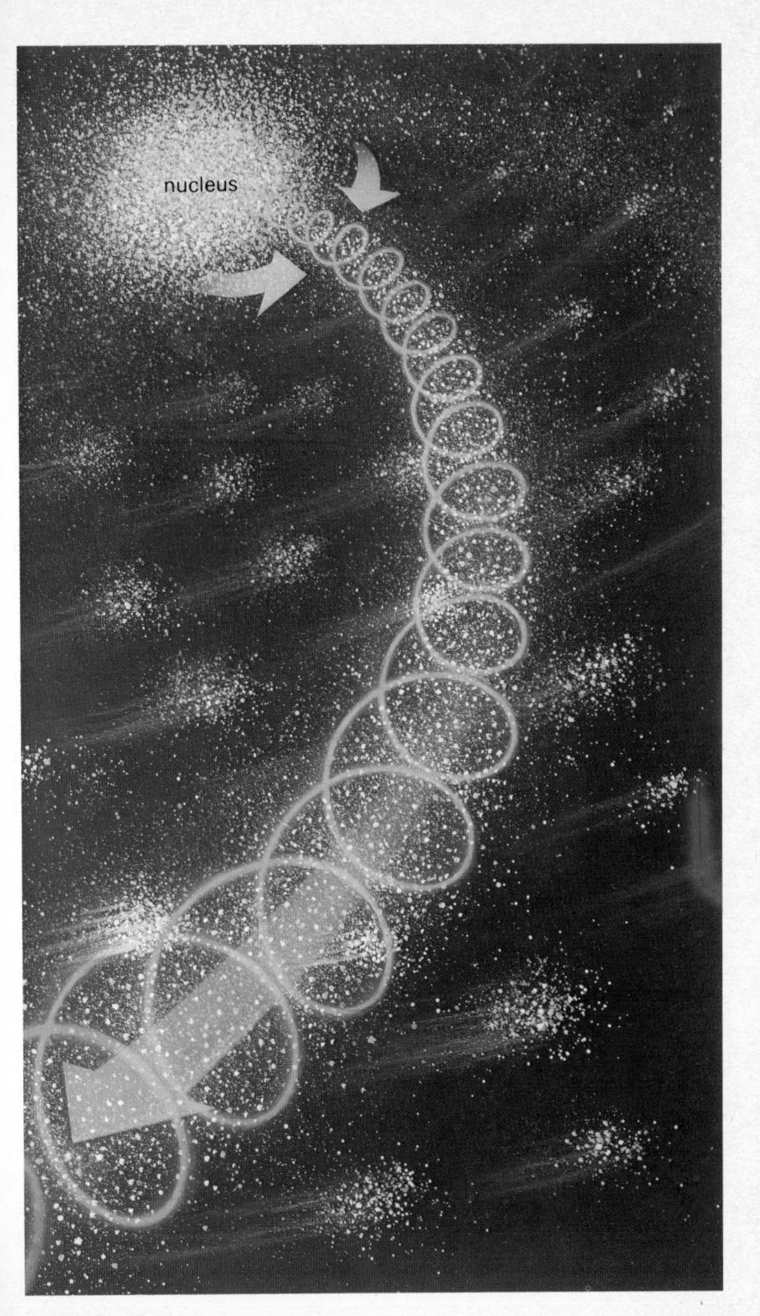
nucleus

solar system. In spite of the rapid motion of the Sun (250 km/s) the span of time bridging the intelligent observation of the galaxy is far too brief to have carried the solar system a significant distance round and through the spiral arms; in the 5 million years or so of human existence the solar system has traversed less than 8° of the 360° in one galactic orbit. Nevertheless, it is possible to apply non-optical techniques to improve the 'view' of the important central regions.

The use of radio observations for probing the dense accumulations of gas was advanced by the work of van de Hulst, who calculated that the 21-cm line emitted by hydrogen could be effectively used to map portions of the galaxy invisible to an optical telescope due to interstellar absorption. Research in this area has confirmed that the bulk of the galactic hydrogen lies along the spiral arms and that accumulations at the centre of the galaxy exhibit random rates of motion. From this it has been observed that arm-like filaments are approaching the Sun at a speed of more than 50 km/s; a qualitative indication of violent events at the centre of the galaxy, the source and magnitude of which is left to speculation.

In the central region of the galaxy, most stars are contained within a flattened ellipsoid 1·5 kpc in the major axis and 1 kpc in the minor axis with about 10 per cent of the galactic mass locked up in this area. The nucleus of the galaxy was discovered in 1959 by Drake, Pariiski and Lequeux and is seen to be centred on Sagittarius A radio source, and is about 10 pc in diameter with a mass of around $1{\cdot}5 \times 10^7$ solar masses.

In summary, the central portion of the galaxy surrounding the nucleus is a dense agglomeration of stars, gas and dust with a mass density, at a distance of 10 pc from the nucleus, $2{\cdot}4 \times 10^4$ times that in the vicinity of the Sun. In this region, unlike the nucleus, stars account for 99·75 per cent of the total mass **(Fig. 119)**.

It is important to examine the problem of invisible objects, and in particular the abundance of inert objects, when considering the galaxy. There is circumstantial evidence suggesting that a massive black hole exists at its centre, and therefore, the mass of the galaxy will be somewhat greater than if total collapse is avoided in the cosmology of the universe. Despite the suspected expansion of the galaxy, in time the energy sources of the galaxy will be exhausted and the formation of stellar objects will cease. Knowledge of the galaxy is not yet enough to derive a model for its final death, and it is a moot point whether the universe will itself come to an end before a significant change is observed.

Extragalactic nebulae and classes

In practical terms it is convenient to assume that all matter has inertia, and physicists in general accept the inevitability of this conclusion, but the cosmologist must ask why this is so. Mach's principle goes

Fig. 118 (top) The Andromeda galaxy, comparable to that containing our solar system.
Fig. 119 (above) Sagittarius; one of the richest star fields visible in the telescope.

some way towards a solution in stating that matter has inertia because of its interaction with the rest of the universe. Disciples of Einstein would go further and say that the value of the gravitational constant, $6{\cdot}67 \times 10^{-8}$ cm^3/g/s^2 (see p. 52) is determined by the amount of mass to be found in the universe.

For this theoretical consideration alone, it is obvious that a great deal of matter must exist beyond the galaxy to which the solar system belongs. However, accepting that the galaxy is flattened and disc-like, observational details indicate that numerous extragalactic nebulae are also present in the universe.

In the early part of the 20th century, it was thought that external nebulae were located mainly above and below the central galactic plane with only a few seen to be in plane with the disc. However, this is considered to be an apparent anomaly brought about by visual obscuration of many nebulae by dust and gas in the arms and central plane of the galaxy.

It is now recognized that galaxies (or external nebulae) can be found throughout the universe grouped in clusters of between 5–10^4 individual components. Our own galaxy belongs to a cluster known as the 'local group', a distribution of 21 separate stellar islands of which only one is larger than ours. All are less than 3×10^6 light years distant with the nearest components barely $1{\cdot}7 \times 10^5$ light years away (remember the diameter of our own galaxy is about 10^5 light years).

Only three of the 21 galaxies in the local group are spiral in structure, our own galaxy, M31 (Andromeda) and M33, with the remainder of irregular and elliptical shape. Only two other galaxies are more than 10^4 light years in diameter and were it not for their comparatively close proximity to our own galaxy they would be impossible to resolve. Andromeda **(Fig. 118)** is a magnificent example of a spiral galaxy and at a distance of 2×10^6 light years is about twice the size of our own galaxy.

Beyond our local group of galaxies, nebulae can be observed which collectively grouped in a range of type classes, permit categorization of local and external cluster components. These type classes, known as the Hubble classification are not meant to imply either evolutionary or sequential steps in the history of the galaxies, but rather are intended to provide a series of type groups to which all the known configurations can be appended.

The Hubble classification provides four general categories of galactic configuration: elliptical, lenticular, spiral and irregular **(Fig. 120)**. Elliptical galaxies are devoid of a pronounced structure and possess no arms or spiral sections. The degree of ellipticity is defined

Fig. 120 Four types of galaxy were identified by Hubble: (a) elliptical, (b) lenticular, (c) spiral and (d) irregular.

a
b
c
d

on a scale E0–E7 and if a and b represent apparent major and minor axes respectively, the index 10 (a–b)/a defines the specific shape. The range E0–E7 represents shapes between b/a = 1 and b/a = 0·3 respectively, with E7 presenting the more pronounced ellipticity. In general, elliptical galaxies contain a bright central nucleus with no absorbing dust and several thousand globular clusters. Within this classification, subclasses D and N are seen to possess faint halos of varying luminosity.

Lenticular galaxies are disc-like structures without arms but with a pronounced central condensation and extended envelope. Three subtypes (SOa, SOb and SOc) represent graduation from an elliptical (E7 type) structure through to a spiral-like configuration devoid of arms or a disc.

The spiral galaxies, like their lenticular neighbours on the class scale, are flattened due to high rotational velocities but unlike other classes they contain visible arms of varying structural form. Spiral galaxies are divided into subclasses Sa, Sb and Sc, each representing a unique set of features **(Fig. 121)**. In particular, arm arrangement determines the specific subclass; the Sa types exhibit closed arms of dense light-absorbing material, and the Sc types exhibit broad, open arms invariably branching into two or more sections outside the central regions. The intermediate Sb class contains more dust than the Sa, presents open and well-resolved arms with a central condensation of moderate size and a bright well-illuminated disc. Both the galaxy to which our solar system belongs and the Andromeda galaxy are of the Sb type, perhaps the most familiar of all galactic classes.

A separate class of spirals, the barred spirals, is similarly divided into subclasses according to the degree of separation between arms. These are: SBa spirals, supporting a prominent but comparatively unresolved bar and tightly wound arms issuing from a central nucleus; SBb types, presenting a more prominent bar and well-defined arms less tightly wound than in the SBa class; and SBc galaxies, exhibiting a well-defined bar and open, unresolved arms **(Fig. 122)**.

The fourth major class in Hubble's classification is the irregular type which is characterized by a pronounced lack of symmetry or apparent defined structure. The Ir I types are rich in interstellar material, young stars and scattered bright regions of uncondensed stars. The Ir II types are less common and present a flattened accumulation of unresolved stars in dense clouds of light-absorbing matter. **Figures 123–127** illustrate some of these spectacular galaxies.

Fig. 121 (above right) Three primary types of spiral galaxy – Sa, Sb and Sc are seen here; type Sa exhibits closely wound arms and type Sc has an open structure.
Fig. 122 (right) The barred spirals – SBa, SBb and SBc follow the same criteria for classification as the spiral types; SBa is a tightly wound structure while SBc is an open one.

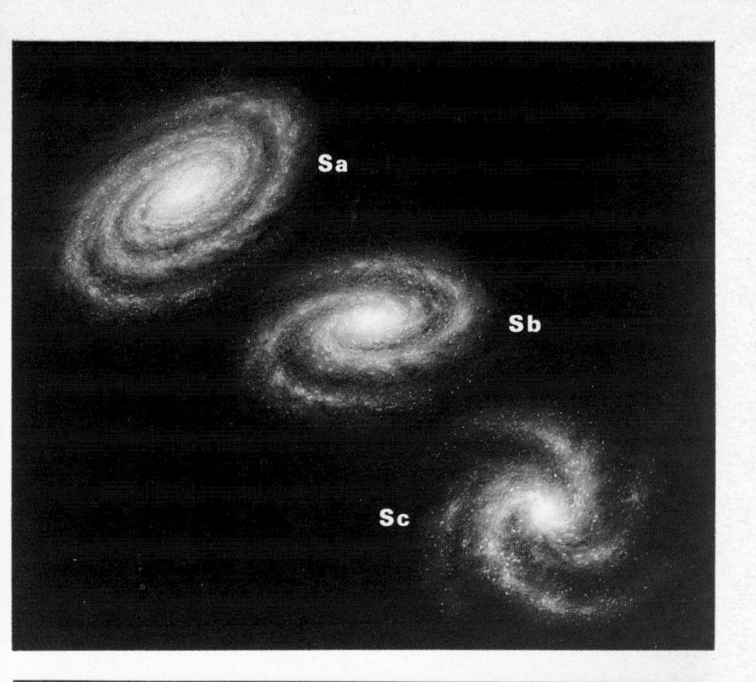
Sa
Sb
Sc

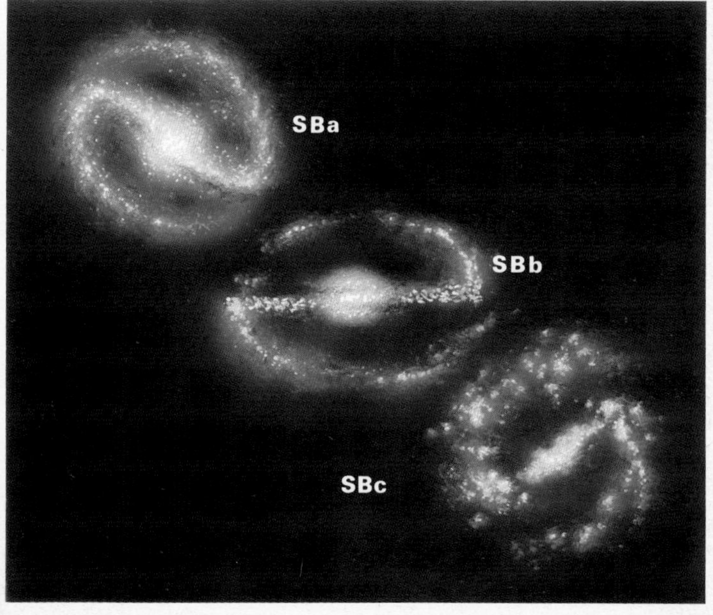
SBa
SBb
SBc

Fig. 123 (left) Galaxy in Sculptor. **Fig. 124 (top)** Galaxy in Triangulum. **Fig. 125 (above)** Irregular galaxy in Ursa Major.

Fig. 126 Whirlpool galaxy (M51)

Fig. 127 Spiral galaxy in Pegasus.

Galactic evolution

It is very difficult to obtain sufficiently accurate measures of the class and abundance of galaxies to permit qualitative judgement on population levels and class ratios. More than 60 per cent of the bright nebulae are of the S or SB type (although Hubble believed there to be only 15 per cent in the SB class) with 22 per cent lenticular, 13 per cent irregular and 4·5 per cent undefined. It was originally believed that galaxies evolved along a sequence of classes and casual observation seemed to indicate that galactic evolution began with ellipticals and progressed to spirals, a type represented by our own galaxy.

Careful analysis of stellar content in nearby examples revealed that the irregulars contain a considerable amount of gas, the proto-material of new stars, whereas ellipticals are devoid of any major process of star formation. This seemed to indicate a sequence whereby galaxies began life as irregulars and evolved through a sequence of spirals to end up as ellipticals. A third possibility, equally valid, assumed that the types and subclasses evolved from either end of the sequence but at different rates, thereby accounting for a variety of types at any one specific time.

In these hypotheses, galactic formation can take place at any period in the evolution of the universe, a factor which allows the acceptance of many different cosmological models. However, various aspects of the physical properties applicable to galaxies permit a more stringent set of criteria to be applied. The only successful way to determine the age of a galaxy is to measure the age of its globular clusters, the oldest groups of stars in the nebula. For all nearby galaxies the ages seem to converge at 10^{10} years and since the measured group consists of a variety of different types, it would seem to be an inevitable conclusion that galaxies may be of different classes irrespective of age.

The only surviving hypothesis implies that galaxies evolve at different speeds and that evolution takes place either from ellipticals to irregulars or vice versa. However, consideration of the masses of the different types of galaxy prevents acceptance of such an evolution.

The bright elliptical galaxies contain, on average, about 30 times the mass of the bright spirals. To accommodate an evolutionary path from spirals to ellipticals would require an unacceptably high rate of accretion of intergalactic material; in any event there would be transitional varieties of intermediate mass but these are not observed. Conversely, an evolution from ellipticals to spirals is equally improbable due to the lack of a satisfactory mechanism for shedding 97 per cent of the initial mass; evaporation from stars and loss to intergalactic space is unlikely to be occurring due to the general lack of interstellar material in elliptical galaxies.

Fig. 128 Classification of galaxy types according to mass and angular momentum.

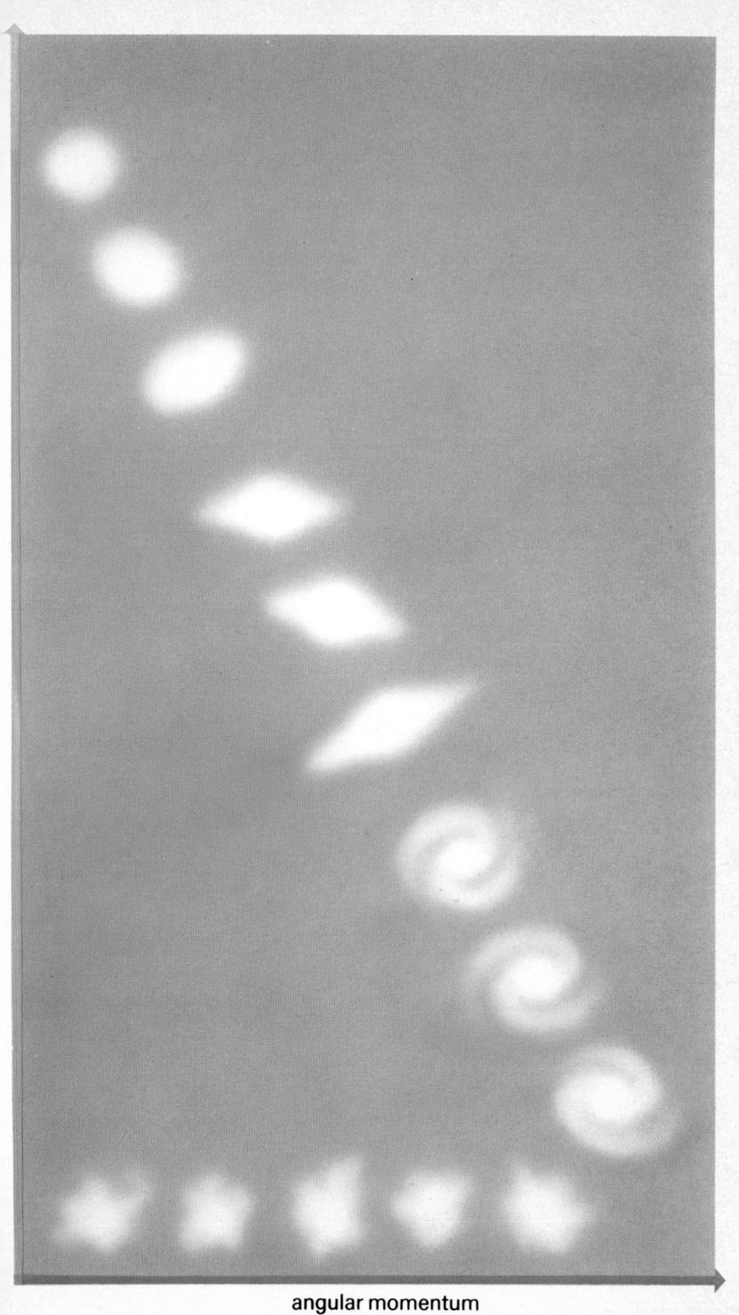
angular momentum

Spiral galaxies possess very much more angular momentum per unit mass than do the ellipticals, and in the absence of any good scientific explanation for either the rapid acquisition or dumping of angular momentum, it is conceptually impossible for elliptical galaxies to become spirals or spirals to evolve into ellipticals. In any case, it would be necessary for galaxies to interact with external material for any appreciable change in angular momentum to occur and such interactions are impossible to observe or envisage theoretically.

Although it seems likely that most of the spiral, elliptical and lenticular galaxies formed at the same time, certain irregular galaxies appear very much younger. An example of the relationship between spiral and irregular galaxies can be illustrated with our galaxy and two attendant irregulars known as the Large and Small Magellanic Clouds. Separated from our own galaxy by about $1{\cdot}7 \times 10^5$ light years, they are between 9–12 kpc across and are linked to the outer region of our galaxy by filaments with a large percentage of gas.

As discussed earlier, matter is continually flowing from the centre of the galaxy into the disc and along the spiral arms and it would seem that, in this situation, irregular attendants to spiral types are the product of matter gradually bleeding away from the outer edge of the spiral. Matter is found outside the spiral galaxies which endorses the assumption that the spirals are losing mass through rotation. Confirmation that irregulars are the product of discharged matter from spirals is obtained, circumstantially, from the fact that spiral and barred spiral galaxies are nearly always accompanied by secondary irregulars.

The general distribution of spiral galaxies, occupying mean background fields in the universe, confirms the tendency for this type to disperse its content as a result of the high rates of rotation. The elliptical and lenticular classes are seen to group in dense clusters avoiding the homogeneous distribution of the spirals.

The redshift

An important observation for all galaxies is that all the lines in their spectra are moved to longer wavelengths than those which are produced in the laboratory. The general movement, or redshift, has become accepted as evidence that galaxies are receding from each other and the Doppler effect is used to determine the distance of very remote nebulae which are too far away for classic methods of measurement. Before 1930, Hubble showed that velocities of recession increase with distance; subsequent measurement of both red shift and distance of galaxies up to 2×10^9 light years away permits extrapolation of more remote objects by the red shift alone.

Fig. 129 Hubble set up the scale of galactic distance versus recessional velocity in the third decade of this century which has since been considerably modified. Here, the accepted scale is shown indicating the apparent velocity of different galaxies, more properly interpreted as an indication of the speed of respective galaxies at increasingly distant points in time.

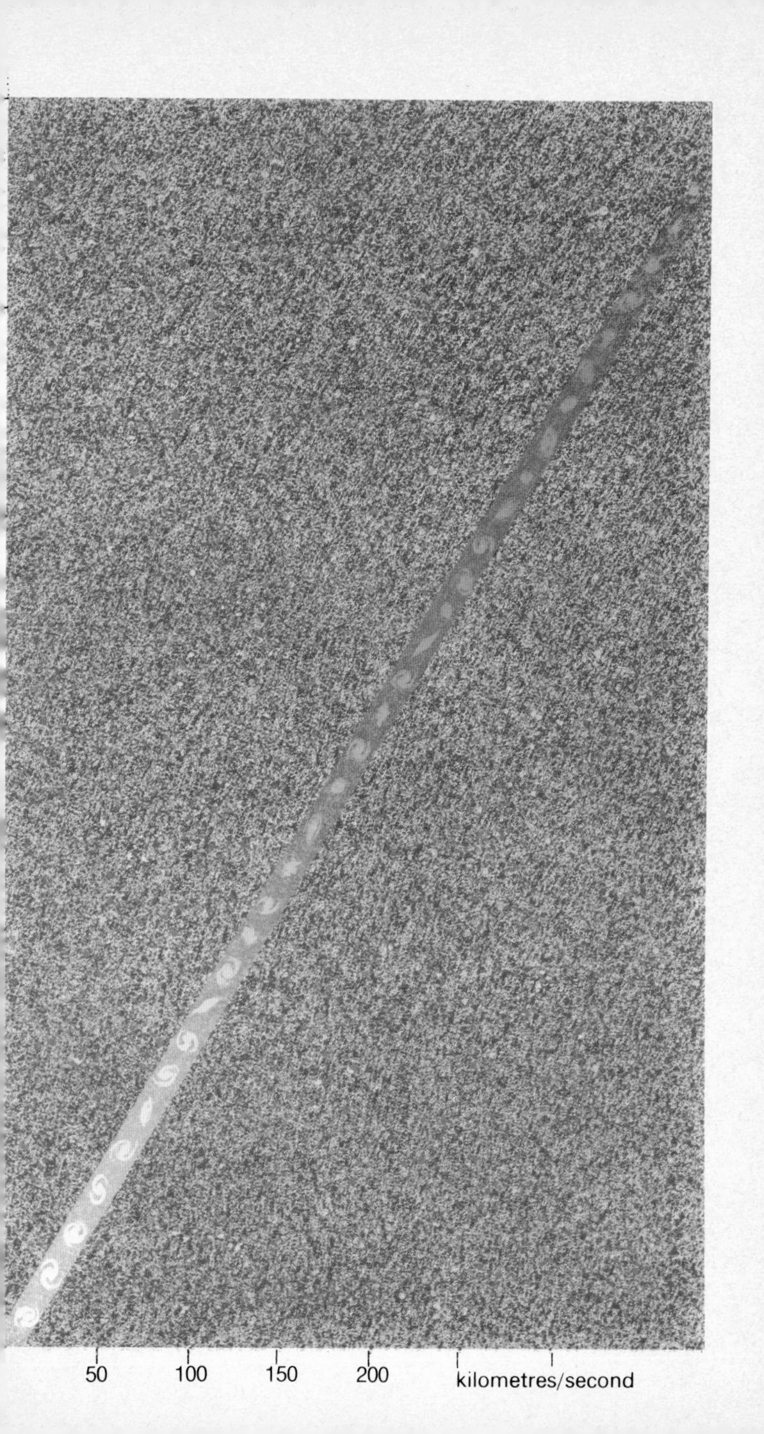

50
100
150
200
kilometres/second

It is obviously of the greatest scientific significance that the amount of redshift corresponding to a particular distance should be correctly determined. Unfortunately this estimation is not easy and the accepted value of the ratio of velocity to distance, known as the Hubble constant, and expressed in km per second per Mpc (Megaparsec = 10^6 parsec) has frequently been altered over the years. For a long time 55 km/s/Mpc was generally used, but more recent investigations seem to suggest that the value is closer to 90 km/s/Mpc. Using the correct value of the Hubble constant is of the utmost importance for cosmology, because as we shall see, it does not apply to galaxies alone, and it is essential in calculations of both the size and the age of the universe itself **(Fig. 129)**.

Formation of galaxies

As we have seen, once formed as a specific type, a galaxy will possess properties that prevent evolution into any other class. In the early stages of galaxy formation the condensation had a simpler chemistry due to the primitive nature of the gas cloud; because most galaxies formed at the beginning of the universe there was little elemental build-up until stars formed from the mechanism which produced the galaxies.

During the stage of the entire cloud's gravitational collapse, the decreasing radius of the envelope increased the mass per unit volume and induced rotation at higher and higher velocities. However, magnetic coupling transferred angular momentum from the centre of the condensation to outer regions preventing complete collapse. Also, because of the lesser resistance to gravitational attraction at angles perpendicular to the plane of rotation, the overall configuration becomes a flattened condensation.

During this process of condensation, magnetic transfer of angular momentum induced local regions of star formation, only becoming effective when the separate subregions consisted of clouds of several thousand solar masses. In fact, the magnetic forces would have prevented local condensations during the collapse of the galaxy if the subregional mass was any less than this value. Thus, massive clusters of stars formed at distances representing the dimensions of the condensing galactic cloud, at that time while the remainder of the galactic material continued to contract and flatten to its present configuration. These condensations, or globular clusters, retained their respective positions to the centre of galactic mass and are seen today as a visible indication of early galactic development.

This scenario applies to spiral and barred spiral galaxies while galactic condensations with less inherent angular momentum become elliptical or lenticular types. The factors responsible for determining the ultimate configuration of a galaxy are, therefore, the result of the mass and angular momentum of the particular condensation in question.

The continued process of stellar formation demands multiple star clusters but the number of stars per cluster becomes very much less than in the early stages when globular cluster formation began to regulate the evolution of the galaxy. Because of this, recent star birth is observed to occur in clouds of several hundred stellar components rather than the more prolific abundance of the older globular clusters.

The H-R diagram was first interpreted as showing an evolutionary cycle for star development which moved from large massive O types down to small, low mass K and M examples. It is now recognized that stars develop from clouds of varying temperature and mass to occupy only one position on the main sequence. Similarly, galaxies were thought to evolve along the Hubble classification spectrum but are now seen to occupy only one type class during their lifetime.

Quasars

By 1960, suspicion was growing that an anomalous class of objects to be known as quasi-stellar objects (QSOs) or quasars **(Fig. 130)** were lying at enormous distances from our local region of the universe.

The QSOs typically show as just tiny star-like points without any

Fig. 130 Quasars are one of the most enigmatic groups in the universe, releasing enormous levels of energy from an apparently small volume.

associated nebulosity or other features (although a few exceptions are now recognized). The majority are radio-quiet, but about one per cent have been found to be active in the radio region, and these are sometimes known as QSSs (Quasi-stellar sources). As a group the quasars were first discovered as a result of surveys to try to identify optical objects corresponding to known radio sources. When their strange nature became evident, intensive investigations added many examples, and it is now estimated that about one million are probably detectable. At first the spectra appeared featureless, and it was some time before absorption lines could be detected.

Careful measurement of the Doppler shift in emission lines in the spectra revealed very great redshifts and consequent velocities of recession. On the assumption that the Hubble relationship still holds, the distances of these objects may be inferred, and these show that some of them must be far out towards the boundaries of the visible universe. Ever more distant quasars are discovered, but at the moment the record is held by an object with an apparent velocity of about 270000 km/s (90 per cent of the speed of light). The actual distance must depend on the accepted value of the Hubble constant (page 248) but at 90 km/s/Mpc, the implied distance of 3000 Mpc equals some $9{\cdot}8 \times 10^9$ light years. Such distances pose problems for cosmology, because a universe which was isotropic (or of uniform distribution) would appear substantially unaltered with changes in location or time. The anisotropy implied by the remote situation of quasars certainly does not agree with a uniform distribution of matter and objects. The universe must therefore have undergone drastic evolutionary changes during its existence. In observing QSOs we are looking far back into the very early stages of the universe.

Their observation over such exceptionally large distances implies that QSOs are extremely energetic and it has been found that their emission may be as high as a thousand times the amount of energy emitted by even the very largest optical galaxies. However, that is not all. Many of them show variations in their emission over very short periods – of the order of days. This implies that the emitting regions themselves must be very small, being limited to the distance that light can travel in a few days. It is quite impossible for any disturbance to trigger any variation at another place in a time that is less than that required for light to traverse the distance between the two points. This means that quasars (or more correctly, the regions that are emitting the energy) are just a few light days across, say about 100 AU at the most. So a region only slightly larger than the size of the Solar System is emitting as much energy as some one thousand galaxies, each consisting of at least one hundred thousand million stars.

The problem now becomes one of accounting for the production of such vast amounts of energy in such a tiny region of space. The only conceivable mechanism that could release so much energy involves processes occurring round a very massive black hole. If this is able to 'swallow' material from a surrounding galaxy, the infalling material reaches very high temperatures just outside the black hole. Although nothing can escape from a black hole itself, outside its boundary the inflowing matter can release a large fraction of its potential energy in the form of radiation. This gives rise to the exceptionally high luminosities of the quasars.

Although at first all that could be seen of quasars at optical wavelengths were tiny, brilliant point sources, this was primarily because they were so distant that any surrounding material was too faint to be detected. With advances in equipment and techniques, more and more QSOs are being found to be embedded in galaxies of some sort. Many of these galaxies appear, as far as we can tell, to be fairly normal galaxies, with no particularly notable features other than the presence of a quasar within them. So quasars do, in fact, appear to be the extremely active central regions of very distant galaxies that existed in the early universe.

Certain astronomers have argued against the distant, and highly energetic, nature of quasars. They have pointed out that if they were exotic, nearby objects, both their luminosities and sizes would be far less extreme, and thus more easily explained. Such a solution would, however, mean that the red shift was not related to the expansion of the

Fig. 131 Pulsar NP 0531 in the heart of the Crab Nebula; a pulsating radio source which gives off emissions at the rate of 30 per second.

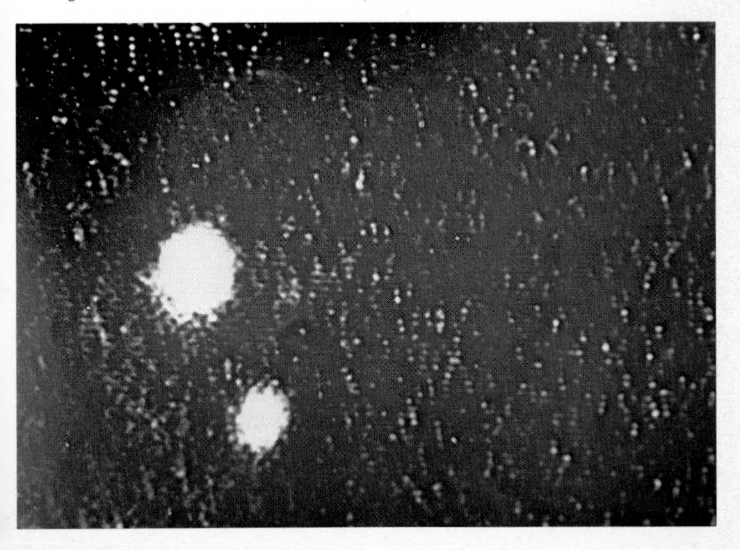

universe. These suggestions regarding the nature of quasars have become very improbable with the discovery that quasars occur within galaxies. Another argument for quasars being distant objects comes from the study of their spectral lines. Apart from characteristic emission lines, many quasars exhibit absorption lines. Some show sets of absorption lines, with each set having a different red shift. These may arise in clouds of gas ejected by the quasars, or in otherwise as yet invisible, intervening galaxies. In a few cases, distorted or apparently doubled images of quasars have been discovered, where the gravitational fields of intervening galaxies have bent the paths of light from the distant quasars in the so-called 'gravitational lens' effect.

Radio galaxies are another form of energetic object, although not emitting as much energy as quasars. They often show two extended lobes of material far outside the parent galaxies and from which most of the radio emission arises. It is beginning to seem as if there may be a progression of objects from ordinary, 'quiet' galaxies, such as our own – which does, however, have energetic processes occurring in its nucleus – right through to the highly energetic quasars. In between the extremes are radio galaxies and other objects, known as Seyfert galaxies, most of which show little radio emission, but which have brilliant point-like central nuclei.

The universe – evolutionary or steady state?

During the first half of this century several important discoveries were made having future implications for both cosmology and astronomy. So far, an examination of the factual content of the universe has provided background information to cosmology and it is that philosophical expression of astronomy which will be considered in this subsection.

As stated above, the universe is expanding **(Fig. 132)** at a rate proportional to the distance from our own galaxy. This implies that the universe is moving out from a central point, and that if the events are theoretically back-tracked to when the material content of the universe was in one location it would be possible to derive its age. The very acceptance of this suggests a single event which started the universe on its current expansion – the 'Big Bang' theory.

Supporters of this theory calculate the age of the universe as between $1{\cdot}1 \times 10^{10}$ and $1{\cdot}8 \times 10^{10}$ years depending upon the value of Hubble's constant which is used. Such an age agrees with that

Fig. 132 The expanding universe. This view is from 'outside' and approaching galaxies appear blue while receding ones show a red shift. However, from a central position all galaxies will exhibit red shift, and from any given galaxy all others will appear to be moving away at a greater or slower rate. The galaxies here are shown in roughly one plane, although expansion is of course in all directions from the central point.

found for the oldest stars in the galaxy and the derived origin of isotopic abundances of chemical elements.

The steady-state theory implies that the density distribution per unit volume in the universe today is essentially the same as it was at any time in the past and will continue to remain the same at any time in the future **(Fig. 133)**. Accepting the uniform separation of any one cluster from another, the theory goes on to say that matter is in a state of continuous creation at a rate sufficient to accommodate a constant density value within the portion of space vacated by overall expansion. The mechanics of this theory are unable to explain the method by which material is created to maintain mass density distribution. However, this anomaly should not be taken as proof of inadequacy because Big Bang theorists are at equal loss to explain the creation of the material content of the primordial 'sphere' preceding the expansion of the universe.

A significant factor in determining the accuracy of the theories came in 1965 when scientists at the Bell Laboratories in the USA were perplexed by a background radio noise present in all areas of the sky. Elsewhere it had been calculated that if the universe really did begin with a cosmic fireball (about $1{\cdot}1 \times 10^{10}$ years ago) as postulated by the Big Bang theory, the residual energy of the fireball should be detected today as a uniform background temperature of about 3° above absolute zero. Since the energy will radiate microwaves at wavelengths >1 mm, this background temperature should reveal a brightness curve exhibiting a specific profile at wavelengths between 100 cm–0·5 mm.

Careful measurements down to about 5 mm construct a profile identical with that set up by a background temperature of 2·7° above absolute zero. This is used to substantiate the theory that the observed microwave radiation is the cooled remnant of the primeval fireball, calculated to be $1{\cdot}5 \times 10^{10}$°C just 1s after the initial explosion.

Therefore, the universe is the volume occupied by the total amount of all the matter in stars, galaxies, clusters of galaxies, gas, dust and radiation observed to have separated from the primeval fireball, or cosmic egg. However, it is necessary to ask whether this 'single' universe actually consists of a number of space-time environments with differing mathematical and physical parameters. Is in fact the universe in which our galaxy resides the same as that billions of light years away? If not, it must be assumed that Big Bang or steady state models allow a multiplicity of universes.

It has already been shown that the steady state theory demands a high isotropy and any imbalance in the space-time relationship will

Fig. 133 In the steady state theory, as galaxies recede from one another, more galaxies are spontaneously created to maintain uniform density. In this hypothetical reconstruction, the four galaxies in the upper view are observed to have moved away from each other in the lower view while four more galaxies have appeared to preserve frequency distribution per unit volume.

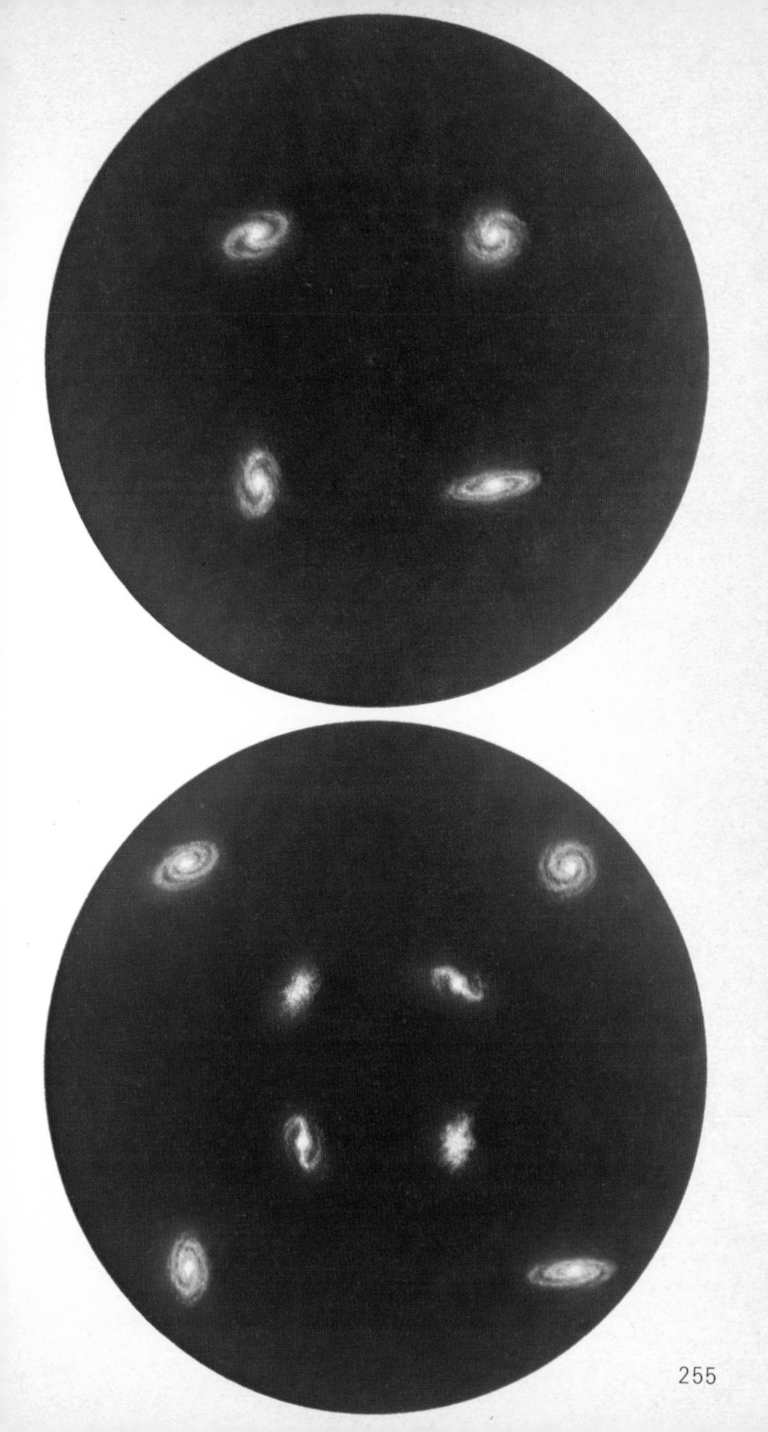

eliminate this theory as a possible explanation of the phenomena observed. If it is assumed that the observed redshifts of QSO sources do in fact derive from cosmological recession, determination of their distribution should make it possible to decide whether the Hubble constant does actually vary in very distant regions of the universe, beyond those in which galaxies themselves can be distinguished.

QSOs represent a special class of object because of the unique nature of their mass energy interchange. Also, if they really are as remote as they seem, their position in the universe implies a material change in the physical content of the universe with time. In other words, a QSO measured at a distance of 5×10^9 light years is seen now as it was 5×10^9 years ago simply because its light has taken 5×10^9 years to reach our position in the universe.

Any material, or mass, change in the isotropy of the universe with time will have a profound effect on the Hubble constant. In other words, the gravitational constant, believed to vary according to the relationship of any specific mass with the rest of the universe, will be different far out in space if QSO sources were present only in the past. Consequently, the linear relationship of distance to velocity (galaxies twice as far out recede at twice the speed) becomes invalid at extreme distances or, as expressed above, varies with the passage of time. The universe must have been evolving throughout the period spanned by the time separating us from the QSO era if QSO red shift is cosmological in origin.

Leaving this hypothethis for a moment, it is useful to consider a possible mechanism for the unparalleled release of energy from the small, low mass QSO sources. Total collapse of stars greater than 2·2 solar masses was described on p. 57 where space-time curvature completely isolated events within the collapsar from the universe outside. Efforts to determine what may occur at and beyond the event horizon, or Schwarzschild radius, while speculative, lead to some interesting possibilities.

The Einstein field equations can be interpreted as having two solutions which are symmetrical about the event horizon. One applies to this universe, and the other can be thought of as describing another universe remote in space and time. If the body which formed the singularity is ignored, it is even possible to conceive of a 'wormhole' connecting the space-time region in which we exist to some other universe **(Fig. 134)**.

In theory therefore, it might be possible for matter and energy to be transferred from 'this' universe into another which, from the form of the equations, must be at a different place and time – or indeed, from some other space-time continuum into our own. By analogy, the regions where matter and energy might emerge into a continuum have been described as 'white holes'. But the solutions do not preclude 'wormholes' which connect regions within the same universe, provided that they are distant in space and time.

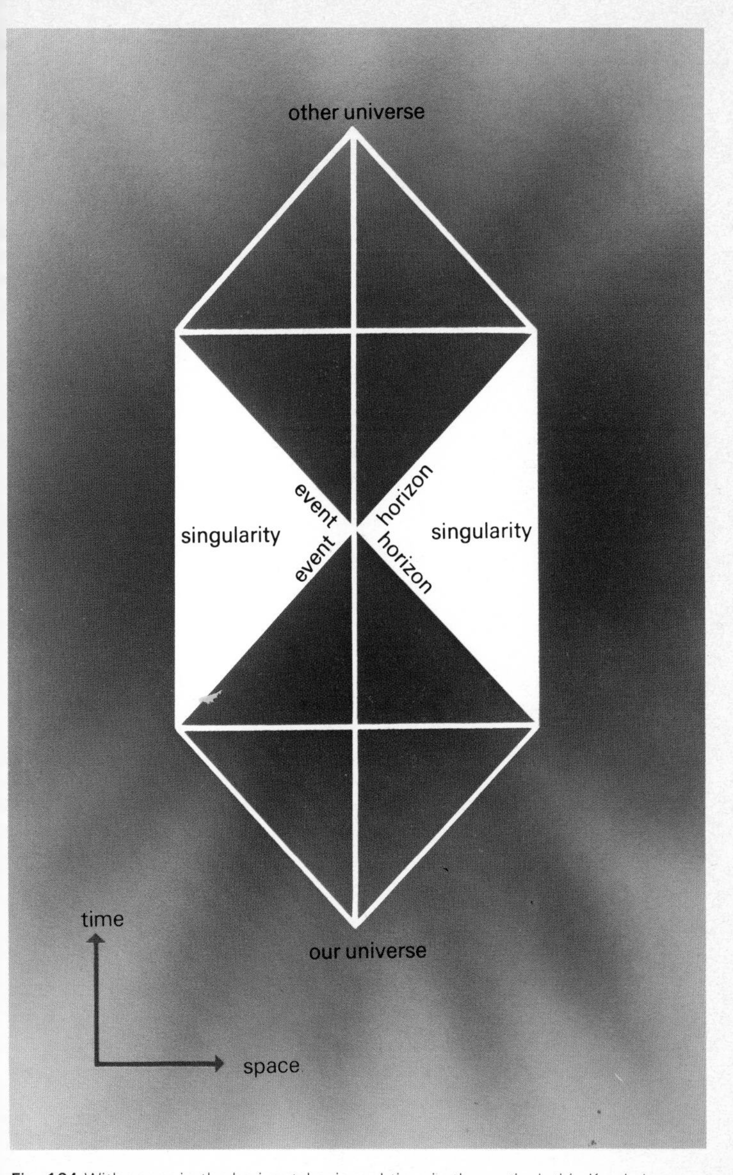

Fig. 134 With space in the horizontal axis and time in the vertical, this Kruskal diagram of the Schwarzschild solution suggests that it may be possible to move matter from our universe to another, or vice versa. This becomes impossible due to the singularity, and matter moving from our universe to another would have to pass through the space between the event horizon and the singularity.

As a result the suggestion has been made that a black hole at one epoch could conceivably manifest itself as a white hole at another time and in a different region of space, and feed material and energy from one point to the other. To some persons this has seemed to be the ideal answer to the problems posed by the highly energetic, tiny sources of radiation which are required to explain the phenomena observed in QSOs. These might, on this speculative theory, therefore be the sites of white holes.

However there are additional complications. In examining the Schwarzschild model we neglected the existence of the body which formed the singularity, something which it is hardly permissible to do. This body effectively 'blocks' the wormhole, and prevents the transfer of energy from one side to the other. But another effect comes into play – that of rotation. The unmodified Einstein field equations cannot incorporate the complication of rotation of the collapsing body, yet calculations indicate that a non-rotating black hole can never be formed. This failing led to the more sophisticated, Kerr solution of the field equations, which does incorporate rotation. Here the rotating collapsar is surrounded by the event horizon as in the Schwarzschild model, but additionally, there is an outer surface known as the 'stationary (or static) limit' where a particle must move at the speed of light to appear stationary. In theory, material objects and energy can escape from the region between the stationary limit and the event horizon.

We now have a rather more complicated model **(Fig. 135)** which does appear to permit exchange of material between universes. However the story is by no means over, as the latest calculations seem to show that even rotating black holes may not be the answer, and that perhaps it is just not possible for material or energy to escape.

It should be pointed out that all this discussion is built upon rather shaky foundations. We really have no justification at all for assuming the existence of any other space-time continuum, or 'universe'. We can certainly never obtain any evidence that it is really there, and although the solutions of the space–time field equations may permit other universes to exist, it is not obligatory for them to do so. Although present theories do predict the existence of black holes, all the other phenomena – white holes, wormholes, other universes, and space–time bridges – could well be pure figments of the imagination – in fact they probably are! It is almost certainly better to seek elsewhere for an explanation of the sources of the energy of QSOs, although at the moment a satisfactory mechanism has yet to be found.

One final point may be made before leaving this subject. If energy could be transferred between universes, or between regions of this single universe, this fact implies an asymmetry which is unacceptable to steady state theories. Similarly, mass and energy transfer would completely invalidate the Hubble distance/velocity relationship for distant regions of the universe.

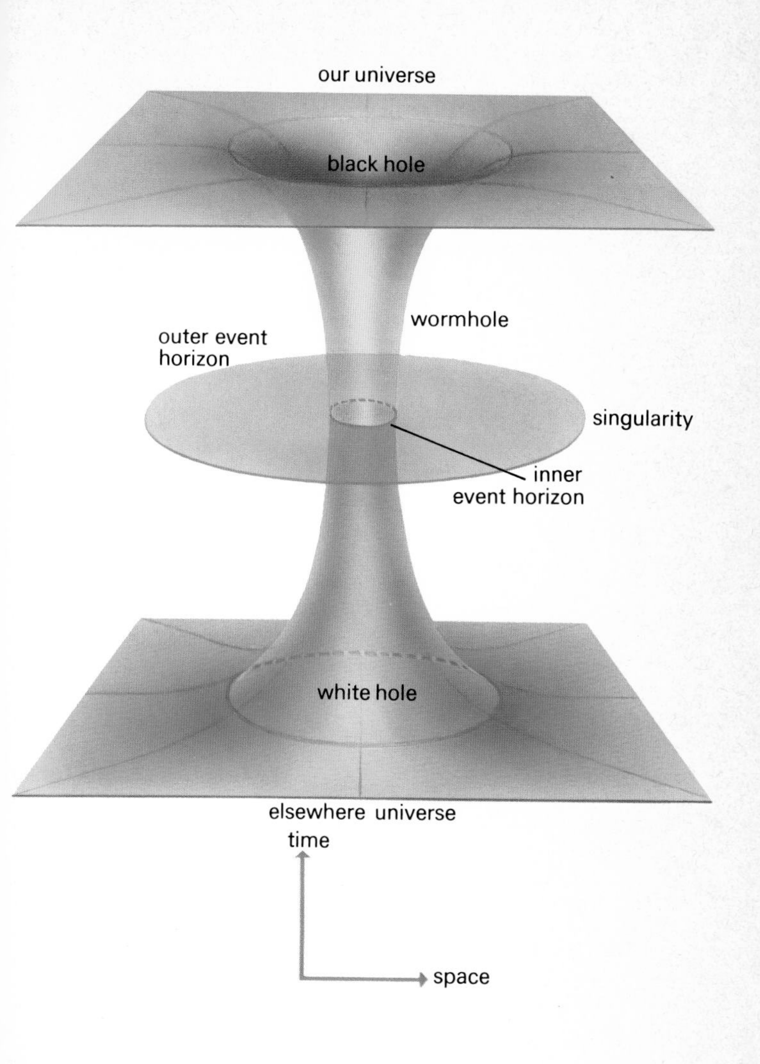

Fig. 135 Application of the Kerr solution to Einstein's field equations reveals the possibility (in theory) of moving matter from one universe to another down a black hole, along a wormhole and out from a white hole. This assumes an asymmetric, rotating collapsar.

Curvature of the universe

Clearly it is necessary to obtain an appreciation of the shape of the universe before finally deciding whether it is evolutionary or steady state and, if it is evolutionary, whether it will continue to expand for ever, reach a point of static equilibrium or eventually collapse back to a cosmic egg. Three models of geometric relationship to space and time are possible. The first adopts the classic assumption of Euclidean geometry that space is flat with straight line coordinates, where the surface of a sphere with radius r is $4\pi r^2$ and the sum angles at the corners of a triangle always equal 180°. The second postulate would assume a positive curvature to space, and in this situation the summed corners of a triangle would be greater than 180°. In the third consideration, a universe of negative curvature, the summed angles of a triangle always equal less than 180°. In other words, the three divergent geometries exhibit shapes reminiscent of a flat surface, a sphere and a saddle respectively.

The implications for cosmology are that a Euclidean universe would display a uniform distribution of galaxies per unit volume irrespective of the area observed, a universe with positive curvature would show a depleted number of galaxies with increasing distance while a negatively curved universe would show an increasing number of galaxies with increasing distance. Radio surveys of distant parts of the universe, conducted within the past decade, appear to indicate that there are fewer galaxies at remote distance than in our own region of space (after corrective factors are applied to accommodate the lost detection of intrinsically dim objects). This indicates a positive curvature to the universe and has a profound significance for the ultimate fate of the universe.

It has already been established that the curvature of the universe in space and time is determined by the distribution of matter and because of this the shape is directly related to the expansion of the universe. Investigations of very distant objects, for example by source counts, indicate that the linear relationship between distance and velocity assumed in the Hubble 'constant' does in fact seem to begin to break down at positions which are very remote. The value has changed with the passage of time since the primordial fireball. This obviously implies that the universe is truly an evolving one, and is additional evidence against the steady state theory.

Recessional velocities alone are not sufficient to determine whether the universe will fly apart for ever, will reach a static equilibrium, or will eventually collapse. This is dependent upon the total mass in the universe, and can be related to the mean density of the universe. Unfortunately this is very difficult indeed to determine. Various estimates can be made of the mass within galaxies and clusters of galaxies, but the amount of intergalactic matter and the presence of non-luminous mass is well-nigh impossible to determine. The best

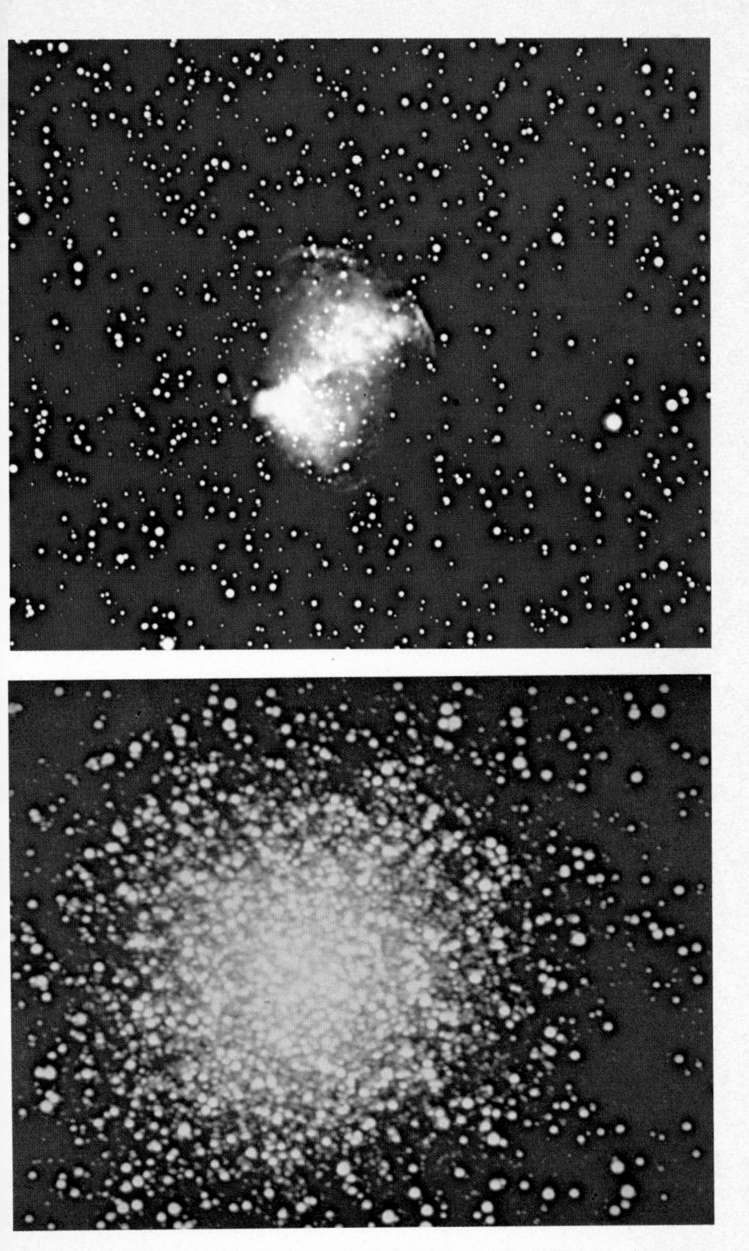

Fig. 136 (top) The 'Dumb-bell' planetary nebula M27, and **Fig. 137 (above)** the globular cluster M13.

estimates which have been made up to the present time seem to suggest that the mean density of the universe is not sufficient to 'close' the universe. It may therefore be expected to expand for ever. However the uncertainties involved, especially with regard to the mass held within black holes and as intergalactic matter, mean that there still remains the possibility that the universe does indeed contain sufficient mass for it to suffer an eventual collapse.

The presence of the background microwave radiation indicating the presence of the 2·7° thermal residue from a cosmic fireball, the mass/density change in space and time, the positive curvature of space and the change in linear relationship of distance and velocity on the Hubble law, all point towards an evolving universe that began about $1{\cdot}4 \times 10^{10}$ years ago which will, at some point in the future, spend its expansion and fall back to a dense cosmic egg.

The theory of an oscillating universe gains support from nearly every new astronomical discovery, and while retaining an open mind on the possibility of theoretical reversal it seems impossible to envision a situation where the steady state view can ever again receive credibility.

It is difficult to speculate on the situation which will emerge after the ultimate collapse of the universe. In a black hole, mass, charge and momentum are conserved because the phenomenon is attached to the universe. In the state of final collapse not even these properties will remain for there will be no field equation (universe) in which to place the solution. Whatever happens beyond the final singularity will cause the 'next' universe to be built from particles, atoms and molecules of profoundly different structure.

Perhaps the universe is destined to turn itself down its own black hole into the white hole of another universe to which it is attached, and in this case there would be a broader field equation to which the solution could be attached. Alternatively, the collapse of matter into existing black holes in the universe may flow through the wormhole connected to the universe yet to emerge from the residual debris of our own.

The use of satellites carrying instruments that explore parts of the electromagnetic spectrum that are not accessible from the Earth's surface has brought about a vast increase in our knowledge of the highly energetic phenomena occurring inside galaxies and quasars, and has given us a far more detailed picture of the processes occurring within them and of how they evolve. This, in turn, gives us a few more clues as to their formation and to the overall structure of the universe.

Some early satellites carried out general surveys of the whole sky. More recent, and forthcoming, satellites are designed to investigate individual objects in detail. The giant telescope of the Hubble Observatory is an example, and will give pictures of astronomical objects at optical wavelengths with an amount of detail unobtainable from the ground.

Fig. 138 The highly successful, Netherlands/UK/USA Infrared Astronomy Satellite (IRAS), surveyed the whole sky at infrared wavelengths, detecting many new and interesting objects to be examined in detail by future missions. Such observations are impossible from the ground.

The Development of Celestial Observation

Aubrey holes

It is impossible to know for certain when Man first began to observe the passage of Sun, Moon and planets across the background of stars, but it is generally accepted that some of these were deified, and that certain prehistoric structures had ritualistic significance. Among these, one of the principal sites is Stonehenge, some 4700 years old. Archaeological and astronomical study has shown that complex alignments predated the major stone assemblies, which were erected about 4000 years ago. The site is one of the earliest to show signs of being more complex than necessary for purely religious purposes.

Although conjecture still surrounds the level of complexity of the original layout it seems likely that the location of the site, the design of the first rings and the setting up of principal stone markers all point towards the considered preferences of priest-astronomers.

The first significant point about Stonehenge is that the four primary stone markers that form a broad rectangle **(Fig. 139)** known as stones 91, 92, 93 and 94, are so arranged that midwinter moonset, midsummer moonrise, midsummer sunrise and midwinter sunset positions are placed at 90° to each other; this is the only point of latitude where such an alignment could be positioned. Moreover, additional stone markers point to the extreme positions of midwinter moonrise as it oscillates through its 18·6-year cycle.

An almost perfect ring of 56 holes, named after the noted antiquarian Aubrey, may have provided a means of measuring three such metonic cycles (18·6 + 18·6 + 18·6 = 55·8). The prime function of the observatory seems to have been to mark the midsummer sunrise position; the orientation of ditches, banks, entrance, approach avenue and (much later) sarsen stones emphasized this alignment.

During a flurry of building some 700–800 years after the initial development of the site, five trilithon 'arches' were set up in an inverted 'U' configuration **(Fig. 140)**. Each stone, weighing between 30–50 tonnes was set up so that prime Sun and Moon rise and set positions could be observed through respective gaps in the trilithon uprights. An outer circle of 30 sarsen stones directed the lines of sight towards the horizon. One member of the sarsen circle is only half the width of the rest making 29·5 standard sarsen widths – the number of days in one lunation.

About 100 years later, two sets of holes were dug encircling the outer sarsen ring with 30 on the outer and 29 on the inner circle. The sum total equals two lunations with an error of only 52 min/year. Finally, a set of 19 smaller bluestones arranged in a horseshoe within

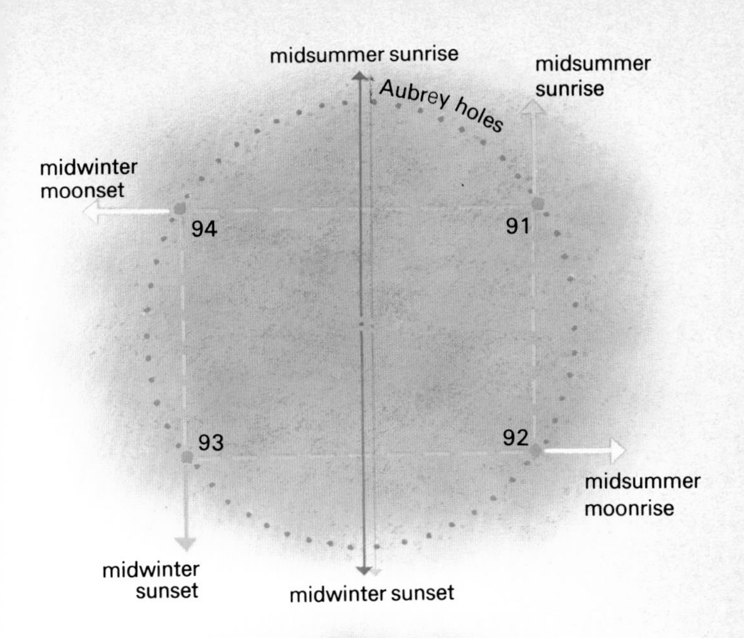

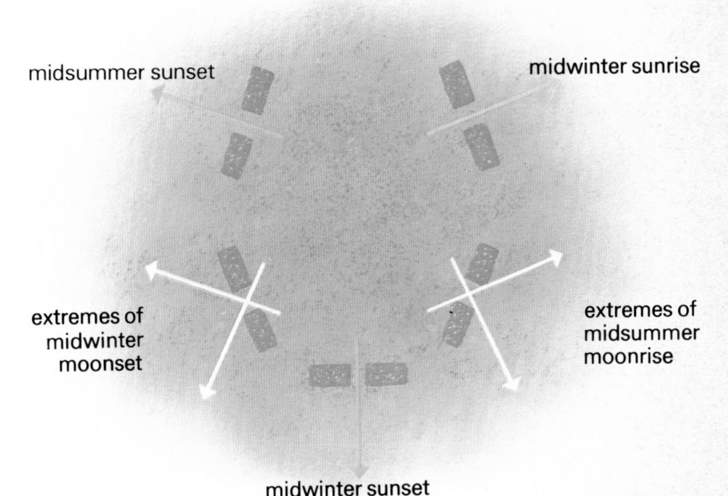

Fig. 139 (top) Nearly 4700 years ago, the first workings at Stonehenge, Wiltshire, England, provided early man with a means of marking lunar and solar rise and set positions. Here, the four markers are shown with the 56 Aubrey holes. **Fig. 140 (above)** Later additions to Stonehenge, in the form of massive sarsen stone trilithons, could have been used to align prime Sun and Moon rise and set positions as viewed from the inner face of each pair. (The diagram only shows the general arrangement of respective angles.)

the trilithon position and a set of 59 bluestones forming a circle outside the trilithons, could have been used to count intervals between extremes in midwinter moonrise positions (18·6 years) and two lunations (59 days) respectively.

Contemporary with the earliest workings at Stonehenge (2700 BC), Egypt displays significant artifacts from an era when priest-astronomers held influential positions in the cultural hierarchy of the day. The Great Pyramid of Cheops is aligned with the celestial north pole and there is evidence to support the view that widespread preoccupation with the movement of Sun, Moon and stars dominated pre-dynastic philosophy. Not until the Greeks, 24 centuries later, did observation lend itself to recorded theories on solar system geometry.

Aristarchus proposed a system with the Sun at the centre although the idea was lost until the 16th century AD. It was left to Ptolemy, in the 2nd century AD, to bequeath the geocentric theory that the Earth was at the centre of the universe and for 1400 years this belief reigned supreme, carried forward by early Christian dogma which effectively stifled reason and retarded the development of astronomical science. In the Ptolemaic system, planets moved round Earth with movement accounted for by epicycles **(Fig. 141)** and stars occupied a shell beyond Saturn (believed then to be the outermost planet).

In 1530, Copernicus rejected the accepted ideas and wrote a treatise

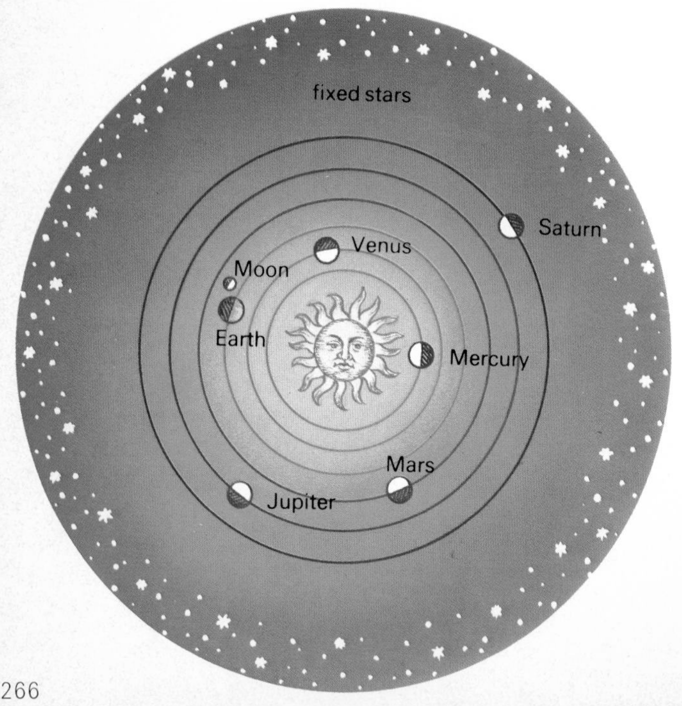

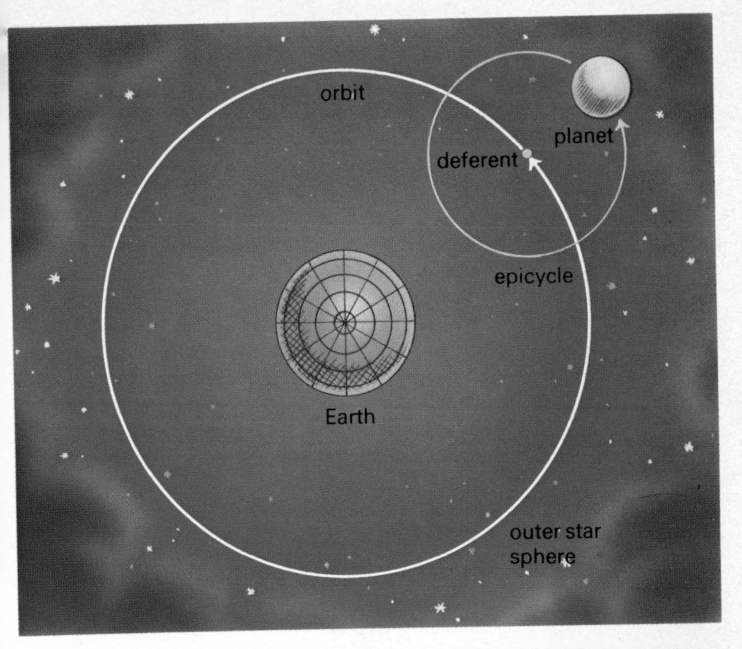

Fig. 141 (top) The Ptolemaic concept of the solar system. (For clarity only one planet is shown moving in epicycles: also, the Sun and the Moon move in the same way.) **Fig. 142 (left)** The Copernican system placed the Sun at the centre of the solar system with the planets, including Earth, moving round in separate orbits. **Fig. 143 (above)** Amateur observatories such as this are used by many enthusiasts.

stating that the Sun was at the centre of the universe with the planets, including Earth, orbiting it **(Fig. 142)**. Building on this theory Brahe, Kepler and Galileo refined the theory to one in which the planets moved in elliptical orbits about the Sun, and by the 17th century it was clear that the Church's teaching would have to change radically.

Finally, in the late 17th century, Newton laid down the fundamental laws of gravitation, but it was not until 1835 that the then Pope allowed Roman Catholic acceptance of the Copernican theory. Such ideological whims of religious and philosophical leaders have certainly influenced and indeed controlled the development of astronomical theory. From the times when astronomical objects were deified, and regarded as having direct influence over human affairs, through the period when preferred religious beliefs dictated which ideas could be advanced, we have now reached the stage when theoretical astronomy is itself a philosophically satisfying and worthwhile pursuit, which is uncovering basic truths about the Earth and the universe. Actual observation of the heavens has obviously been of primary importance in bringing this situation about, and all forms of telescopic observation, amateur or professional, have their own value.

Telescopes – on Earth and in space

The 17th-century invention of the optical telescope gave the amateur observer greater scope and application **(Fig. 143)**, and the most commonly used ones have been described on p. 6. However, within the last 100 years the camera has been married to the telescope to reveal images quite invisible to the human eye. Unlike the eye, prolonged exposure of a suitable photographic plate gives an increasing amount of detail through the resolution of fainter and fainter objects. Also, the exposed plate, when developed, provides a useful and permanent record of the sky at a specific time, and the image can embrace a wider range of wavelengths than those interpreted by the human eye.

Most modern, large telescopes are primarily designed for use with a photographic plate but small telescopes can be put to important use for recording stellar or planetary objects and for the discovery of comets and novae. For wide-field photography the Schmidt telescope is most effective (see p. 14).

In the past half-century a new tool for astronomy, the radio telescope, has been made available so that radio waves can be collected and analysed to provide insight into many of the energetic activities developing within distant sources **(Fig. 144)**. The radio telescope sometimes makes use of a steerable dish for collecting the faint energy waves and focussing them on to a horn that routes the converging rays through a wave guide to the receiver. Others, like the

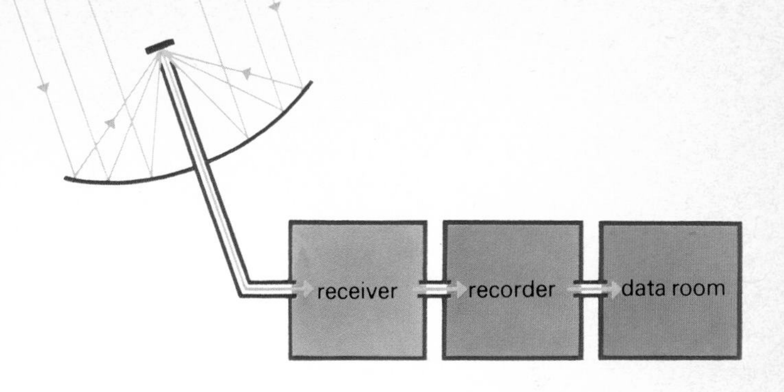

Fig. 144 (top) Radio astronomy has added a new and important dimension to observation of the universe. The basic operating principle is shown here, whereby radio waves are received and focussed by a concave dish and then processed through a receiver, a recorder and a data room. **Fig. 145 (above)** The Arecibo radio telescope effectively makes use of a natural bowl in the earth with a moveable horn/receiver hung between three towers.

Arecibo array in **Fig. 145**, are massive, fixed antennae, and here the 304-metre-diameter telescope has a moveable horn/receiver assembly which seeks out radiation reflected from different angles.

The ultimate use of the telescope for planetary observation is where a camera is transported to the vicinity of another planet so that close imaging of the surface can reveal small-scale features **(Fig. 146)**. A typical example is the type of camera system employed on Mariner missions to Mars, Venus and Mercury in the 1960s and 1970s. This was brought to perfection in the orbiter of Project Viking in 1976.

The orbiter carried two cameras, each one consisting of a 38 mm selenium vidicon tube and a catadioptric Cassegrain lens with a focal length of 475 mm. A photoelectrically formed image was scanned by an electron beam which converted the image into electrical signals. Each scan line carried 1182 pixels and each frame was made up from 1056 lines for a total pixel count per picture of $1{\cdot}248 \times 10^6$. In transmission, each pixel was made up from 7 bits with each picture providing a resolution of 50–100 m from an altitude of 1500 km.

As an extension of the investigation of the universe with unmanned satellites, two Orbiting Astronomical Observatories were launched, in 1968 and 1972, carrying ultraviolet telescopes and X-ray spectrometers **(Fig. 147)**. These are typical of the many astronomy satellites now orbiting the Earth. As already stated on p. 248, distance in the universe depends upon the Hubble constant, and the Space Telescope **(Fig. 148)** has the best opportunity for moving the radius

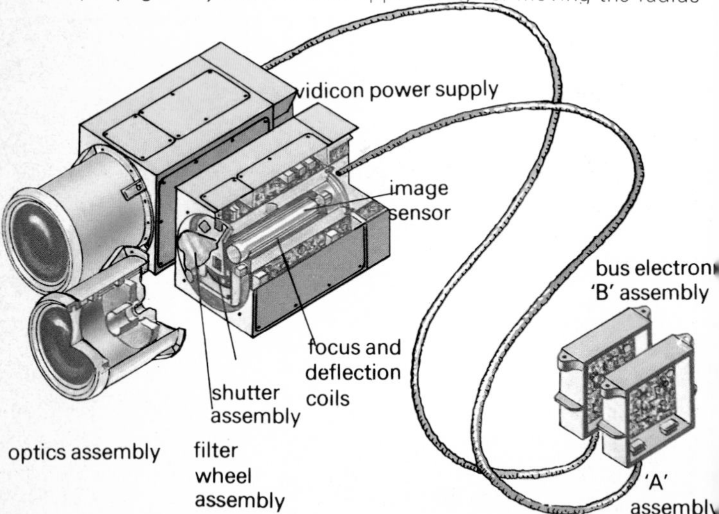

Fig. 146 Cameras carried by unmanned space vehicles provide much valuable information about the topography of planets in the solar system. Without the information sent back by radio signals our knowledge of the evolution of neighbouring worlds would be largely speculative.

Fig. 147 (top) The Orbiting Astronomy Observatory spacecraft was the first major step towards probing portions of the electromagnetic spectrum denied to surface observers. **Fig. 148 (above)** The Space Telescope will be launched by Shuttle, a reusable delivery system that reduces the cost of sending payloads into orbit.

of observation out to the limits of the universe. From

$$\frac{299\,790}{90} = 3331$$

where the velocity of light in km/s is divided by the Hubble constant (see p. 248), the limits to the observable universe are set at a radius of 3331×10^9 pc or $1{\cdot}1 \times 10^{10}$ light years. The Space Telescope is capable of resolving objects of magnitude 27 (see p. 33) and thus carries the science of celestial observation to limits undreamt of by Galileo less than four centuries ago **(Figs 149 and 150)**.

Meanwhile, there is considerable work for the amateur to do and **Fig. 151** shows the kind of result which might be expected with modest photographic equipment. Although many stellar phenomena are of interest to the amateur, notably variables, within the solar system important contributions can still be made.

Among these are the continued surveillance for new comets, observation of the Moon for Transient Lunar Phenomena (see p. 161). observation of the atmospheric changes in Mars' atmosphere and induced visual changes in the cloud structure of Saturn from the magnetospheric tail of Jupiter. Also, careful observation and recording of meteor showers contributes to a well-ordered understanding of the way cometary debris is collected by the orbital passage of these remarkable emissaries from the edge of the solar system.

Today, as in the day of Galileo, astronomy advances on the strength of amateur and professional support, and the mysteries which are so profound today are corner stones of tomorrow's unwritten science.

Fig. 149 (left) The shuttle opens the possibility of the wider application of space activity. **Fig. 150 (top)** One of the highly successful Voyager spacecraft that visited Jupiter, Saturn, Uranus and Neptune. **Fig. 151 (above)** The Orion Nebula taken by an amateur through a 215 mm reflecting telescope.

Tables

Table 1 The 20 brightest stars

Name of star	Designation	Apparent Mag	Absolute Mag
Sirius	α Canis Majoris	−1·43	+1·5
Canopus	α Carinae	−0·73	−7·4
—	α Centauri	−0·27	+4·1
Arcturus	α Boötis	−0·06	−0·3
Vega	α Lyrae	0·04	+0·5
Capella	α Aurigae	0·09	−0·6
Rigel	β Orionis	0·15	−8·2
Procyon	α Canis Minoris	0·37	+2·7
Achernar	α Eridani	0·58	−1·3
Betelgeuse	α Orionis	0·4	−5·9
Hadar	β Centauri	0·66	−4·3
Altair	α Aquilae	0·8	+2·4
Aldebaran	α Tauri	0·85	−0·6
Acrux	α Crucis	0·87	−3·4, −2·9
Antares	α Scorpii	0·98	−5·0
Spica	α Virginis	1·00	−2·9
Formalhaut	α Piscis Austrini	1·16	+2·0
Pollux	β Geminorium	1·16	+1·0
Deneb	α Cygni	1·26	−6·2
—	β Crucis	1·31	−4·5

Table 2 Planetary data

	Mercury	Venus	Earth
Mean distance from Sun (million km)	57·9	108·2	149·6
Mean distance from Sun (Au)	0·39	0·72	1·0
Period of revolution	88 days	224·7 days	365·3 days
Eccentricity of orbit	0·206	0·007	0·017
Inclination of orbit	7°	3·4°	0°
Inclination of axis	<28°	3°	23·45°
Rotation period	59 days	−243 days	23 h 56 min
Orbital velocity (km/s)	47.9	35	29·8
Diameter at equator (km)	4880	12 104	12 756
Oblateness	0	0	0·003
Density (g/cm³)	5·4	5·2	5·5
Volume (× Earth)	0·06	0·88	1
Mass (× Earth)	0·055	0·815	1
Atmospheric pressure (bars)	10^{-6}	90	1
Satellites discovered and named	—	—	1

Microgram	γ	Pi	π
Micrometre	μm	Radius	r
Millibar	mb	Right Ascension	RA
Millimetre	mm	Second	s
Minute	min	Volt	V
Parsec	pc	Watt	W
Parts per million	ppm		

Glossary

Absolute magnitude: The amount of light received from a star at a distance of 32·6 light years (10 parsecs)
Achromatic lens: Lens free from chromatic aberration due to two glass lenses placed together compensating for the dispersion
Ångström (Å): Unit of length equal to 10^{-10} metres
Angular momentum: The product of the moment of inertia and the angular velocity
Aphelion: The most distant point on an orbit about the Sun
Apparent magnitude: The amount of light received from a star as observed from the Earth
Autumnal equinox: A point on the celestial sphere at which the ecliptic crosses the equator
Bits: Binary digits
Bolometric magnitude: The total amount of energy emitted by a star
Chandrasekhar limit: A value of 1·4 solar masses above which the degeneracy pressure in a star is unable to counter the effect of gravitational attraction
Chromatic aberration: Light dispersed into coloured bands due to refraction through a lens
Collimating lens: The optical element that brings light rays to a parallel path before they strike a prism or a spectroscope
Eccentricity: A measure of the degree of ellipticity in an orbit, defined as the ratio between one half of the major axis and the distance between the centre of the ellipse and one focus
Ecliptic: The plane of the Earth's orbit about the Sun which, when presented on the celestial sphere, becomes the path of the Sun about the Earth prescribing an imaginary line inclined 23° to the Earth's equator
Electromagnetic radiation: Radiation comprising waves of energy travelling at the speed of light ($2{\cdot}9979 \times 10^{10}$ cm/s)
Electron: An elementary particle having a negative charge and occupying valence shells around an atomic nucleus
Event horizon: A theoretical sphere surrounding a black hole where the velocity of escape from the collapsar exceeds the speed of light

Eyepiece: Lens or system of lenses used to observe an image from the objective
Focal plane: The point at which light is made to converge from the objective; the distance between the two is the focal length
Frequency: Wave motion equal to the velocity divided by the wavelength
Heliosphere: The region dominated by the solar wind
H–R diagram: The Hertzsprung–Russell chart relating all stars to a scale measuring mass versus luminosity
Inverse square law: A law which states that the intensity of an effect varies inversely as the square of the distance
Ionization: The formation of ions, electrically charged atoms, produced by removing or adding electrons to the valence shells surrounding a nucleus
Isotopes: Atoms of the same element which have differing numbers of neutrons and, hence, a different mass
Isotropy: Uniform distribution of matter in all directions
Libration: An apparent oscillation in the visible hemisphere of the Moon caused by the elliptical path of the Moon about the Earth
Light year: The distance travelled by light in one year
Luminosity: The amount of light emitted by a star, also known as absolute magnitude
Magma: Molten material usually found beneath the outer crust of a planet
Neutron: An elementary particle found in the nucleus of every atom except hydrogen having neither a positive nor a negative charge
Nova: A form of binary star where transfer of material between the stars produces sudden outbursts in luminosity
Nuclear fission: A nuclear reaction whereby an atomic nucleus splits into two equal parts, emitting nuclear energy
Nuclear fusion: A nuclear reaction between two atomic nuclei as a result of which a heavy nucleus is formed and nuclear energy is released
Oblateness: The degree of 'flattening' in any body
Objective: Lens (in a refractor) or mirror (in a reflector) focussing light to a focal plane
Parallax: A shift in the apparent position of a body due to the motion of the observer
Parsec: Unit of distance equal to a parallax of one arc second (3·26 light years)
Perihelion: The closest point to the primary of an orbit about the Sun

Photon: A quantum of electromagnetic radiation sometimes regarded as an elementary particle
Planetismals: Large asteroidal-sized objects present in the early phase of the solar system's evolution
Precession: The result of a coupled motion imparted to a spinning body which causes rotation around the axis of spin prescribing a cone in the process
Proton: An elementary particle with a positive charge equal in magnitude to the negative charge of the electron and found in the atomic nucleus
Reflector: A telescope in which light is reflected from a primary mirror back to an eyepiece
Refractor: A telescope which refracts light through a main lens to a specific point (the focal plane) from where a magnifying lens presents the image to the eye
Resolution: The ability to distinguish between two separate objects of equal brightness
Seismic activity: The movement of solids or fluids within a terrestrial-type planet which causes measurable shock waves
Sidereal period: The actual period of one 360° revolution of a planet about the Sun
Singularity: The hypothetical centre of a black hole into which all incoming matter is drawn
Spectroheliograph: An instrument for observing the Sun in the light of one spectral line
Spectroscope: Instrument for observing spectra
Spectrum: Separation of light into its component colours when passed through a prism
Spherical aberration: Distortion caused by different rays of light being brought to different focal planes by a lens or mirror
Supernova: The disruption of a star's equilibrium so that a high fraction of its mass is released in an explosion
Synodic period: The period between successive conjunctions of a planet as observed from the Earth
Transient Lunar Phenomena (TLP): The obscuration of lunar surface features by some endogenic event, usually on a localized scale
Vernal equinox: The primary point on the celestial sphere at which the ecliptic crosses the equator; reached by the Sun about 21 March each year
Wavelength: Determined by dividing the velocity of light by the appropriate frequency and expressed in metres
Widmanstätten: A characteristic pattern observed in a cut and polished iron meteorite suggesting slow cooling from a molten state

Greek alphabet

Greek Alphabet

Alpha	α	Iota	ι	Rho	ρ
Beta	β	Kappa	κ	Sigma	σ
Gamma	γ	Lambda	λ	Tau	τ
Delta	δ	Mu	μ	Upsilon	υ
Epsilon	ε	Nu	ν	Phi	ϕ
Zeta	ζ	Xi	ξ	Chi	χ
Eta	η	Omicron	o	Psi	ψ
Theta	θ	Pi	π	Omega	ω

Further reading

Audouze, J. & Israël, G., *The Cambridge Atlas of Astronomy*, 2nd edn. Cambridge University Press, Cambridge, 1988.

de la Cotardière, P., (ed), *Larousse Astronomy*. Hamlyn, London, 1987.

Dunlop, S.R., *Astronomy: A Step-by-step Guide to the Night Sky*. Hamlyn, London, 1985.

Kippenhahn, R., *100 Billion Suns* (trans. J. Steinberg). Weidenfeld & Nicholson, London, 1983.

Kippenhahn, R., *Light from the Depths of Time* (trans. S. Dunlop). Springer-Verlag, New York, 1987.

Menzel, D.H. & Pasachoff, J.M., *A Field Guide to the Stars and Planets*, 2nd edn. Houghton Mifflin, Boston, 1983.

Norton, A.P., *Star Atlas and Reference Handbook*, 17th edn. Gall and Inglis, Edinburgh, 1978 (*not* later Longman edn).

Peltier, L., *Guide to the Stars*. Astromedia, Milwaukee, and Cambridge University Press, Cambridge, 1985.

Ronan, C.A.R. (ed.), *Amateur Astronomy*, rev'd edn. Hamlyn, London, 1988.

Verschuur, G.L., *The Invisible Universe Revealed*. Springer-Verlag, New York, 1986.

Whipple, F.L., *The Mystery of Comets*. Smithsonian Institution Press, Washington, D.C., 1985.

Index

Page numbers in bold type refer to illustrations

Amalthea 196
ammonia 192, 204, 208, 217
Ångström, A. J. 18
  unit 18
Anubis 94
Apollo space mission 161
Argonauts 90
Ariel 204
Aristarchus 266
asteroids (*see also* minor planets) 212, **212**, **213**, 226
astronomical unit (AU) 49, 146
Atlas 200
Aubrey, J. 264
Australia 226
Baade, W. 32, 46
Barnard, E. E. 145
Barlow lens 14, **15**
Bayer, J. 86, 98, 106, 112, 114, 122, 126, 138
Bell Telephone Laboratories 253
Berenice 100
Big Bang theory 252, 253
binary system 61, 62, 229
black dwarf 40
black hole 41, **43**, 57, 60, 61, 62, 104, 230, 250, 258, 262
blue shift 49
Brahe, T. 100, 214, 268
British Astronomical Association 5
British Interplanetary Society 5

Cacciatore, N. 106
calcium 44
Callisto 196, 198
Calypso 204
carbon 55
carbon dioxide 176, 190, 217
carbon-nitrogen cycle 40, 41, 44
Cassegrain, N. 270
  reflector 10, **13**, 21, 270
Cassini division 200
Chandrasekhar, S. 57
Charon 210
Cheops, Great Pyramid of 268
chromosphere 36
comets 214, **215**, **217**, **218**, **219**, **220**, **221**, **223**
  Arend-Roland 215
  Encke 222, 223
  Halley's 216
  Kohoutek 220
  orbits 218–222
  radio emission 218
  Seki 215
  structure 215–218
conjunction, inferior 150
  superior 150
constellations 76–145, **78–85**
  Andromeda 29, 32, 46, 86, **87**, 114, 126, 128
  Antlia 86, **87**, 112, 130, 140
  Apus 86, **87**, 88, 98, 100, 122, 124, 138
  Aquarius 88, **89**, 96, 106, 108, 126, 128, 134
  Aquila 88, **89**, 96, 106, 110, 132
  Ara 86, 88, **89**, 102, 122, 124, 132, 136, 138
  Argo Navis 96, 112, 130, 140
  Aries 90, **91**, 126, 128
    First Point in 74, 128
  Auriga 90, **91**, 92, 110, 118, 126
  Boötes 90, **91**, 94, 100, 102, 110, 138, 142
  Caelum 92, **93**, 100, 106, 108, 112, 128
  Camelopardus 92, **93**, 118, 126, 138
  Cancer 92, **93**, 110, 112, 114, 118
  Canes Venatici 90, 94, **95**, 100, 138
  Canis Major 94, **95**, 100, 116, 120, 130
  Canis Minor 92, 94, **95**, 110, 120
  Capricornus 88, 96, **97**, 106, 120, 128, 132
  Carina 96, **97**, 98, 112, 122, 128, 130, 140, 142
  Cassiopeia 86, 92, 96, **97**, 114
  Centaurus 86, 96, **97**, 98, 100, 104, 112, 118, 122, 140
  Cepheus 29, 98, **99**, 114, 140
  Cetus 88, 90, 98, **99**, 108, 128, 134
  Chamaeleon 86, 96, 98, **99**, 120, 121, 122, 142
  Circinus 86, 100, **101**, 116, 122, 138
  Columba 94, 100, **101**, 116, 128, 130
    Noae 100
  Coma Berenices 90, 94, 100, **101**, 114, 138, 142
  Corona Australis 102, **103**, 132, 136
    Borealis 90, 102, **103**, 110
  Corvus 102, **103**, 104, 112, 142

Crater 102, 104, **105**, 112, 114, 132, 134, 142
Crux 98, 104, **105**, 122
Cygnus **66**, **67**, 104, **105**, 114, 118, 126, 142
Delphinus 88, 106, **107**, 108, 126, 132, 142
Dorado 106, **107**, 112, 120, 128, 130, 138, 142
Draco 90, 92, 106, **107**, 110, 118, 138, 140
Equuleus 88, 106, 108, **109**, 126
Eridanus 108, **109**, 112, 116, 124, 126
Fornax 108, **109**, 126, 134
Gemini 92, 110, **111**, 118, 120, 124
Grus 110, **111**, 114, 120, 126, 128, 134
Hercules 88, 90, 102, 110, **111**, 118, 122
Horologium 106, 108, 112, **113**, 130
Hydra 86, 92, 102, 104, **112**, **113**, 114, 116, 120, 130, 134, 142
Hydrus 106, 108, 112, **113**, 120, 122, 130
Indus 110, 114, **115**, 120, 122, 124, 136
Lacerta 86, 114, **115**, 126
Leo 92, 100, 104, 112, 114, **115**, 116, 138, 142
Leo Minor 114, 116, **117**, 118, 134, 138
Lepus 94, 100, 108, 116, **117**, 124
Libra 112, 116, **117**, 118, 132, 142
Lupus 100, 116, 118, **119**, 120, 122. 132
Lynx 92, 110, 116, 118, **119**, 138
Lyra 110, 118, **119**, 120, 142
Mensa 98, 106, 112, 120, **121**, 122, 142
Microscopium 96, 110, 114, 120, **121**, 128, 132
Monoceros 94, 110, 120, **121**, 124, 234
Musca 86, 96, 98, 100, 104, 122, **123**
Norma 88, 100, 118, 122, **123**, 132, 138
Octans 86, 98, 112, 114, 120, 122, **123**, 124
Ophiuchus 88, 110, 116, 122, **123**, 124, 132, 134
Orion 108, 110, 116, 120, 124, **125**, 234
Pavo 86, 114, 122, 124, **125**, 136
Pegasus 86, 88, 106, 108, 114, 126, **127**, 128, 142
Perseus 86, 90, 92, 126, **127**
Phoenix 108, 110, 126, **127**, 134
Pictor 96, 100, 106, 128, **129**, 130
Pisces 86, 88, 126, 128, **129**
Piscis Austrinus 88, 96, 110, 120, 128, **129**, 134
Puppis 94, 96, 100, 112, 120, 128, 130, **131**, 140
Pyxis 86, 112, 130, **131**, 140
Reticulum 106, 112, 130, **131**
Sagitta 88, 132, **133**, 142
Sagittarius 88, 96, 102, 120, 122, 132, **133**, 136, **235**
Scorpius 88, 102, 116, 118, 122, 132, 134, **135**
Sculptor 88, 108, 110, 126, 128, 134, **135**
Scutum 88, 132, 134, **135**
Serpens 88, 90, 102, 110, 116, 122, 124, 132, 134, **135**, 142
Sextans 104, 112, 114, 134, **135**
Taurus 90, 108, 110, 124, 126, 136, **137**, 234
Telescopium 88, 102, 114, 124, 132, 136, **137**
Triangulum 86, 90, 126, 128, 136, **137**
Triangulum Australe 86, 88, 100, 122, 138, **139**
Tucana 110, 112, 114, 122, 126, 138, **139**
Ursa Major 90, 92, 94, 96, 100, 114, 116, 118, 138, **139**
Ursa Minor 92, 140, **141**
Vela 86, 96, 112, 130, 140, **141**
Virgo 90. 100, 102, 104, 112, 114, 116, 142, **143**
Volans 96, 98, 106, 120, 128, 142, **143**
Vulpecula 106, 118, 126, 132, 142, **143**
coordinates 72, 74
Copernicus, N. 148, **266**, 268
cosmic rays 262
coudé reflector 10, **12**, 21

Darwin, C. 248
declination 74, **75**
Deimos 184, 190

Dione 204
Doppler, J. C. 224, 246, 250
Drake, F. D. 234

Earth 72, **73**, **75**, **77**, 192, 200, 213, 214, 248, 277
  magnetic field 196
  orbit 146, 150
  velocity 224
Earth-Moon system 148, 166
earthshine 162
ecliptic 72, 76, 132
Einstein, A. 236, 256, 258
Enceladus 200
Epimetheus 200
equator 72, 76
equinox 72, 74
  vernal 128, 136
Eratosthenes 100
Euclid 260
Euphrates 108
Europa 196, 198

Fraunhofer, J. von. 18
  lines 18, 36

galaxies 228–263, **237**, **240**, **241**, **242**, **243**, **245**, **261**
  Andromeda 228, **235**, 236, 238
  elliptical 236, 244
  evolution 244–252
  irregular 236, 244, 246
  lenticular 236, 238
    radio 250, 252
    Seyfert 252
    spiral 236, 238, **239**, 244, 246, 249
      structure 236
Galaxy, The 228–263
  centre 230
  disc 230
  mass 230
  spiral arms 230–232, **233**
  stars 228, **229**, 230
  structure 230, 231, 236
Galilei, Galileo 7, 16, 198, 268, 272
Galle, J. G. 208
gamma rays 17, 262
Ganymede 196, 198
Ganymedes 88
Golden Fleece 90
gravitational collapse 248
  constant 60, 236, 260
gravity 56, 57, 58, 236

Habrecht 130
Halley, E. 90, 214
Helene 204
helium 40, 43, 44, 46, 54, 55, 192, 204, 208
helium-carbon cycle 54
Herschel, W. **12**, 204
Hertz, H. R. 17
Hertzsprung, E. 38
Hertzsprung-Russell diagram 38–46, **39**, **41**, 54, 55, 249, 250
Hevelius, J. 92, 114, 116, 118, 134, 142
Holbrook (USA) 224
Hubble, E. P. 236, 244, 248, 252, 258, 262
  classification 236, 238, **247**, 248, 250
  constant 248, 256, 260, 272
hydrogen 40, 43, 44, **45**, 46, 50, 54, 55, 145, 192, 193, 204, 208, 217, 234
hydrogen cyanide 218
hydroxyl 217
Hyperion 204

Iapetus 204
Io 196, 198
iron 44, 58

Janus 200, 204
Jupiter 36, 43, 44, 148, 190–198, **191**, **193**, **194**, **195**, 200, 204, 212, 213, 216, 222, 272
  atmosphere 192
  magnetosphere 193, 196, **197**, 204
  mass 148, 190, 198
  orbit 190
  red spot 192
  revolution 192
  rotation 192
  satellites 190, 196, **197**, 198, **199**, 204

Kepler, J. 26, 148, 268
  laws of 148
Kerr, F. J. 258, **259**
Kirin Province (China) 225

Lacaille, N. L. de. 92, 100, 108, 112, 120, 122, 128, 130, 134
Large Magellanic Cloud (LMC) 106, 112, 120, 138
Lequeux, J. 234
Le Verrier, U. J. J. 208
life (extraterrestrial) 62
light 8, **9**, 10
  chromatic aberration 10, 11

dispersion 8
gathering power 15
light year 28
local group of nebulae 236
luminosity 28, 29, 38, 41, 50, 55, 56

Magellan, F. 106
Mach, E. 235
magnesium 44, 54
magnification 14
magnitude 28, 29, 32, 34, 38
Mariner space mission 172, 270
Mars 180–190, **181**, **182**, **183**, **186**, **187**, **188**, **189**, 212, 213, 222, 270
atmosphere 184, 272
density 180
distance 180
gravity 180
Hellas region 184
mass, 148, 180
Olympus Mons 184, **185**
orbit 180
polar caps 184
rotation 180
satellites 184, 190, **191**
surface 184
temperature 184
Tharsis region 184, **185**
Mercury 171–179, 196, 212, 213, 270
composition 172, **173**, **174**, 175
density 172
mass 148
orbit 171, 172
revolution 172
rotation 172
temperature 172
Messier, C. 138, 146, 228
Messier Object
M3 (globular cluster) 94
M4 (globular cluster) 132
M6 (open cluster) 134
M7 (open cluster) 134
M9 (globular cluster) 124
M10 (globular cluster) 124
M11 (open cluster) 134
M12 (globular cluster) 124
M13 (globular cluster) 110
M14 (globular cluster) 124
M18 (open cluster) 132
M21 (open cluster) 132
M22 (globular cluster) 132
M23 (open cluster) 132
M24 (star cloud) 132
M25 (open cluster) 132
M26 (open cluster) 134
M27 (Dumbell Nebula) 142
M28 (globular cluster) 132
M29 (open cluster) 144
M30 (globular cluster) 96
M31 (Andromeda Nebula) 86, 228, 236
M33 (spiral galaxy) 138, 236
M34 (open cluster) 126
M42 (Orion Nebula) 145
M43 (nebula) 145
M44 (open cluster, Praesepe) 92
M48 (open cluster) 112
M49 (globular galaxy) 142
M51 (Whirlpool galaxy) 94
M52 (open cluster) 96
M54 (globular cluster) 132
M55 (globular cluster) 132
M56 (globular cluster) 120
M57 (Ring Nebula) 118
M58 (spiral galaxy) 142
M59 (elliptical galaxy) 142
M60 (globular galaxy) 142
M61 (spiral galaxy) 142
M63 (spiral galaxy) 94
M67 (open cluster) 92
M68 (globular cluster) 112
M69 (globular cluster) 132
M70 (globular cluster) 132
M71 (globular cluster) 132
M74 (spiral galaxy) 128
M75 (globular cluster) 132
M79 (globular cluster) 116
M80 (globular cluster) 132
M81 (spiral galaxy) 138
M82 (irregular galaxy) 138
M83 (spiral galaxy) 112
M87 (globular cluster) 142
M89 (elliptical galaxy) 142
M90 (spiral galaxy) 142
M97 (Owl Nebula) 138
M101 (spiral galaxy) 138
M103 (open cluster) 96
M104 (Sombrero Nebula) 142
M108 (spiral galaxy) 138
M109 (spiral galaxy) 138
meteorite 224–226
Hoba (South West Africa) 225
impacts 224
types 224, 225, **227**
meteoroid 222–224
bolide/fireball 224
meteors 222–224, 272
crater (Arizona) 224, **225**
showers, Beta Taurids, Leonids, Orionids 223
methane 204, 210, 217

methyl cyanide 218
Milky Way 88, 96, 106, 120, 134, 142, 144, 228–235
Mimas 200
minor planets 212, **212**, **213**
Ceres, Juno, Pallas 212
Miranda 204
mirror 10
Moon 62, 150, 161–175, **162**, **165**, 214
age 166
atmosphere 162
composition 166, **167**, **168**, **169**, **170**
craters 164, 166, **167**, 171
Clavius 164
Copernicus 164
Tycho 164, 171
density 166
distance 161
libration 161
maps **152**, **153**, **154**, **155**, **156**, **157**, **158**, **159**
index to 160, 161
mare 164
Mare Imbrium 164
Nectaris 164
Nubium 164
Serenitatis 164
Tranquillitatis 164
maria 171
mass 162
Montes Apenninus 164
Oceanus Procellarium 164
orbit 162, **163**
radius 162
rotation 161
seismic activity 171
surface 164, 166

nebulae 236, 238, 244
Coalsack (Northern) 144
(Southern) 104, 233
Crab 60, **64**, **65**, 136, 233
Horsehead **70**, **71**, 145
Horseshoe 132
Lagoon 132
North American 144
Owl 138, 142
Ring 120
Saturn 88
Trifid 132
Veil 144
Whirlpool 94
neon 54
Neptune 110, 146, 208, **209**, 210
satellites **208**, 210
Nereid 210
neutron star (collapsar) 40, 41, 58, 60, 230, 256
Newton, I. 6, 214, 268
NGC Object 2506 120
2244 120
2323 (= M50) 120
2808 96
5253 112
6067 122
6720 (= M57) 120
6779 (= M56) 120
6853 (= M27) 142
nickel 54
Nile 108
novae, 44, 46, 50, 52, **52**, 60, 120, 136, 230

Oberon 204
Observatory
Hale 21, **24**
High Energy Astronomy **263**
Orbiting Astronomical 270, **271**
Palomar 14
Paris 92
Odyssey 90
Olympus 88
Orion 145
Great Nebula in 234, **273**
oxygen 54, 55, 145, 217

Pandora 200
parallax 26
Pariiski, N. N. 234
parsec 28
period, sidereal 150
synodic 150
Phobos 184, 190
Phoebe 204
Piazzi, G. 106
planets, density 146, **147**
exterior 150
interior 150
mass 148
orbits 148, **149**, 150, **151**
revolution 148
rotation 148
Pluto 110, 146, 172, 175, 176, 210, **211**
Population I 32, 46, 47, **47**, 48, 228, **229**,
II 29, 32, 46, 47, **47**, 48, 228, **229**
precession 76, **77**
proton-proton reaction 40, **40**, 42, 43
Prometheus 200

proper motion 26
Ptolemy, C. 76, 86, 92, 102, 104, 116, 142, 148, 226, 266, **267**
Puck 204
pulsar 58, 60, **251**
Pultusk (Poland) 224

quasar **249**, 250, 251, 252, 256, 262

radio source 104, 250–252, 262
  Cygnus X-1 104
red giant 39, 52, 90, 126
Red Planet 180
red shift 49, 248
resolving power 15, 16
Rhea 204
Right Ascension (RA) 74, **75**
Röntgen, W. C. 17
Rosen, N. 256, 258
Russell, H. N. 38

satellites 148
  (orbiting) 57, 262
Saturn 198–204, **201**, **203**, 210, 222, 266
  appearance 200, 272
  brightness 200
  constitution 200
  density 200
  mass 148, 200
  orbit 198
  revolution 200
  rings 200, **202**, **205**
  rotation 200
Schmidt camera **13**, 14
Schwarzschild, K. 256, **257**, 258
  radius 256
silicon 44, 54

Small Magellanic Cloud (SMC) 112, 138
solar flares 36
solar system 146–150, 266, **266**, **267**
  evolution of 213
  (*see also* universe)
solar wind 36, 196, 216, 217
solstice 72
South China Sea 226
spectrograph 18, 270
spectroscopy 16–20, **19**, 49
stars, binary 48, **49**, **51**, 94, 96, 98, 100, 106, 110, 116, 118, 126, 128, 130, 132, 134, 138, 140, 142, 144
  Cepheid variables 28, 29, **29**, 32, 88, 124, 140
  eclipsing binary 50, **51**, 90, 96, 102, 118, 124
  RR Lyrae stars 32
  spectroscopic binary 90, 106, 110, 138, 140
  T-Tauri variables 54
  variable 96, 98, 116, 118, 122, 124, 132, 136, 142, 144
Achernar 108, 126
Acrux 104
Albireo 104, 144
Alcyone 136
Aldebaran 136
Alderamin 98
Al Dhanab 110
Alfirk 98
Algieba 114
Algol 126
Alhena 110
Alioth 138
Alkaid 138
Almaak 86
Alnair 110
Alnilam 124
Alnitak 124, 145
Alphecca 102
Alpheratz 86, 128
Al Rischa 128
Alshain 88
Altair 88
Alya 134
α, β, γ, Andromedae 86
Ankaa 126
Antares 132
α, θ, Apodis 86
α, β, γ, λ, Aquarii 88, 108
η, ι, Aquilae 88
α, β, Arae 88
Arcturus 90
α, γ, Arietis 90
Arneb 116
Asad Australis 114
Ascella 132
α, β, ε, ζ, θ, Aurigae 90
Bellatrix 124
Betelgeuse 34, 124
α, γ, ε, η, Boötis 90
γ, δ, Cancri 92
α Canis Majoris 94
Canopus 96, 128, 130
α Canum Venaticorum 94
α, β, Capricorni
β, ε, η, ι, Carinae 96
α, β, δ, ε, Cassiopeiae 96
Castor 110
α Centauri 98
α, γ, δ, ζ, Cephei 98